"Know Thyself." Such was the advice constantly offered over 2,000 years ago by the famed Greek Oracle of Apollo at Delphi. It was given in response to those who sought her counsel regarding the course their destiny was likely to take. It is still sound advice for most of us in the modern world.

To come to *really* know oneself—discover one's distinctive temperament and character—requires frequent self-scrutiny. It is well nigh impossible to know what makes one "tick" without recognizing the nature of one's attitudes and responses to life in the outside world, while also acknowledging the highly personal inner psychological drives of feeling, thought and imagination. The consciousness that impels us is psychologically deep and wide-ranging. The search for the essential Self requires a "Sherlock Holmes" mentality and discipline: it's a hell of a job to unify outer and inner "consciousnesses."

This book should help. Every chapter can be seen and read as its own "story" describing an especially significant aspect of consciousness. Cumulatively, they are meant to help readers attain a sense of their own body-mind-spirit complexes and *who* they are as entities unto themselves. And then to ask the question as to where "reality" is to be found: in the mental life of thoughts and feelings . . . or in physical encounters with the material world of time and space?

By the same author:

Form, Space and Vision (4 editions)

Art and the Creative Consciousness

War Night Berlin

Antarctic Odyssey

WHAT THE HELL ARE THE NEURONS UP TO?

The Wire-Dangled Human Race

Graham Collier

AuthorHouse™
1663 Liberty Drive
Bloomington, IN 47403
www.authorhouse.com
Phone: 1-800-839-8640

© *2011 Graham Collier. All Rights reserved.*

No part of this book may be reproduced, stored in a retrieval system, or transmitted by any means without the written permission of the author.

First published by AuthorHouse 3/23/2011

ISBN: 978-1-4567-0177-2 (sc)
ISBN: 978-1-4567-0178-9 (hc)
ISBN: 978-1-4567-0179-6 (e)

Library of Congress Control Number: 2010917738

Printed in the United States of America

This book is printed on acid-free paper.

Any people depicted in stock imagery provided by Thinkstock are models, and such images are being used for illustrative purposes only.
Certain stock imagery © *Thinkstock.*

Because of the dynamic nature of the Internet, any Web addresses or links contained in this book may have changed since publication and may no longer be valid. The views expressed in this work are solely those of the author and do not necessarily reflect the views of the publisher, and the publisher hereby disclaims any responsibility for them.

FOR

Patricia and Mary; Wendy, Andrew and Ruth

LIST OF CHAPTERS

INTRODUCTION xiii

FIRST THOUGHTS xvii

1. 'EXCUSE ME: AREN'T YOU FORGETTING SOMETHING?'. 1

2. WHAT THE HELL ARE THE NEURONS UP TO? 19

3. THE NEURONS – ORDER & ENIGMA: THE STRANGE AFFAIR ON EASTER ISLAND 42

4. THE COMPASS POINTS OF CONSCIOUSNESS 49

5. SOLITUDE and the SYMBOLIC ENCOUNTER. 71

6. TIME and the DREAM 87

7. TIME and the WHIRLPOOL 102

8. TWO SIDES OF LIFE'S COIN: THE FLYING BISHOP & HIS 'SERMON' ON OPPOSITES 113

9. LEADING A DOUBLE LIFE: MATERIAL NATURE: ENLIVENING SPIRIT 127

10. THE BRAIN'S PRODIGIOUS CEREBRUM & CORTEX: THE LATERAL SPECIALIZATION OF THE TWO HEMISPHERES . . 147

11. THE SNAKE and the GREAT ROUND 168

12. CONSCIENCE & MORALITY: CHARACTER - FOR GOOD OR ILL 180

13. WHAT PRICE A 'BRAVE NEW WORLD'? 221

14. WAR and KILLING: QUESTIONS OF EVIL, RELIGION, SPIRITUALITY, INNOCENCE 241

15. THE MORAL FIBER of NATIONS: GREECE and ROME 273

16. MIRROR, MIRROR, ON THE WALL... 282

17.	MIND and BRAIN	316
18.	THE AGE-OLD CONCEPT OF SOUL: ITS VITAL ROLE IN CONSCIOUSNESS	326
19.	A PYRAMID OF SOULS? A SENSE of the HOLY: MUSIC of the SPHERES	375
20.	SOUL and the WANDERING SPIRIT	407
21.	CONSCIOUSNESS WITHOUT A BRAIN?	411
22.	HUMANITY AT LARGE: DECLINE and FALL?	423
23.	ON THE SIDE OF THE ANGELS: LOVE and COMPASSION	445
24.	LIFE WITHOUT MYSTERY	463
	BIBLIOGRAPHY	469
	ABOUT THE AUTHOR	473
	INDEX	475

Acknowledgments

Throughout the book I have drawn constantly on the writings of those whose wisdom - and whose lives-as-lived - testify to the extraordinary range of human thought and sensibility. Their words have set me thinking and wondering from those early days of following *Alice* through her *Wonderland*.

My gratitude then to the shades of those Greek mentors: Aeschylus and Euripides; Socrates, Democritus, Plato, Aristotle... And to those of the Roman philosophers and statesmen who possessed a Greek turn of mind: particularly Cicero, Ovid, and Pliny the Elder... and that wisest of all emperors, Marcus Aurelius, whose *Meditations* 'say it all' when it comes to expressing the ideals and moral principles to which we can aspire. Their philosophical insights are quoted freely throughout these pages.

Nowadays, the statements made by scientists, psychologists and philosophers often complement each other, allowing 'reality' to be defined in several ways throughout these pages. I have drawn heavily on *The Amazing Brain* - both the extraordinarily lucid text written by Robert Ornstein and Richard F. Thompson, and the wonderfully expressive drawings by David Macaulay. The book makes neuroscience a thrilling read, and I am indebted to Dr. Thompson for offering advice when it came to delivering philosophical and psychological conjectures from neurological facts.... and for allowing me to reproduce passages - written by Dr. Ornstein and himself - from *The Amazing Brain*.

Gretel Ehrlich's moving story - *A Match to the Heart* - describes the neural mayhem suffered by her brain as a result of a lightning strike. It is spellbinding read and vividly illuminates the astonishing complexity of brain-structure and function described by Ornstein and Thompson. I am grateful for permission to quote from her powerful testimony.

Loren Eiseley died in the summer of 1977. I cannot over emphasize my indebtedness to him and wish to acknowledge him here. Whether writing as scientist, philosopher or poet... his words - always evoking his wonder at the functioning complexity of the natural world; the ineffable mystery of our human journey through it - both clarify and inspire. I quote from several of his books - particularly *The Star Thrower*. And thank him *in absentia.*

The phenomenon we call 'electromagnetic energy' is discussed in its many manifestations throughout the following pages - electrical forces playing the preeminent role of vitalizing our bioelectric selves. Nowhere is this most comprehensively and vividly described than in *The Body Electric*, by Robert O. Becker, M.D. and Gary Seldon. The passages I cite from this book provide some of the essential science which make credible the complex range of our consciousness. Finally, I must recognize the significance of two recently published books which illuminate the general metaphysical conclusions pursued throughout this volume. *The Horse Boy* by Rupert Isaacson (telling of his Mongolian shaman-seeking journey), and *Born to Run* by Christopher McDougall. They both reveal the ability of the human spirit to surmount the normal limits of physical and psychological endurance. Their publication came at the right moment.... providing illustrations of the kind of moral and spiritual powers which are evidenced in my last three chapters.

Carl Gustav Jung, Swiss doctor and psychologist, died in 1961. Even now, so many years later, it would be remiss of me if I were not to acknowledge here his great influence in bringing me to form a meaningful philosophy of life. Six Chapters of this book contain extracts from his own writings - many taken from *Psychological Reflections* edited by Jolande Jacobi. His teachings, and indeed his 'presence', pervade these pages.

The visionary insights of the great poets can often convey the profound nature of our thoughts and feelings more intensely than the most logical prose. Consequently verse which serves to further illuminate - render more celebratory - the conclusions reached in the text appears constantly. Most of the poets are no longer with us, but I would pay my respects to them all, but especially the two 'Wills' who dominate these chapters: William Shakespeare and William Wordsworth.

I must also acknowledge the book *The New Quotable Einstein* by Alice Calaprice from which I have extracted many of Einstein's statements and casual remarks.

I have received support and advice from many wise friends over the seven years this book has been in the making. Elizabeth Gaythorpe, whose book *Somewhere In Loving* affirms the psychical power of human love after the death of her brother in the Second World War. Her letters and her life of devotion were inspirational when it came to writing this book. Professor Elizabeth Renk's advice ensured that my pages on Mozart, as well as my comments on the anti-Nazi 'White Rose' members of the University of Munich who were executed, were historically sound and philosophically acceptable. Dr. John Adams was constantly available to instruct on the holistic view that body, mind, and spirit work together for our total wellbeing. Dr. Joan Poultney read parts of the manuscript at several stages in the writing, and confirmed or modified statements concerning the goals of contemporary psychotherapy. And to Carol Haralson who designed the dust jacket, my appreciation for her ability to create just what I had in mind.

Closer to home I have sought the counsel of some wise and learned friends. Amala and Eric Levine have always been there to 'put me right' when my own 'uncertainties' were hard at work. And Austin Chinn - whose head seems to be seldom out of a book - has, from time to time, given me the nod to continue. My dear and wise departed friend John Hilberg was always there to encourage when spirit was flagging. Michael McClellan will not be aware that he provided critical support when I - book included - was at a low ebb. While my good friend Professor Traugott Lawler was a great help in suggesting revisions involving both fact and style.

And my long-standing friend Richard Broakes-Carter has made many suggestions which have been invaluable in the editing.

My wife, Patricia, has been very patient during the many rewrites of this book; apart from which her encyclopedic knowledge and critical acuity have been - and still are - constantly called upon to come to the rescue....

Finally, I must record the debt I owe to my daughter Ruth. Without her proficiency in utilizing the electronic means by which texts are prepared for publication nowadays, and without her dedication to the work, the constant reorganizing of lengthy sections of this manuscript might not have been accomplished. I owe its ultimate appearance, in large part, to her efforts on my behalf.

Graham Collier
August 2010
Sharon Connecticut

WHAT THE HELL ARE THE NEURONS UP TO?

INTRODUCTION

'Whence Come We? What Are We? Whither Do We Go?'

Such was the title given by the great French painter, Paul Gauguin, to one of his later works painted in Tahiti in 1897. His words could well serve as the title of this book, for they ask the unfathomable questions that we have been asking ourselves throughout our recorded history: questions born of uncertainty concerning our human status in the natural world, niggling away in consciousness, and provoked by the idea that some mysterious truth lies lurking behind the surface of life as we know it. Questions which we pursue throughout the book.

Human consciousness does indeed present us mentally with two fundamental modes of awareness - a 'two track' way to knowledge. One is concerned with apprehending the physical realities of the *outside* world in time and space, courtesy of the five senses; the other with conveying the psychological reality of our *inner* world of thought and feeling, courtesy of intuition and imagination. If we make full use of this range of consciousness, then we cannot avoid living what can only be described as a 'double life'.

Among all of the masterpieces Gauguin painted during his years of self-imposed exile in Tahiti, this painting in particular demonstrates the artist's ability to fuse together both aspects of this double life. He reveals the natural lushness and beauty of the 'outside' tropical world *and* the physical grace of its inhabitants - while also managing to convey the fact that, individually, the Tahitians depicted are caught up in an inner world of reverie, seemingly entranced.... *'Perchancing to dream'* as Shakespeare might have put it. Thus the artist shows how we are *in* the world; yet not necessarily *of it*. For most of his life Gauguin was troubled by this dual nature of our existence. The title of this masterpiece alone indicates how he was haunted by the three questions which inevitably accompany this human condition - and which many of us also quietly ponder.... wondering about our origins, what kind of creature are we, and where are we on our way to - if anywhere?

Throughout the four to five thousand years of our recorded history we find evidence that such wonderings as to the *purpose, meaning,* and *ultimate destiny* of human existence have been universally present. Even in prehistoric times the archeaological evidence of the rituals practised suggests that it was so. I would describe such introspective questioning as a form of 'metaphysical anxiety'…. generated by a 'built-in' and unconscious echo of a spirit-state-of-being - a psychical element that lies beyond the self of flesh and bone inhabiting a timebound material world.

It could be said that one significant result of our creative abilities in both the sciences and the arts - together with the mental discipline we call philosophy - is to bring about some reconciliation between our sensory involvement with the outside world as *fact*… and the more intuitive, internal reveries of mind and spirit emanating from the realm of *imagination*.

Fact and imagination. It is my hope that the contents of this book will help bridge the divide. And in so doing reveal how the mental exchanges that go on between these two levels of consciousness determine not only our behavior, but the form and nature of our character and personality. Such an exploration demands that we look into the workings of that amazing physiological structure, that springboard of consciousness we call the *brain* - and ask the question how on earth does this living organism manage to serve those two masters of consciousness: the objective senses and intellect vis-à-vis the world; the subjective imagination fuelling our creativity and pushing us through a maze of moral and spiritual values.

The human brain has evolved over some 500 million years to become the 100 billion neuron-strong powerhouse responsible for the ultimate emergence of *homo sapiens* - a development which some anthropologists and neuroscientists believe occurred some 100 thousand years ago. And with it came a release from the mindless tyranny of instinct. No longer were we at the mercy of a mechanical 'stimulus-response' reaction to life's events - leaving the brain free to make intelligent decisions as to how to deal efficiently and practically with the challenges presented by the environment, while at the same time conjure up 'felt-thoughts' about the meaning of it all. (Which brings one to wonder whether or not the chimpanzees from whom we evolved ever wondered 'who' and 'what' they were, and 'where' they went from 'there'.)

This is not a 'technical' book, even though I draw heavily on neuroscientific explanations as to how the brain works. After a long and varied life, I have found that memory and hindsight come to the fore and allow one glimpses of the essential qualities that constitute *humanness*: insights that become more revelatory the longer one lives - providing one indulges a persistent curiosity about the 'how' and 'why' of everything in a state of existence: from the cosmos surrounding us to the small planet we inhabit; the elephant to the ant; the chimp to the man; the human to the angelic.

Yet one of the most significant questions to emerge from these writings is whether any form of awareness can exist without a functioning brain. The reader will draw his or her own conclusions after reading Chapter 21. And then there is the enigma we call 'Mind': is this merely another word for 'Brain'? And what, if any, is the difference between the traditional concepts of 'Spirit' and 'Soul'?

I wrote this book in an attempt to shed some light on such human issues as these, and especially to try and reply to Gauguin by disclosing the amazing complexity of consciousness and its metaphysical overtones: an enigma, a mystery, that cannot be dismissed. And I like to think that if Gauguin had been around to read it, he might have found some answers to assuage his heartfelt appeals.

I find it relatively easy to comprehend an evolutionary process that, over many millions of years, will make of an ape a *super* ape - a creature showing the transformations of physique and brain in order to attain the bodily and practical skills required to adapt to changing environmental circumstances. But it is surely more difficult to understand an evolutionary movement that can, on the one hand, produce creative and moral human beings such as Mozart, Florence Nightingale, Einstein, Mother Teresa.... while on the other put together such destructive and basely immoral creatures as 'Jack-the-Ripper', the evil Adolf Hitler and members of his Gestapo, the equally vicious Joseph Stalin, Pol Pot, Saddam Hussein.... Evolution has brought about many enigmas and variations on this theme of humanness, moving from the extremes of those who seek to enhance and further the cause of human life, to those who work to diminish life and destroy their fellow human beings wholesale.... and more efficiently and callously than any ape from whom we have evolved. How are we to account for such a range of individual differences? Well, we do our best throughout the pages of this book. We discuss the role played by neurological function, the startling mathematics of genetic inheritance, and the traditional belief in the part played by non-biological forces such as soul and spirit.... all of which go to influence the psychological identity of every individual; all contributing to the great mélange we think of as the human race.

It is perhaps revealing that Sir Julian Huxley - brother of Aldous, and distinguished biologist who died in 1975 - should have written in 1964:

> *Though undoubtedly man's genetic nature changed a great deal during the long proto-human stage, there is no evidence that it has in any important way improved since the time of the Aurignacian cave man... Indeed during this period it is probable that man's nature has degenerated and is still doing so.*

The significance of this statement is discussed in Chapter 22: *Humanity At Large: Decline and Fall?*

In the dream, the ultimate question is asked of the Wise One: 'Does evolution have a goal?' He replied, 'Only if you can imagine the perfect hybrid: the man-angel'.

FIRST THOUGHTS

Consciousness: Where Inner and Outer Worlds Meet

Go far; come near
You must still be
The center of your own small mystery.
Walter de la Mare: from the poem
Go Far; Come Near

When Benjamin Disraeli - Queen Victoria's favorite Prime Minister and highly regarded author in his own right - was addressing a meeting of the Oxford Union in 1864, he was asked by an undergraduate what he thought of Charles Darwin's recently published *Origin Of Species*. The Prime Minister pondered for a moment, and then replied: *'I suppose you're asking if I think that man is ape or angel. Well… I'm on the side of the angels.'* His reply may well have been tongue-in-cheek - many of his fellow members in Parliament certainly thought so: gales of laughter echoed through the Chamber when news of his 'angel' pronouncement reached London.

I wonder how many of the Prime Minister's audience took him seriously when he introduced angel-like qualities into evolutionary theory? Perhaps there were a few among them who would have been somewhat bemused…. having read the French philosopher and mathematician Blaise Pascal's thoughts on mankind in his *Pensées* - a work which appeared in its entirety only after Pascal's death in 1662 - and represents a less optimistic view of the human condition than that expressed by Disraeli. Here is Pascal: *'What a chimera, then, is man! What a novelty! What a monster, what a chaos, what a contradiction, what a prodigy! Judge of all things, feeble worm of the earth, depositary of truth, a sink of uncertainty and error, the glory and the shame of the universe.'*

And yet…. when Pascal describes man as a *'….depository of truth….the glory and shame of the universe'* and *'… 'Judge of all things…'* he is drifting into Disraeli's camp. For abstract concepts such as 'truth', 'glory', 'judgement'…. relate more to the spirit-territory of Disraeli's 'angels' than to the earthbound world where the five senses dominate, and reason holds sway.

So, despite his exuberant rhetoric, Pascal nevertheless implies that man is not totally *'a feeble worm of the earth'* - *(an intelligence in servitude to his organs'* as the novelist and philosopher Aldous Huxley put it.) Also, in using the word *'chimera'*, Pascal is pointing out that we are harried from pillar to post by a consciousness leading us up more than one garden path to serve contradictory ends - from the highest or the most base of motives.

Evolution has not, as yet, placed many of us in the angel category. But it has provided us with a brain capacity and range of awareness far exceeding that possessed by our chimpanzee ancestors. Five hundred million years of brain evolution have long ensured that in responding to life's happenings, we are no longer controlled solely by the mechanical processes of instinct. Instead, we have developed complex systems of thought and feeling that determine our reactions. Over countless millennia we have mentally advanced to possess an incredibly comprehensive consciousness. So much so that we have evolved to live in *two* mental worlds. On the one hand we are governed by an *inner* and subjective imaginative life, intuitively prompted, creative, contemplative, self-examining. On the other, we are - courtesy of the five senses and a questing intellect - objectively bound to the *outside* world of time and space: seeking to know how nature works, pursuing the sensual and psychological satisfactions such worldliness offers, and beset by the painful vulnerability of body and mind to accident, disease and age.

The following twenty-four chapters tell of the most significant neurological and psychological insights that nowadays shed some light on this paradoxical dual life…. and suggests that in trying to eventually reconcile our *inner* and *outer* selves we grow in humanness and wisdom; become 'whole' as human beings. They tell of how the *five senses* serve the powers of *reason;* how the independent mental forces of *intuition* serve the inventiveness of *imagination*. But they also address the presence of one other vital factor in consciousness: namely the psychical *force majeure* we traditionally think of as the *soul* and its attendant emissary the *human spirit* - the means by which we mobilize the will to act inspirationally, heroically, in human situations demanding the deployment of such powers…. and in so doing reveal the spiritual element which Disraeli likely had in mind when he placed us *'on the side of the angels'*.

In the late 1940's, the great Welsh poet, Dylan Thomas, described us as

The Wire Dangled Human Race - a particularly apt description for it vividly indicates the psychological stresses of the 'dangling' to which we are subjected by such a diverse stream of urges and 'directives' emanating from consciousness. Yet Thomas' evocative 'one-liner' is also prophetic, for it was written at a time when neuroscience was in its infancy. Nowadays, contemporary neuroscientists actually use the word 'wiring' to denote that the brain's 100 billion neurons make trillions of electrical and chemical connections with each other... numbers that evoke a sense of awe - if not mystery - compelling us to ask.... just what the hell *are* the neurons up to?

A major problem confronting us today is that the outward-looking worldly-self, ever hungry for new sensations and experiences is dominant. The inner realm is in retreat. Personal success in life is measured only in material terms. Education also serves this philosophy by emphasizing the gathering of *facts*, without demanding that students follow up and think for themselves - personally assess the meaning, value, and potential of the information they receive or access. We are becoming progressively psychologically unbalanced as a society: in the words of the great American paleontologist Loren Eiseley, '*Unconsciously, the human realm is denied in favor of the world of pure technics.*'

Mind and Brain

Nowadays we turn to the scientific discipline of neurology, and the medical studies of psychiatry, to help explain consciousness in general and the range of individual psychological characteristics in particular. Yet the question as to whether the brain is the sole instigator of every thought and every action is stirring debate among the philosophers and psychologists. Chapter 21 for example, *Consciousness Without A Brain?*, tells of the case of a patient, 'watching' from 'outside' herself the progress of an operation being performed to save her life. She had been deliberately rendered 'brain dead' for the 48 minutes or so required to complete the surgery, but was nevertheless able to question the surgeon later on what she had 'seen' - quite accurately - going on. While this report may raise questions concerning the absolute and ultimate authority of the brain in terms of human awareness, it is not prejudicial to the fact that this amazing organic structure - part building on part over 500 million years - has performed the key role in the development of this complex and discerning consciousness to which we have become heir. Even so, the neurologists and psychiatrists still have to explain one highly puzzling mental phenomenon.

Namely, how can the brain, as a *bio-physical* entity, bring an inner and *abstract* mental life of thoughts, feelings, dreams and imaginings…. into being - and this by means of the machinelike, chemical and electrical activity of 100 billion brain neurons and their trillions of interconnections?

It is a question that pervades every chapter, but in Chapter 17, *Brain and Mind,* the concept of 'Mind' is proposed as a possible answer. Seen as a psychical, *extra*-biological medium *informing* the brain, Mind's appearance on the evolutionary scene might be regarded as the ultimate extrasensory faculty responsible for the mental range of our inner life: for its inspired imaginative leaps, and the range of its dreamings. The expression, 'mind over matter' suggests which is the master here. Thomas Hewitt Key's witty epigram in *Punch* of 1855 certainly makes the distinction between the mental and material worlds compellingly absolute.

> *What is mind?*
> *No matter.*
> *What is matter?*
> *Never mind.*

If Disraeli had read his *Punch,* and been called upon to defend his suggestion that at some point in the millions of years through which we have passed, some trace of the 'angel spirit' could be responsible for Mind's role in the human psyche, he might have quoted Thomas Hewitt Key. Tongue-in-cheek again, you think? Who knows! It was difficult to tell with Disraeli, but he was a very wise man. Yet I think he would have felt himself to be on firmer spiritual ground had he been around to read, and quote from, the evocative and intellectually challenging thoughts of one of my scientific heroes who - and more significantly - was greatly admired by the late and great philosopher-poet W.H. Auden. I am referring to the distinguished American anthropologist, paleontologist and poet, Loren Eiseley, who died in 1977, and of whom I speak several times in the following pages. In Chapter 19, sub-titled *A Sense of the Holy,* I quote the following from his book, *The Star Thrower:* '… *without the sense of the holy, without compassion, his* (man's) *brain can become a gray stalking horror - the deviser of Belsen.* And then, suggesting how such a spiritual sensibility may have intruded itself into the processes of human evolution, he writes: '... *it was something happening in the brain, some blinding irradiating thing. Until the quantity of the gray matter reached the threshold of human proportions no one could be sure whether the creature saw with human eye or looked upon life with even the faint stirrings of some kind of religious compassion...*'

Incidentally, in *Webster's Collegiate Dictionary* the Greek word *menos* (spirit), appears as one definition of Mind.

The Double Life

It was just after finishing Eiseley's *The Star Thrower* and commencing these short descriptive essays that I chanced upon a sentence written by that strangely mystical 19th century English poet, Francis Thompson. He was a great admirer of the always youthfully enthusiastic poet Percy Bysshe Shelley, and writing an essay about him in 1909 entitled *The Beautiful Years*, Thompson described Shelley's opposition to the narrow, conventional world of his time in a single lyrical and highly symbolic line: '*So beset, the child fled into the tower of his own soul, and raised the drawbridge.*'

We all pull up the drawbridge from time to time - but some contemporary psychiatrists would likely say, more prosaically, that Shelley was prone to experiencing 'psychotic retreats from the world.' Yet, as a poet, he was a man *of* the world, sense perceptions and intellect keenly focused on the forms and forces shaping the physical reality of *outside* events. While at the same time he was also his own man, possessed of strong feelings and a powerful imagination - thoughts, hopes, fears.... representing an *inner* psychological reality. This is how consciousness works for most of us - if at a less intense outer and inner level than for the highly creative individual - but ensuring, nevertheless, that as human beings we are destined to live a double life... knowing two forms of 'reality'.

The *world* provides the facts. *We* are the provocateurs of the dreaming beyond the facts. In Chapter 4, *The Compass Points of Consciousness*, the psychological interplay between these two modes of awareness is described in some detail - both diagrammatically and verbally. Therefore a person can only be described as psychotic when either one mode dominates

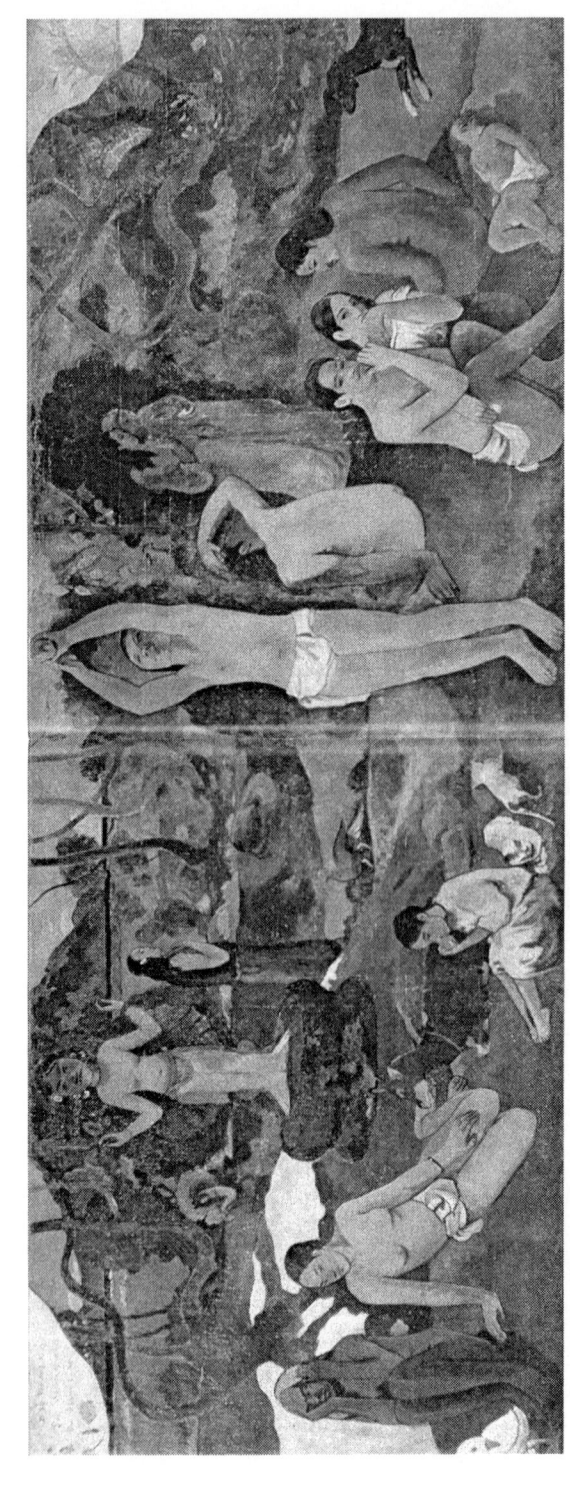

Paul Gauguin *Whence Come We? What Are We? Whither Do We Go?* M.F.A. Boston

consciousness to the exclusion of the other. There was nothing psychotic about Shelley - like all great visionaries, poets, philosophical and scientific thinkers, the drawbridge could never be always up. Only after having had his fill of the world's wonders would he sound the retreat - withdraw into himself and mull things over: allow reflection and contemplation, the whisperings of spirit or soul…. to deliver their moments of *in*sight, their intimations of life's meaning. The scientists amongst us ask of themselves, '*How* does nature and the cosmos work?'; the philosophers, '*Why,* and to what end, does everything exist?'

When Walter de la Mare, in his poem *Go Far; Come Near,* talks of '….*your own small mystery*' he is referring to the puzzling, if not disconcerting fact, that the face one sees in the mirror is the self of flesh and bone, whereas the quintessential Self is the invisible one who is at home in *'…the tower of…'* the *'soul'*. It would be difficult to find a better example than Paul Gauguin, the great French painter, of the demands this double life can make on our physical and psychical wellbeing. He was a man compulsively driven to discover his own true center, or quintessential Self, by what is often called the 'power of spirit', And he did indeed *'Go far…'* to try and find it, venturing to Tahiti where in 1887 - towards the end of his sad exiled years in the South Pacific - he painted one of his most powerful works entitled: *'Whence Come We? What Are We? Whither Do We Go?'* A sense of bewilderment and loneliness pervades this masterpiece, conveying more powerfully than words the anxieties and uncertainties that beset him: the need for meaning and identity that had been chasing him throughout his life. We are all in the painting, standing there alone among the still and silent figures, caught up in wondering about the 'why's' and 'wherefore's' of existence. It is part and parcel of every human life to live - consciously or relatively unconsciously - with undercurrents of 'built-in' psychological anxiety: the most pervasive being the shadow of mortality dogging our footsteps, causing us to question who and why we are.

The ape has travelled a great distance to achieve a Gauguin-like consciousness. And we have moved but a few years since Gauguin's day in understanding how neurologically we travel this dualistic labyrinthine journey. Yet from a philosophical point of view we are still in the dark as to what, if any, deeper and more transcendent truths might lie beyond the span of our transitory, material life - wondering whether consciousness leads us on a wild goose chase to nowhere.... or is breached from time to time by angel-inspired 'voices' encouraging us in our Odyssey to push on to distant and transfiguring shores.

The Wire Dangled Human Race

The brilliant Welsh wordsmith and poet, Dylan Thomas, wrote this evocative and memorable line. A great humanist, philosopher, metaphysician... he was at the height of his powers in the late 1940's, writing moving and vividly perceptive torrents of verse that expressed the general uncertainty of human life - the 'wire dangled' life as he called it. He died in 1953 at the age of 39. There is frequently an undercurrent of despair running through his verse as if, in the final analysis, he saw no ultimate victory for the powers of the human spirit when it comes to overcoming physical and psychological stress - the trials and tribulations described by Hamlet as '.... *the slings and arrows of outrageous fortune'.*

But then life is like that. We have never had any guaranteed *physical* tenure in terms of years. Nature has dozens of ways to rid herself of our presence. In addition to which, we are our own worst enemy. The violence we continually practise against each other takes on many forms, mercilessly destroying lives and leaving intense suffering in its wake. Our own body too, has a limited 'shelf life' - a built-in 'sell by' date: an obvious fact that subliminally gnaws away at our general morale.

When it comes to *psychological* well-being, few escape the personal tragedies that beset us: sickness, accident, ill fortune... the ultimate bereavement. Our emotional and intellectual lives are generally a round of conflicting feelings and sentiments, wherein we wrestle with competing - often contradictory - thoughts, ideas, judgments... in coming to make any kind of decision. In addition, consciousness has us living, sometimes simultaneously in three time-zones: the factual *present* as presented by the five senses at any given moment; together with memories of the *past,* and dreams of the *future*. Neither should we forget the subtle effect the dream-life of sleep can have on our day-to-day sense of reality - sabotaging the authenticity of clock-time the following morning; sometimes bringing us to wonder just *who* we are. And a further complication: the primitive brain's instinctual 'fight or flight' mechanism - a rudimentary form of consciousness where neither conscience or reason have ever lodged - is always poised to act blindly in tense and unfamiliar situations. All in all, we are not 'built' to sail through life on an even keel of surety and equanimity. The American philosopher and writer Henry David Thoreau had considerable justification for observing that we endure '.... *lives of quiet desperation*'.

Yet there are those who, while travelling this maze of consciousness, find ways to avoid the dead end 'desperation' route. Some are curious and adventure in the mind - embarking on imaginative voyages of scientific, philosophical, or artistic exploration. Others feel compassion for those less fortunate than themselves and work to help them - moved to act out of conscience, often in situations requiring exceptional courage. The wise ones throughout the ages have always said that to pursue a Cause - creative, compassionate, searching in one form or another.... brings meaning and fulfillment to life; is the way to become psychically complete by discovering the Self while so doing; attain peace of mind and reach a mental plateau where an egocentric and *angst*-ridden existence is surpassed. Viktor E. Frankl, the distinguished Viennese doctor and psychiatrist, who was imprisoned in Auschwitz and Dachau during World War II, came to believe that mental healing occurred when a patient understood that the *purpose* of life is to be mentally engaged in the search for some *meaning*, some *truth*, that brings one to face life serenely, stoically, hopeful of some positive resolution at the end. Even the role of suffering in enduring the most terrible of life's disasters can play a part in bringing one to know the Self that lives in spirit. In Chapter 21, I recount several instances - taken from Dr. Frankl's book, *Man's Search for Meaning* which was published in Austria in 1946 and is still in print - which tell of his extraordinary visionary experiences while incarcerated. And Chapter 4, T*he Compass Points of Consciousness*, provides a basic introduction - by both diagram and text - as to how the interactions between consciousness' various mental 'departments' can result in the kind of endurance and illumination-through-suffering of which Frankl speaks. The diagram in particular indicates how great 'Einsteinian' discoveries explaining the material nature of the world, owe much to a 'mix' of rationally gained objective information, and feats of imagination and intuition. While on the other hand, the music of a Mozart springs from *within* Mozart and owes little - if indeed anything - to his objective experience of 'sounds' in nature. Consequently, his musical 'discoveries' can be said to be more abstract and spiritual than the materialistic findings of science.

Some years ago, Professor George Steiner, Distinguished Fellow of Churchill College, Cambridge, made the following statement in a public lecture. It is an observation which captures the essence of this book. He was discussing the puzzlingly wide range of human awareness - of a mental life that can venture beyond the world of hard facts into the realms of an imagination where creative insights concerning the 'why's' and 'wherefore's' of ourselves - and of our world - come unbidden to mind.

> *There is too much of our cortex. We could do with far fewer cells and synapses and still have an excellent information system. Something much deeper is going on. Man has a marvellous excess of invention. He can say 'No' to reality.*

'Curiouser and Curiouser...' murmured Alice in her Wonderland as she watched the Cheshire Cat in the tree disappear and reappear, leaving only its enigmatic grin to linger momentarily on the air as it vanished. My first introduction to Alice was on my tenth birthday when I received a copy of *Alice's Adventures in Wonderland* - and being a lad immoderately driven by curiosity, I identified with Alice immediately as she pursued her way through the dream world of the unconscious. Even now, so many years later, I find myself chanting 'curiouser and curiouser' when encountering events or theories that defy comprehension or belief - as, for example, when neuroscientists tell us that there are perhaps 100 billion neurons (nerve cells) working away chemically and electrically in the brain, making a trillion or so interconnections with each other; or when geneticists inform us of the 3.1 billion letters of the DNA code that comprise the human genome; and when physicists and cosmologists discuss, seemingly quite casually... the trillions of stars occupying billions of galaxies.

Mystery breeds curiosity. The cosmos is mysterious. The natural world is mysterious. We, collectively as a species and singularly as individuals, are compellingly mysterious. Six and a half billion of us resident on this planet, everyone behaving according to his or her exclusively personal psychological drive: a veritable gamut of human beings ranging from base sadists and egomaniacs devoid of compassion and conscience....

to 'decent', well-meaning people occupying a moral middle ground.... to those selfless 'saints' (sung and unsung) who inhabit the high plateau and put their lives totally at the service of others. And all exercising their own personal agenda... for good or ill.

This ability to make choices in life - in terms of both attitude and action - would seem to result from an expansion of consciousness that brought about an early recognition of a 'raw' and unrestrained sense of self - a freewheeling awareness of 'I' that occurred at the higher end of brain evolution between two and four million years ago. You will read in Chapter 8 about the Flying Bishop - my companion on a flight in an old pre-World War II DeHavilland Rapide - who declared that as long as such primal egocentrism is abroad in the world, unmodified by the humanness-inducing factor of a developing *natural* spiritual sensibility - we cannot expect to see any significantly improved levels of wise and compassionate behavior between either nations or individuals. So the levels of violence, and the mix of prejudice and intolerance to be found in every society, seem likely to remain with us until the voice of reason and a universal enlightenment of spirit.... (transcending the grossness of ego and the dogmas of regional and institutional religions) - takes over mankind.

Despite constant progress in the sciences to ameliorate our common lot, we seem to be a far cry yet from seeing that we are all in the same boat, share a common destiny.... whatever religious faith we follow. We live for a brief instant of time, unable to share our common humanness, and then we die. A significant aspect of the tragic nature of human life is that we cannot universally apply the one psychologically ascendant force most of us experience: that mysterious psychological phenomenon we call 'love'.

For is it not a curious psychological phenomenon that to love deeply is to be transported to an altered state of consciousness - a state where one is temporarily freed from mundane worries; even from an over zealous ego.... for one is walking on air, released from any search for meaning - subliminal or conscious. *'True'* love, as the more romantic or poetic amongst us have long described it, is characterized as bringing us to identify with, and dedicate one's life to, the object of one's love. And when such love is lost - especially on the death of a lover, husband, wife, child or beloved dog - bringing us to suffer the debilitating suffering and anguish that can cause the world to become so bleak, then the will to live in it begins to wane, and even the primal instinct to survive at all costs is surpassed. That such suffering is universal is apparent when viewing the victims of natural and man-made disasters worldwide.

The Brain & Beyond

In their wonderfully informative book, *The Amazing Brain*, the neuroscientists Robert Ornstein and Richard F. Thompson describe the brain as *'... unique in the universe, and unlike anything that man has ever made.'* In Dylan Thomas' day neuroscientists knew much less than they do now about the prodigious structure and activity of the brain - about the chemical and electrical activity of those hundred billion neurons.... all able to link up with each other to create the trillions of neural interconnections that comprise, shape, and govern each individual mentality and character the world over - this *'wild inextricable maze ...'* as the English novelist Richard Blackmore put it: a maze which can be seen as wondrous at the *micro* level of human physiology, as in the clustering of a hundred billion stars within the Milky Way galaxy at the *macro* level of the Universe.

Consequently, in writing chapters where aspects of 'humanness' are discussed, some basic information concerning the brain's extraordinary biological intricacy and *modus operandi* should be part of the picture. And this I have tried to provide, particularly in Chapters 2, 3 and 21. Some neuroscientists see the brain's 'architectural' structure as evolving over a span of some 500 million years, finally attaining its present highly sophisticated and fully operational state between two to four million years ago, resulting ultimately in *homo sapiens*. This development represents an astonishing biological event which is described in some detail in Chapter 10. We moved, over countless millennia, from being creatures acting solely on the dictates of instinct delivered from the old brainstem, to largely self-governing human beings empowered to think and feel individually - due in large part to the ultimate arrival of the cerebra and cortex. Even so, this incredible maturation provides no guarantee that the decisions and judgments we make reflect any absolute truth about our own existence, or that of things in general. And even if we talk about being responsible for our own destiny, and act in accordance with our evolved free will, we still cannot control, or even know, what our actual fate will be. So near and yet so far... inasmuch as the brain is concerned.

In contemplating the issue as to whether or not the brain is responsible for the initiation of every aspect of our psychological life, every type of human behavior, one is inevitably taken into the realm of the 'metaphysical' - a word coined by Aristotle to denote concepts, intuitions, and behavioral life-forces of which we seem to become aware and driven through *extra*sensory means. By a *sense*less brain? It is a supposition which brings us to consider the possibility that non-biological intelligence-systems operate within the human psyche - particularly the psychical powers long referred to as 'Soul', 'Spirit' and 'Mind': forces that operate beyond the range of those known to the natural sciences, bringing visionary ideas and supportive feelings of spiritual, moral, and aesthetic persuasion to consciousness. And that *employ* the brain to this end. Such metaphysical propositions have been argued throughout the written records of human history - as they continue to be debated here throughout the book. Questions such as whether Soul and Spirit represent the same kind of immaterial force; or if Mind can be seen as both independent of Brain, yet *of* Brain (a paradox similar to that describing the phenomenon of *light* as both particle and wave). Wilder Penfield - Canada's late preeminent neuroscientist and brain surgeon - said that he could only describe Mind as a *'... non-temporal, non-spatial entity.'*

Plato in *Phaedrus*, records a plea to the Gods made by Socrates to allow him *'...to become beautiful in the inner man'*. Central to the healing ministrations of shamans, priests and doctors throughout history has been the supposition that an inner and spirit-self coexists with one's temporal and physical existence - a self with which one must become familiar if any breakthroughs to *'... the centre of (one's) own small mystery'* are to occur. Throughout many of the following pages we shall see how this idea of an 'alternate self' can be supported. In Chapter 4, the abbreviated account ('borrowed' from my book *Antarctic Odyssey*) of how Sir Ernest Shackleton - the famed British Antarctic explorer - used his visionary powers to accomplish the remarkable 800-mile small-boat journey from Elephant Island to South Georgia, provides a convincing example of such insights at work. Without such guidance and inspired leadership it is unlikely that any member of the ill-fated Endurance expedition would have seen home again.

We have travelled some distance since Plato's day in understanding how, neurologically at least, we travel the day-to-day labyrinth of consciousness. But we are still in the dark as to how Plato's *'inner man'* and Shackleton's insightful powers may relate to what, if any, deeper and transcendent truths might lie behind the fact of our transitory life.

People, Music, and Places…

From time to time one meets fellow human beings who radiate an aura of moral and spiritual strength, wisdom, and equanimity - men and women who have moved far beyond the 'aspiring ape' level of evolution; even beyond the 'faltering angel' stage…. to attain a truly virtuous state of discernment and compassion. In the years following the Second World War, when working as both practising artist and teacher, it was my good fortune to encounter a few of these advanced souls. My years of friendship with Herbert Read (later to become Sir Herbert) were - and remain - the most influential years of my life. His books on the philosophy and psychology of the arts revealed the significance for mankind of the poetic and artistic imagination. While his novel, *The Green Child*, established the yardstick by which I could make some metaphysical sense out of human existence. His poetry, flowing from the crucible of the trenches on the Western Front during World War I, gives us the most moving and expressive revelations of the depths of human stoicism, heroism, and suffering to come out of those tragic four years. He inspired one to believe in the creative worth of one's own work; simply to be in his presence assuaged anxiety of whatever nature: he was a supremely compassionate and wise man. The Distinguished Service Order and Military Cross he was awarded were decorations given for saving the lives of his own men when caught in dire situations behind enemy lines after a German advance had overrun the British positions. I would have followed him anywhere.

The power of music is often referred to throughout these pages. From the age of 10 until my voice broke in adolescence, I sang in the choir of a church with a great musical tradition: Handel, Haydn, Bach…. led me to Mozart…. and from then on I was hooked, and began my Mozart record collection. Curiously enough, and with no planning on my part, I found myself - on demobilization from the Royal Air Force after World War II

- commissioned to make pen, ink, and wash drawings of well known musicians. Challenging subjects to portray: men and women whose features were individually distinctive, the outward mark of those able to live deeply within themselves.

Sir John Barbirolli, conductor of the famed Hallé Orchestra, was the most compelling of them all. I spent many hours in his company making line sketches for the B.B.C.'s *Radio Times,* before moving on to a finished portrait for the Hallé Concerts Society. Sigmund Freud once said that *'Music is the royal road to the soul'* - and if ever one needed confirmation of that observation.... it was only necessary to spend a little time with John Barbirolli. His head was rarely out of a musical score - which made drawing difficult - and on the occasions we had lunch together at his flat in Rusholme he would bring out a green tin and we would munch on soggy cheese biscuits. He survived largely on air and music, and curiously enough I never felt particularly hungry on those occasions. It seems that one doesn't in the presence of a great soul. A similar ethereality encompassed Kathleen Ferrier, the most wonderful mezzo-soprano singer of my generation. Her unwavering, pure and perfectly modulated tone - whether she was singing in the low contralto range or in the soprano - convinced me that any 'guardian angel' I might encounter would be of womanly persuasion and Ferrier-like voice. She died young, at the height of her fame; I was unable to complete the drawing.

Barbirolli was an inspiring presence whose spirit took wing in the making of music: Gustav Mahler was his greatest source of revelation. But the most saintly man of the Church I encountered was Father Trevor Huddleston C.R. whose book, *Naught for your Comfort,* condemning apartheid in South Africa was widely read. He was a priest in whose presence one could literally 'feel' the intensity of a spiritual force. It was not possible to dissemble in any way when in his company: no pretensions, no self-serving surges of ego, no embellishing of facts... One knew oneself to be psychologically stripped bare - capable only of speaking whatever truth resided at one's core.

There have been occasions when both music and *place* have come together.... when one seems to be released from both the physical bonds of time and one's own bodily presence. Driving through Normandy in the summer of 1949, making line and wash drawings of France's superb

Romanesque and Gothic churches for exhibition later in the year, I stopped first at the breathtaking Cathedral of Beauvais. Standing in the choir - the loftiest Gothic structure in Europe at 157 feet from floor to stone-vaulted roof - the vertical uprush of stone columns created a cosmos of space that left one weightless and lost to time. And when the sound of the great organ swelled through this lantern of stone and glass I was abruptly 'translated' (as is said) into a strange and completely indwelling reverie: a state of absolute mental tranquility and elation.

Travelling much further afield - longitudinally from the Arctic to the Antarctic; latitudinally from the Gobi desert to Easter Island in mid-Pacific - the vast and empty reaches of ocean and desert overcome the dominance of the senses and move consciousness into neutral: a retreat of a Beauvais-like nature, but without the spirit-confirming power of music. One is pushed to reflect on the brevity and purpose of a human life when set against such awesome backdrops. Absolute silence possessed of a brooding and supranatural potency - an elemental 'atmospheric' force that had never been fully exorcised by a persistent human presence. Walking over the fearsome volcanic terrain of Easter Island one travels back in time seemingly waiting for some supernormal happening to occur at any moment. An account of one such mystifying event that took place while I was there is given in Chapter 3.

Over the course of several years I have participated in a complete circumnavigation of Antarctica's western South Atlantic side, and a partial one of its Indian Ocean eastern coastline. On the first of these voyages in the western reaches of the Southern Ocean, we were able to reach the remote ice-girt island known as *Peter the First* - named after the Tsar of Russia when discovered by the Russian Admiral and explorer, Thaddeus von Bellingshausen during his remarkable Antarctic voyages of 1819-1821. Our landing was only the ninth recorded since the island's discovery. Basically a volcanic peak some 5750 feet high, the island lies some 600 miles south and west of the Antarctic peninsula, and is inaccessible most of the time due to packed sea ice, terrible weather and very high seas. *Peter I* far outdoes Easter Island in remoteness and hostility, threatening one's very existence for having dared to invade its brooding solitude. The enormous span of time and distance between the geological birth of this bleak pile of volcanic rock and one's own sojourn on the planet, brings a stomach-tightening unease when walking the narrow strip of black sand beach. If ever there was an entrance on this earth to the ancient Greek underworld of Hades, then it must have been here on *Peter I*.

On such long voyages through pack ice around a large Continent that was never inhabited by man until the explorers and scientists moved in towards the end if the 19th century - where days follow days of high-rolling ocean swells bearing along errant icebergs of monumental, sculptural grandeur - the doctrines of a conventional western cultural heritage no longer provide convincing philosophical answers to the spiritual issues that beset us. And yet it is here that the senses - caught up in the white glistening world of ice and ocean - release consciousness to go on walkabout.... listening-in on 'voices' sounding from some supersensible level of one's being.

Commander Frank Wild, who was Sir Ernest's Shackleton's second-in-command on his Antarctic explorations, when asked why he kept going down to the bottom of the world, replied: *'Because of the little voices...'.* Shackleton himself, in his book *South*, reveals how he was inspired at moments of extreme crisis by an inner intelligence - more of spirit than of sense - guiding his actions. I know a little of what he and Wild were talking about. In the book *Antarctic Odyssey* I describe sitting beneath an icefall on Mount Erebus in McMurdo Sound. Nerve-breaking silence. Limitless vistas of ice and water. Absolute aloneness. The 'little voices' off and running. The French explorer Jean-Baptiste Charcot - who made his last voyage in 1906 - wrote: *'Where does the strange attraction of the polar regions lie, so powerful, so gripping that on one's return from them one forgets all weariness of body and soul and one dreams only of going back?'* Had Charcot read William Wordsworth's *A Poet's Epitaph*, written some eighty years earlier, he would have found the answer in the lines: *'Impulses of deeper birth/ Have come to him in solitude'*. The power of natural phenomena to magnetically transfix the senses, and then go further to quicken some inner intelligence of spirit, is difficult to accept for those who are never deeply moved by events - never experience feelings that allow consciousness to overstep the boundaries of reason. Yet one of the most objectively existential and renowned writers in post World War II Europe - the French author Albert Camus - could write about the mysterious power of *places* to affect us. In *The Myth of Sisyphus* he wrote: *'And here are trees and I know their gnarled surface, water and I feel its taste. These scents of grass and star at night, certain evenings when the heart relaxes - how shall I negate this world whose power and strength I feel... The soft lines of these hills and the hand of evening on this troubled heart teach me...that if through science I can seize phenomena and enumerate them, I cannot, for all that, apprehend the world.*

Finally, I should conclude these *First Thoughts* with the words of André Malraux - a renowned French Resistance leader of World War II and distinguished writer and historian - who wrote the following lines in his book *The Walnut Trees of Altenburg*. If I had been restricted to making a very brief Introduction to this book…. these words of his are the ones I would have used.

> *The greatest mystery is not that we have been flung at random between the profusion of matter and of the stars, but that within this prison we can draw from within ourselves images powerful enough to deny our own nothingness.*

I

'EXCUSE ME: AREN'T YOU FORGETTING SOMETHING?'

We are of such stuff
As dreams are made on; and our little life
Is rounded with a sleep.
Shakespeare: *The Tempest*, c. 1610

THE PROSPERO ENIGMA

Is that it, then? Has Will Shakespeare—displaying his customary enigmatic wisdom — got it right? That our whole life is as insubstantial and ephemeral as the dreams that attend our nights; that the sleep of death is a final state of nothingness, knowing no dream? Of course, there are those who may well be disinterested in the whole question of what, if anything, lies behind living and dying; who pay but little attention to how significantly alive they may be from time to time and, consequently, have scant knowledge of how relevant such moments of profound living can be when facing the unassailable fact that one day they will be dead. I would urge them to bear with me for a while.

"So what" is frequently the standard response when the subject of life's brevity crops up—accompanied by a shrug of the shoulders and "there's nothing I can do about it, anyway." Such is the typical and resigned dismissal of a tantalizing, and at the same time, vitalizing human quandary—the need to reconcile, intellectually and emotionally, the knowledge that the face you see in the mirror will one day no longer exist: a fate not thought about in advance—or so we believe—by non-human creatures, although neural research has established that animals such as cows and sheep experience severe emotional responses to witnessing the death of one of their kind. For myself, I find that a brief but tingling shockwave attends each unpredictable time—usually while regarding myself in the glass during the morning shave—when this quick realization of mortality strikes home. And I venture to suggest that the abrupt starkness of such an attack is not relieved even when one possesses a faith in God and personal continuity.

The knowledge that not everyone is curious about those formidable questions, 'Who am I,' 'Why am I,' and 'How long will I be around,' always comes as a bit of a surprise. An open mind is one thing, but a *blank* one … Which reminds me of the time when I was introduced to the chief executive officer of a large national corporation. He ambled up to the corner of the large room where I was leaning on the grand piano, sipping my wine and taking a breather from the press of bodies down by the fireplace. A big, barrel-chested man with a big booming voice and a big handshake, he put his glass down on the piano and started in.

"Well, Professor, what exactly do you profess?"

This is always a difficult moment: the history of human thought and how it has been expressed over sixty thousand years does not lend itself to the witty response of a brief one-liner.

Taking a long and sibilant sip of wine, I looked at him thoughtfully and said, "I suppose the best way to describe it is to say that I talk about consciousness and reality."

"Reality? That's easy enough, isn't it? This piano's real. I'm real. You're real …"—giving the piano a thump and doing a Tarzan on his chest, allowing my own substantiality to go untested.

"True enough, but surely you have to go further than the tangibility of your ribcage to determine your own reality? Or are you seriously telling me that you consider yourself to be just that—a walking ribcage? What about your mental life … ideas, feelings? What's going on in your head— your response to love … women, children, dogs, music? Beauty … or killing and war, for example? All just as real as your sternum and ribs. No? Or don't you look at reality from both a Platonic and Aristotelian point of view?" A typical professorial response, I thought. Just what he asked for.

The CEO picked up his glass, finished off his drink, gave me the kind of look reserved for non-members of the country club, and glanced toward the fireplace as if he needed the warmth. Then: "Heavy stuff, man. Heavy stuff." And he walked off. You see what I mean—this "heavy stuff" is not for everyone.

The lines, "*We are of such stuff* …" are spoken by *The Tempest*'s principal character, Prospero, the usurped duke of Milan. They are declaimed after a spectacular display of the speaker's supernatural powers—a lively performance of assorted spooks and spirits put on to trouble the Milanese courtiers, newly shipwrecked on Prospero's lost isle, among whom are some of those who most gravely wronged him in the past.

But how seriously are we meant to take the somewhat nihilistic implications of the Duke's short soliloquy—the suggestion of life's dreamlike drift and barren conclusion? After all, the words are spoken by a man who has more than a nodding acquaintance with a whole other world: one enlivened by a range of extrasensory phenomena not explainable by the known laws of nature. Prospero has not only managed to gain access to mysterious dimensions of time and space, but also learned how to command cohorts of preternatural regions to materialize in our mundane realm and do his bidding.

Would Shakespeare's audience not find something paradoxical here? Would it not seem strange to them (as to us) that a sorcerer of Prospero's caliber—a veritable spirit-master, no less—should dismiss our "little life" so readily without taking into account the mystifying implications of his own flitting about between this world and a supernatural one? It is a feat of mind over matter, space, and time, which reveals a prodigious mental capability to transcend the finite bounds of the physical world. On the evidence of his own psychic power alone, Prospero might well be expected to declare the significant authority of the human mind in the cosmic scheme of things, rather than diminish it as a dreamlike ephemerality, and at least allow that death may be either a sleep *or* an awakening.

These lines from *The Tempest* set the scene for what is to follow in this book. For in exploring the quite extraordinary workings of human consciousness and the deep, often quite unconscious source from which true inspirational breakthroughs spring, it seems to me that any proposition which regards us as biological entities pure and simple, may, quite reasonably, be called into question. Such moments of high consciousness lie behind the greatest achievements in both the arts and the sciences – as well as furthering those heroic and altruistic acts of selflessness when an individual will risk his or her life to save that of another, or brave danger in defense of a vital principle: behavior which flies in the face of the tyrannically strong instinct for survival. These inspired states of mind can be described as a form of insight, an envisioning which brings knowledge and understanding to the forefront of awareness – the prickings of conscience, solutions to seemingly insoluble problems, 'out of the blue' original thoughts concerning the relative truth of this or that... - all leading to moments of acute self-realization; achieved without apparent forethought, and crowning the analytical and reasoning processes by which the objective consciousness reaches its conclusions.

Many of us may have said on occasion, 'Ah, I've just had an inspiration' – meaning that an unusual and surprising idea has come to mind entirely of its own volition and, most likely, when one's conscious attention is directed elsewhere, perhaps when making the bed or mowing the lawn. The world 'inspiration' has long been used to denote a 'brainstorm' of this kind: its Latin roots *in*, and *spirare* signifying that the knowledge so unexpectedly gained results from suddenly partaking of a very rarefied kind of air indeed - the 'breath of spirit', one could say - a non-sensory source of insight empowering one to experience highly imaginative, creative thoughts and comprehensions unconsciously held, and intuitively released. Certainly the stuff of awareness which can spring to mind when the senses are elsewhere. Just up Prospero's street, I would have thought.

Spirit is a term having a number of connotations. I use it here and throughout the book to denote two particular concepts. First, in cosmic terms, that of a preternatural power which brought the very first subatomic particle into being, and provided the initial energy which started the universe off, and which still holds the cosmos in a certain state of being. And then, in terms of the individual human being, that internal force which determines the strength or weakness of temperament and character – the so-called 'mettle' or ability of a person – and which is responsible for the most inspired levels of human thought and action.

Let me reassure you: I have no religious axe to grind here – at least not the sort that would fall specifically within the dogmas of the major world religions. Many of us have grown up with the much-used Christian term, *'The Holy Spirit'* – a phrase charged with mystery. Yet in an increasingly secular society the significance of the words 'holy' and 'spirit' has diminished considerably. But if one withdraws them from a specifically ecclesiastical context I suggest that they are still very relevant in our attempts to comprehend the universe and ourselves. Given the remarkably complex theories and images resulting from advances in the physical and biological sciences over the last fifty years – not to mention the knowledge gained from space programs – there are many instances when the findings inspire awe, evoke a sense of mystery and of the profound. One is left to ponder the workings of dynamic cosmic forces ranging for seemingly infinite distances, just as capable in my view of being attributed to a spirit intelligence as to random mechanical cause and effect.

Those who pride themselves on their rational turn of mind will, no doubt, find this statement old fashioned, well behind the times. Yet reason now finds itself confronting scientific theories which constantly challenge conventional cause and effect logic. Consider the 'action at a distance' happenings in quantum physics: simply put, this is a phenomenon whereby a transference of energy from one molecule to another at a distance can occur with no known transmission link existing between them – the inference drawn being that such an exchange must take place beyond the speed of light, a hitherto seeming impossibility. We might use the language of quantum physics to suggest that inspiration - occurring beyond reach of the senses - is driven by a quantum-like leap of spirit, operating in a similarly unaccountable way: 'at a distance', that is. In my view, nothing of the mystery surrounding everything in a state of being is lost by the revelations of today's science.

The brain constitutes an awesome powerhouse, and one we are as yet far from understanding in its entirety. It is, as Ambrose Bierce put it in *The Devil's Dictionary*, '*An apparatus with which we think that we think.*' It comprises 100 billion neurons and trillions of neuron connections or synapses – (difficult to assimilate numbers such as these) – and is responsible for new theories in theoretical physics which envisage worlds beyond space and time, parallel universes, time warps... as well as adding a tenth dimension to explain such a plethora of universes: all highly imaginative mathematical adventures which, it seems to me, are already taking science into more mystifying metaphysical territory than any proposed by contemporary theologians or philosophers of whatever ilk. How do you respond to this?

> *What's all the hype about hyperspace? Most of us have our hands full dealing with just one universe. But Kaku takes us confidently into another dimension, or ten, to see why physicists think that universes are parallel, plural, and positively fermented with wormholes!*
> John Barrow, author of *Theories of Everything*, in a review of Michio Kaku's book, *Hyperspace*.

When one confronts the possibility of multi-universes, and the incomprehensible magnitude of light-year distances, the thought that an immaterial, unearthly power – best described as spirit – oversees the whole shebang, is hardly less credible than the physicist's theories.

THE GIRAFFE and the ELECTRICAL CONNECTION

What is spirit?
What a laugh!
Said the tall and thin Giraffe.
Here I have my head
Up high,
To snatch each amp
Come pinging by.

Many years ago in the 1930's, when I was about twelve, my stepfather gave me a book called, as near as I can remember, *The Life of a Cell*. It was written – again, so far as I can recall – by a Polish scientist, and made a great impression on my stepfather who was a metallurgical and chemical engineer. Much of the text itself I could not understand, but the general theme of the book was made clear by the diagrams, some of which I can still bring to mind. They were of irregular freehand circles, within which and from which, red and black arrows zipped about all over the place: the red arrows represented positive electrical charges, the black, negative ones. Each circle (or cell) was sustained internally by a positive and negative charge, while externally the red and black arrows linked up to the electrical systems of proximate circles.

There was no doubt but that the author regarded electricity as the very stuff of life – (a Frankensteinian approach, or so it seemed to a boy's imagination at the time) – as he developed the thesis that good health was the result of equilibrium between positive and negative electrical energies within the cell, while ill health came about as a consequence of disequilibrium. My stepfather thought that this made a lot of sense: "We're electro-chemical creatures, my lad," he declared one night as I was about to set off up the stairs for bed, "and without the electro we'd be dead meat. Stands to reason we've got to keep these millions of plus and minus charges in balance, otherwise we could just blow up or shut down." He really had me worried with this last possibility for I had no idea where the electricity came from, and even less as to how one exercised any control over the amps involved. I lay in bed, pyjama-less between the sheets, body parts suddenly nothing more than a myriad pricking pinpoints of electric sparkings – the life of the cells! One could do nothing save pray for a tie between the red and black factions. And when I was stricken with the sudden onset of my first involuntary erection I closed my eyes, prepared for the worst – terrifyingly ready for the Big Bang that was to come. It was obviously a near thing: the combined cellular hosts of the rest of the body must have won out to preserve an electrical balance. Henceforth, for a year or so of puberty, I knew a profound distrust in the anarchy of a water tap that could become such a tingling pillar – a veritable electric organ.

The following Sunday morning I asked to be let off going to Sunday school. My thoughts concerning electricity – this newfound elixir of life – seemed strangely incompatible with the scriptural teachings of the Church of England. My stepfather looked me in the eye; whatever he saw there seemed to convince him it was in my best interests to remain at home. "All right lad, but remember, going to church helps you become a well-balanced chap: body, mind, and spirit, working in equilibrium - encourages the spirit to get its antenna up into the ether. Lots of electricity up there. Alright, do some reading then."

Thus were planted a mixed-up series of thoughts and associations of ideas which, for a number of years, were tinged with religious teachings: the ether (the heavens) was the source of electrical energy; spirit, as the enlivening power, was our link to ethereal regions and therefore the conduit by which the vital electricity was transmitted and distributed. If the spirit was strong then a balanced electrification of the body ensured, maintaining one in good physical shape as well as keeping heaven within hailing distance, as it were. QED – any perception of an ultimate and higher reality results from our link to a sustaining cosmic force of an electromagnetic nature. (Curiously enough, QED is also the abbreviation for *quantum electrodynamics*.) Consequently, for some years after delving into *The Life of a Cell* I fancied myself as a cosmologist, finding God in a naturally religious way. And, in contrast, felt the churches' reliance on the history of human prophets with their revelatory teachings – revealed religion – to be less convincing than the mystery of spirit and cell electrification.

Hence, to my youthful mind came the Giraffe, most exalted and highly charged of creatures. What a design: head, at the end of that long aerial-like neck, almost in the ethereal zone – good reception ensured. No wonder the animal ambled so contentedly through the African grasslands. I made lots of drawings of giraffes, and amused myself by writing a variety of jingles about such an extraordinary creation.

These days of my early teens were brought vividly back to mind just before the end of the Second World War in Europe. With air operations against the Germans winding down, a number of Royal Air Force officers were encouraged by the Air Ministry to transfer to ground duties of one sort or another. To help one make a choice, a lecture was given daily by a top man in some field of administration, intelligence or science.

One morning in October 1944 fifty or sixty of us were gathered in the lecture hall waiting for the arrival of Sir James Hopwood Jeans, noted physicist and astronomer. I was sitting in the front row next to a certain Squadron Leader 'Taff' Jones DFC, a man who had completed two tours of night bombing operations over Germany (60 missions), and who suffered a spasmodic twitching of his right eye – a nervous tic which sometimes caused the whole right-hand side of his face to jump. When this happened, Jones would automatically bring up his right hand and flick at his cheekbone as if to swat a fly.

Sir James commenced his talk with an introduction to the electromagnetic force, pointing out that it takes on a variety of forms comprising electricity, magnetism, and light itself. He spoke of the fundamental nature of all matter as being electrical. After describing how the atom is made up of negative charges (electrons), positive charges (protons), and electrically neutral particles (neutrons), he suddenly and surprisingly declared, "Well, having said all that, gentlemen, I have to admit that I don't really know what electricity is. Do any of you here have any idea?"

At this precise moment 'Taff' Jones – whose tic seemed to be giving him more trouble than usual – brought up his right hand and flicked his cheek.

"Ah," exclaimed our lecturer, pointing directly at the Squadron Leader, "I see that this gentleman on the front row can answer the question for us. Yes Sir, what is electricity?" Jones scratched his cheek as if to satisfy an itch, raised his eyebrows and looked quizzically at Sir James before shaking his head vigorously. "Well, that's really a pity,' responded the physicist and

Astronomer Royal. "You see, gentlemen, there are only two beings around who can tell us what electricity is: one is God... and the other is this gentleman in the front row, and damn it, he's not going to say!"*

Jeans had introduced us to a spectrum of electrical energy – a spread that ranged from the atom at the microcosmic level to planetary and galactic gravitational fields on a macrocosmic scale. But he not only gave us science: he gave us poetry also. For he invested his subject matter with a certain air of the wondrous, a sense of the holistic nature of complete systems. He was the one who said on a later occasion, *'The universe begins to look more and more like a great thought than a great machine.'*

He invited questions, but my raised hand was lost in a seeming forest of arms. I was going to ask – inasmuch as he regarded the fundamental nature of all matter as being electrical – how he would regard a theory that postulates the existence of <u>hyper</u> electromagnetic energy: a force, as yet unidentified, which surpasses the manifest functions of electricity, and could be the means by which not only the incredibly complex mechanics of the universe are explained, but also the imaginative, creative abilities of the human mind and brain.

It took another scientist, many years later, to cause me to realize that I could simply have asked Sir James Jeans if my idea of hyper electromagnetic energy might not explain the force we call spirit.

THE 'EXCUSE ME...' INCIDENT

*I had rather believe all the fables in the
Legend and the Talmud and the Alcoran than
That this universal frame is without a mind.*
Francis Bacon: *Of Atheism*

"Excuse me, aren't you forgetting something?

The question was unexpected; the look on my face, no doubt, uncomprehending. "You're in the Art or Philosophy Department, aren't you? Odd.

---

* This was, of course, a rhetorical question. Jeans knew that electric currents involve the free flow of electrons; that Niels Bohrs' model of the atom, which postulates electrons orbiting the nucleus, provides an ample sufficiency of moving electrons which, with their 'spin,' conduces the phenomenon of magnetism. He was simply drawing attention to origins - to the fact that such an incredible, if not mysterious combination of electrical forces, should be universally present in the first place.

I mean here you are talking to me about the great dynamic systems in the universe and you've never once mentioned spirit."

The challenge came at me gently enough, yet there was no disguising the surprise which the Professor of Botany obviously felt at the need to ask it at all. He was a man of note in the university community – a highly regarded scientist researching into the role played by genes in the mutability of plant species and known for his directness and skill in debate. His query took the wind from my sails. For here I was, talking with him for the first time, and stupidly seeking to impress by going on about strong and weak nuclear forces, electromagnetic and gravitational dynamics, in that earnest way typical of enthusiastic amateurs.

Professor Leon Dure's interruption came some thirty years after the lecture on electricity by Sir James Jeans - thirty years during which my professional interests lay not so much in pursuing the history of the visual arts in a conventional way, but in attempting to discover something of the inner world of truly creative artists – the psychological basis of the compulsion to find the form, through one medium or another, which best expresses or symbolizes their most vividly experienced sensations, thoughts, and feelings. The professor's *"Aren't you forgetting something..."* served as a kindly reminder that scientists are uncovering extraordinary facts concerning the nature and operation of matter and energy, and that it is no less reasonable for any one of them to speculate as to whether or not some sustaining transcendent power - call it spirit, if you like - was responsible trillions of light years ago for bringing the universe as we know it into being, as it is for poet, philosopher and shaman. Had I not been so preoccupied with the need to impress my esteemed colleague I would likely have remembered the adage about 'fools rushing in...'

It was Aristotle of Stagira (384-322 B.C.), one of the western world's greatest thinkers and natural scientists, who believed that the result of any new and significant discovery in the physical and natural sciences was always to leave the researcher confronting an inevitable and fundamental question: namely, why, and by what prime Cause, did anything come into existence at all? For example, writing two thousand years or so after Aristotle, Charles Darwin (1809-1882) the English biologist famed for the theory of evolution set out in his book, *On the Origin of Species by Means of Natural Selection,* did not believe that his ideas could stand by themselves as a complete explanation of how man came to be. And although the clergy of his day regarded his conclusions to be in complete opposition to the Christian faith, Darwin did not regard himself as an atheist. In a letter to the Rev. J. Fordyce, dated July 1879, he wrote: *I feel compelled to look for a first cause... and I deserve to be called a Deist. I have never been an atheist in the sense of denying the existence of a God... an agnostic would be the more correct description of my state of mind.*

So for Aristotle – student of Plato – the logical and proper end of scientific inquiry is to ultimately go beyond the existential facts of the visible world and occupy the mind by pondering the mystery of what he called the 'Prime Mover' – the agency responsible, at the outset, for creating the very first sub-atomic particles leading to the formation of matter; of matter then taking on specific forms disposed in space; and the start of time. If the contemporary 'Big Bang' theory is correct as a 'physics-astronomy' explanation of the beginning of the universe, then Aristotle would go on to ask not only where the chemical, electrical and sub-atomic elements came from before they reached a single point of nearly infinite energy density and exploded, but also why.

It seems that no amount of sophisticated science can manage to eliminate the Prime Mover question. Many contemporary scientists recognize this and subscribe to the Aristotelian position. Others seem to think they have overcome the problem, yet in advancing arguments intended to dispense with the hypothesis of a Prime or First Mover, show that they never understood Aristotle's proposition in the first place. Here is a case in point:

> *Ultimately, we can trace the origin of life itself back to the spontaneous formation of protein molecules in the early earth's oceans without appealing to a higher intelligence. Studies performed by Stanley L. Miller in 1955 have shown that sparks sent through a flask containing methane, ammonia, and other gases found in the early earth's atmosphere can spontaneously create complex hydrocarbon molecules and eventually amino acids (precursors to protein molecules) and other complex organic molecules. Thus a First Designer is not necessary to create the essentials for life, which can apparently emerge naturally out of inorganic chemicals if they are given enough time.**

---

* Michio Kaku, *Hyperspace*: (New York, Oxford University Press, 1994) p. 194.

What is being described here is a process - a complex series of chemical and electrical interactions producing the building blocks of matter. But it cannot be passed off as the origin of life. To do this it is necessary to explain where 'the early earth and its oceans' came from; how methane, ammonia and other gases – not to mention electrical energy – came to be around so fortuitously to spark off the conversions described. Similarly, in the last paragraph, the 'essentials' for life are the inorganic chemicals without which the process the writer describes could not take place – not the process itself. So it is illogical and misleading in this context to eliminate the role of Aristotle's 'Prime Mover' when the origin of the elements necessary for the process Kaku describes to get underway is simply taken for granted....

We are left with the same old question: how do we explain the necessary and convenient presence of oceans and atmospheric gases – everything in place – all systems at 'go' – ready for blast off?

Whatever the terminology – whether one talks about a First Designer, Prime Mover, or Great Spirit... the intention is to recognize the possibility of a prepotent source of creation – one responsible for the appearance of the first bit of *anything*. Furthermore, and especially when using the word 'Designer' to describe such a source, the added implication of purpose is introduced. For to 'design' is to plan – to have in mind a goal, a particular end. For example, if we knew absolutely that built into everything, into every molecule or combination of molecules, is the blueprint for a distinctive, individual morphology and condition-of-being to consistently manifest itself – a schema which determines that each object will evolve to its own complete, predetermined potential – would it not be reasonable to suggest that design is a factor in the origin of such things? If it is the case that the acorn, fixed in its 'oakness,' will only produce an oak tree, or evolve into variations on the oak tree theme, then are we seeing a designed or a chance happening?

When Professor Dure queried the absence of spirit in my list of life's vitalizing forces, he might – had he been an American Indian – have referred me to the Great Spirit. In which case, he would have been alluding to the concept of a preternatural intelligence which not only could stand as an initial creative force (Aristotle's Prime Mover), but which the Iroquois regarded as an ongoing, life-giving and elemental presence which must be respected as the power pervading the life of the planet and circumambient universe – as we see in the following passage:

> *To the Great Spirit, however, the Iroquois ascribed creative power. He created not only the animal and vegetable world, but also adapted the elements, and the whole visible universe to the wants of man.*
>
> *That the Indian, without the aid of revelation, should have arrived at a fixed belief in the existence of one Supreme Being, has ever been matter of surprise and admiration. In the existence of the Great Spirit, an invisible but ever present Deity, the universal red race believed. His personal existence became a first principle, an intuitive belief, which neither the lapse of centuries could efface, nor inventions of man could corrupt.**

It is not commonplace in an outgoing hedonistic age like ours to sit back and consider the mystery inherent in a universe teeming with stars and planets; or in that of an earth such as our own hosting an incredible variety of living things. There is not much time, and even less inclination, to develop a philosophy of life. The frenetic pace of contemporary existence, and the ready availability of round-the-clock entertainment to 'take one out of oneself,' effectively anesthetizes the natural tendency of the mind to inquire into the purpose of one's existence; the 'whys' and wherefores' of multitudes of suns, moons, and stars in galaxies galore.

Yet the time does come, usually in middle age if not before, when many men and women will understand what the Welsh poet and painter David Jones was getting at when he said that he wished *'to discover the forms of which I myself am made.'* He is talking about an internal journey of self-discovery which, shaped by poetic and painterly images, will go on for the rest of his life. In essence, I believe it entails remembering and recognizing what manner of things and happenings one has come to feel strongly about over the years – those to which one has been strongly attracted and welcomed... and those one has disliked and rejected. It is then but a short step to recognizing the nature of one's personality and character, in terms of the personal values and sensibilities that have been nurtured over time.

So it is really important to constantly embark on reflective voyages

---

* Lewis Henry Morgan, League of the Iroquois (New York: Citadel Press – Carol Publishing, 1993), p. 154.

of self-scrutiny - the sort of contemplation which surveys the past to link up with childhood memories, and chart the course one has followed from those early days of wonder and curiosity. For, contrary to much of current opinion which considers that such introspection stifles self-growth and progress, it is a vital self-discipline if one is to come face to face with one's true nature; recognize its good and bad sides, and decide if one likes the self which is revealed. Unfortunately, failure to turn inward and sum up the course of one's existence every so often, diminishes the ability to feel deeply about things and causes thought to direct itself solely outward. And the those who comes to live solely in the external world are likely to suffer a reduced sense of their own individuality becoming, in the end, gray and uncurious characters reduced to living each day at the mechanical level of stimulus and response, unknowing, even of themselves.

In <u>Faust</u>, Part 1, Goethe emphasizes the importance of feeling as a way of knowing. At one point he has Faust, in a grand soliloquy, pondering the vast range of subjects he has studied throughout his life: theology, astronomy, jurisprudence, chemistry, philosophy.... Thoughts which bring him to say to himself, *"Yet here I stand, not one whit wiser than before"* – followed by, *"Unless you <u>feel</u> it, great efforts are in vain."* The 'great efforts' to which Faust refers can be seen as an ironic reference to the limitations of intellectual, purely cerebral learning – the gathering of facts unaccompanied by any particular affinity or feeling for the subject concerned. For it is the way we feel about something - the way random emotions have organized themselves into firm sentiments - that govern not only the intensity of our interest, but also provide the drive, the strength of our motivation to act and pursue that interest to the limit. Without such visceral reactions we would simply be encyclopedic, mental storehouses, capable of little imaginative reaction, making shallow judgments, failing to recognize worth.

Unhappily, as I have mentioned, in some circles nowadays it is considered psychologically non-productive to cultivate the 'felt-thoughts' - as Sir Herbert Read, the great English poet and essayist (who died in 1968) described the results of inward reflection. And one wonders how many disgruntled Fausts are running around out there - folk who have neglected and distrusted their feelings over the years, and wonder why they now stand, *'not one whit wiser than before'*. And their numbers will certainly increase if the pursuit of what is considered 'the good life' continues, with the engine of consumerism ensuring that our predominant interests - together with the values that accompany them - are determined by a preoccupation with the world of fashion and style; with acquiring possessions; with ready mobility, change and the temporary excitement of the new; with entertainment as a panacea for boredom and minds grown lazy; and with a belief in technological innovation as the be-all and end-all of existence. Of course all of the above can bring pleasure and a certain temporary satisfaction to living, for there is always the enjoyment of the moment, and when that is over the expectancy of something alluring turning up down the road. And who would want to deny the great advances in medicine and medical technology which have lengthened and enhanced our lives. In terms of being relatively enlightened and free from ignorance we are light years away from the existence which the 'huddled masses' of past centuries endured; Neither do most of us in the modern world suffer the harsh deprivation of the basic nutrients to sustain life, or the medical services to save it...as do so many men, women, and children living in undeveloped countries; not to mention those existing below the poverty line in affluent nations..

Yet I think we have to constantly remind ourselves that another kind of poverty exists - a poverty of spirit which can take over if we put all our psychological eggs in the basket of materialism; reach a point of craving the 'instant gratification' of appetite or desire, or come to believe that happiness in life is to be gained by moving from one 'new' experience to the next - 'progress by substitution' it is sometimes called. I suspect that this sort of poverty was not common among the Iroquois. The very sparsity of their material resources ensured that they were fortified internally through their intuitive awareness of 'The Great Spirit'.

Obviously, to be human is to possess a consciousness which serves an inner life of feeling and thought, and conveys through the senses the form and life of an outside world. Equally obviously, common sense dictates that some kind of balance between the two sides should be achieved. For to be completely introspective, introverted to a degree that eliminates the significance and presence of external things, can render one either otherworldly and saintly, or psychotically lost to any kind of reality. While to take the opposite tack and become totally extroverted, living only with the sensory, existential happenings taking place outside oneself - at the expense of inner 'felt-thought' responses - will seriously curtail the imagination, the ability to be moved by love, compassion, or the beautiful...; or will silence the 'still, small voice' which tells of conscience, and advises on the quality and worth of this or that experience. To become a total extrovert is likely to result in a despairing vacuity of consciousness in life's later years. We become complete as *human* beings, experiencing the full complement of mental riches available, only by keeping a foot in both camps.

> *Today, civilized man notices a gilded cloud and at best mutters, "cumulus."*
> William Wordsworth, *Intimations of Mortality*, c. 1800

If Wordsworth's statement rang true 200 years ago at the start of England's industrial revolution, think how much more relevant it may well be to our time when finance, commerce and market forces drive the world along, seeming at times to be the only reason we see for our existence.

"Excuse me, haven't you forgotten something...?" Haven't many of us forgotten something? Forgotten that spirit, as an inspirational force, takes us beyond our biological and existential selves, challenging us continually to face the mystery of our brief passage through and in time - indeed, of the mystery of time itself - and look to find an answer to the enigma of being 'human' somewhere along the way.

The French poet and philosopher Paul Valéry would, when pressed, give the following advice to a young writer: *'By all means indulge in a reverie – but please, please, let it be aided by a little accurate information.'* In that one sentence Valéry describes the desired interplay between the two faces of consciousness. His counsel obviously does not only apply to writers. It is the way we should all naturally function. Yet, to use Wordsworth's words, there seem to be a hell of a lot of 'cumulus mutterers' out there.

As I write this I remember the first time I used a dictionary to gain a little factual knowledge in order to express myself on a subject I felt pretty strongly about. I was about six or seven years old, and our class in elementary school had finally learned how to join up the letters in a word and write a decent looking and concise sentence. We were sent home that weekend and told to return on Monday with three lines on 'ordinariness.' I asked my stepfather for advice. He sent me to the dictionary. After I had struggled with the definition, he called me back and said, "Well lad, what do you think is ordinary in your life, then?" "Going to Sunday

School," I promptly replied. "Right. Then go and write three lines about what you mean by 'ordinary' when you think about going to Sunday school." The dictionary had given the word 'boredom' as one connotation of ordinariness; it caught the nature of my feelings exactly. And so began the practice of always seeking a *'little accurate information'* before setting out to express my own feelings and opinions, before indulging in the inevitable *'reverie.'*

> *To see a world in a grain of sand.*
> William Blake: <u>Auguries of Innocence</u>

A FEW ORDINARY MARVELS RECENTLY PERCEIVED in the outside world - memorable vignettes, courtesy the senses of sight and hearing - which triggered boundless reveries.

1. A family of Canada geese – two parents and six goslings – taking off alongside the small pond to which the adults return each year to breed. The fussing of the parents as they line everybody up – a certain tenseness to the moment – and then they take to the air, the youngsters taking longer runs to become airborne, adults circling slowly at low altitude until all are safely aloft, and then, gaining height, the wings of the goslings beating more rapidly than those of their parents, they head north in a line of flight as straight as an arrow, one adult leading, the other bringing up the rear. On course – sure of their heading in that vast region of space we call sky.

2. A ruby-throated hummingbird at the honeysuckle: about 3" long, motionless in the air, seemingly impervious to gravity, wings a filmy blur, long pipe stem of a beak poised as delicately within the flower as the probe of a spacecraft docking miles above the earth. Moves to investigate other blossom – flying *backwards*, I mean; then proceeds to check out the purple blossoms swaying high over the beams at the opposite end of the arbor, but now flying upside down, allowing itself to peer directly at the most promising looking blossoms above. Its sudden, darting, changes of direction – vertical and horizontal flight equally facile – makes any man-made flying machine look like a lumbering

dinosaur. The book says that a hummingbird's wings make 78 beats a second. Think about it – think of the motor mechanisms involved, of the neuron activity in that chip of a brain. Think of 4680 wing beats a minute.

3. Monarch butterflies have arrived in the large field. On time. The milkweed plant is grown and ready for monarch larvae to feed on. The butterflies have flown all the way from Mexico – over 2000 miles – flimsy aviators pitted against the turbulence of wind and weather. They arrive prior to the bursting of the pods when the leaves are most succulent: a feat of navigation and timing. An eminent biologist tells me that the way back to Mexico is imprinted in the creature's neuron system, yet the butterfly which has made the journey north may not survive to make the return; but the offspring which start life in my field will accomplish it successfully. Inherited imprints of navigational intelligence?

4. How is it that my old black Labrador's tail begins to beat on the living room floor the moment I put one foot out of bed in the morning? She is 30 feet away and the door between the rooms is always closed. Furthermore, she is becoming hard of hearing. Yet there it is, thump, thump, thump…

> *The bodies of animals and plants afford 'clear and cogent arguments of the wisdom and design of the Author.'*
> Robert Boyle

2

WHAT THE HELL ARE THE NEURONS UP TO?

Whether viewed as a biological organ, an incredible machine, a supercomputer, or simply a miracle, the human brain is surely one of the most amazing things in the universe - and the more we find out about it, the more amazing it becomes.
The Amazing Brain: Robert Ornstein and
Richard F. Thompson

I have just been reading Gretel Ehrlich's inspiring book, *A Match to the Heart*. It is a finely written, most personal account of her recovery after being struck by lightning – a work at times of poetic grandeur, charged with insight, and replete with the kind of 'accurate information' Paul Valéry insisted should go hand in hand to support contemplation and 'reverie'. Her neurological account of brain function sets the stage for the personal drama she experienced, and is so graphically and scientifically drawn that it plunges one into the maelstrom of electrical and chemical energy that is responsible for consciousness - at the same time bringing one to feel strongly with the author in her plight.

I quote directly from her book because what she has to say about the neurons, and the way she says it, provides the perfect introduction to this chapter. And I would not want to either paraphrase Gretel Ehrlich's account, or to try and find my own words to write about the lightning strike she experienced. Her own description is so compelling that it should not be edited or abridged.

An electrochemical pulse beats in every one of our hundred billion nerve cells. It is the 'life force' referred to in other cultures. Much like the cumulonimbus clouds, where lightning is born, nerve cells are structured with a difference in electric potential between the inside, which is negative, and the outside, which is positive, so that in response to stimuli, polarization can take place. Sudden storms of firing neurons travel on long tendrils that sprout from the brain stem and spinal cord and burrow into every organ and muscle in the body. We are the body electric – or, more precisely, the body electrochemical. I no longer think

> *of the brain as being that hard globe atop our shoulders but a body within a body: a long-limbed and flexible apparatus hunting and gathering messages, parlaying them into tiers and nets of neuronal connections almost unthinkably intricate, in which tiny voltage spikes fire into the burning bush of perception, activity, and consciousness.*
>
> *An electrical impulse travels down an axon – the transmitting arm of a cell – and when it arrives at the axon terminal, certain chemicals called neurotransmitters, are released from a gated surface – like a gated irrigation pipe – into the synapse, a gap across which those chemicals must travel to reach the other shore. The shore is the dendrite, which is the receiving arm of a cell body.*
>
> *The synapse is holy. <u>Apse</u> comes from <u>apsis</u>, whose roots mean, to loop, wheel, arch, orbit, fashion, or copulate, and the apse of a church is a place of honor. The synapse is the gap where nothing and everything happens. Bodies of thought swim in the synaptic lake, sliding over receptors, reaching for the ones that live on the other shore. An interval of between 0.5 and 1 millisecond transpires before an impulse makes its way across the gap, as in the bardo where we pause between life and death, treading water in the oblivion of a gray sea. What is a thought before it registers as memory? Is it a shape, or a shadow, or only unarticulated grayness that can't be held? Is it like unrequited love, or a lover who is spirit only, who has no body?**

This, then, is the phenomenal activity we call consciousness - the means by which we love and hate; will come to the help of others, yet also kill them when circumstances seem to warrant it; create images and concepts through inspired activities in art and science, yet destroy cultures in times of war: behave in courageous or cowardly ways; admire beauty, nobility and elegance yet embrace vulgarity and coarseness... A veritable mishmash of psychological drives and behavioural actions. And all courtesy of 'electrochemical pulse beats' in 100 billion neurons. It might well cause one to raise eyebrows in incredulous disbelief at the extraordinary range of chemical and electrical forces running our lives - all radiating from this neural powerhouse we call the brain. The marvel becomes more intriguing when you consider that ordinarily - save in the case of identical twins, perhaps -- no two of us are, psychologically, completely alike. Can this be due to significant variations in the numbers of neurons with which each person is endowed? Or to levels of efficiency in the neuron 'firing'

---

* Gretel Ehrlich, <u>A Match to the Heart</u> (New York: Penguin Books, 1995), p. 69, p. 70.

processes between batches of neurons? Or to differences in 'hard-wired' circuitry - *inborn* patterns of fixed neural interchanges; while some brains are less adept at developing 'soft-wired' circuits - *new* linkages between neurons brought about by life's experiences?

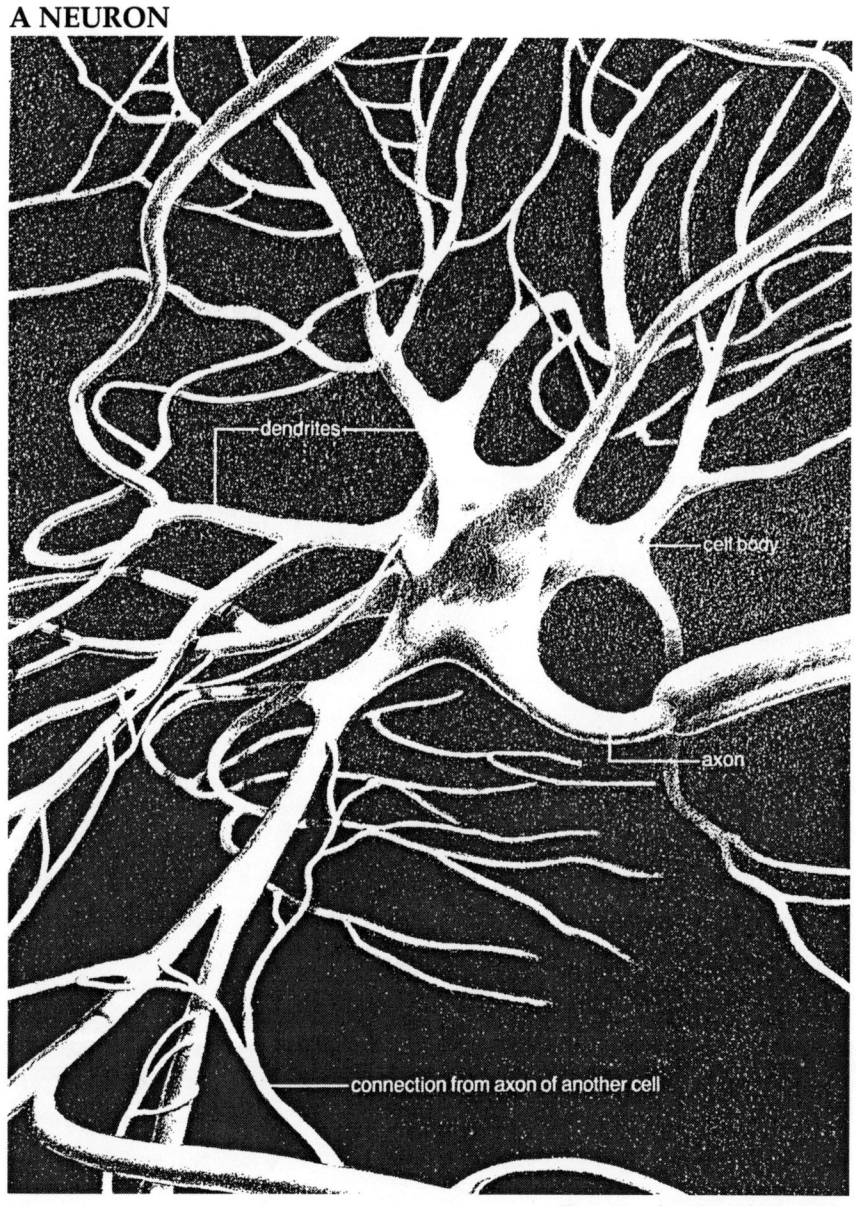

A NEURON

Drawing by David Macauly

It is also recognized that physiological differences exist between brain and brain; that the forms taken by various parts are not consistently the same, resulting in interesting morphological variations. The corpus callosum in women, for example, is larger towards the back of the brain than is the case with that in men. Robert Ornstein, one of the authors of *The Amazing Brain* noticed as he worked in the laboratory day upon day, that each brain he dealt with had its distinct differences - a bulge here, a larger occipital lobe or smaller temporal lobe there.... - and he comments that, ...'*people's brains are as different as their faces.*' Differences also occur within the internal workings of the brain. The brain organization of left-handed people is different from that of those who are right-handed. And the right and left cerebral hemispheres in males are more specialized - less integrated - than those of females.

Also, one must consider how long the human brain has taken to physically become the highly complex organ that governs our lives today. It is some five hundred million years ago that the old brain stem evolved to sit atop the last vertebra of the spine, and become the means by which the mechanism of instinct controlled our behaviour. And over the many thousands of years since then the midbrain, hindbrain, and forebrain have come into being. The crowning glory has been the left and right hemispheres of the cerebrum and cortex (the forebrain) - thought to be fully operational between 2 and 4 million years ago. But the degree to which this final cerebral development has brought us to live intelligently, creatively, and morally.... has not been uniform. The fact that there are some six billion 'modern' brains now inhabiting the planet - all obviously 'doing their own thing', displaying greater or lesser levels of sensibility, sagacity, and moral awareness - would seem to indicate that the range of differences found between human brains is much greater than that which exists between those of other animals. Such human diversity in brain performance provokes the thought that all brains are not born equal.

In Classical Greece the word '*psyche*' signified the animating spirit-force of the soul: a mysterious psychic power acknowledged throughout the annals of both Eastern and Western civilizations. It is seen as an inner and involuntary psychical takeover of normal consciousness - moving awareness and motivation beyond the hold exercised by ego and sensory, material involvement with the world. As such, it brings about an altered state of consciousness possessing a mystical quality - evoking moods in which time slips and the imagination wanders.... moving into realms of spiritual contemplation, moral awareness, and rushes of creative insight - moments of truth, so-called, regarding the 'how' and the 'why' of existence.

Later Greek and Eastern philosophers saw the power of *menos* or mind* to be the soul's great ally: the grand mental conduit by means of which the psyche influences and shapes consciousness. First, in providing and developing the mental skills by which the senses bring us not only to perceive the world, but to order our perceptions into structured categories of form, as well as bringing analytical skills to bear which reveal underlying principles of cause and effect. Second, in being able to shift from the job of seeking hard facts to inducing the more abstract and transcendent modes I have just described. And third, through its reflective powers of recall in bringing us to mull over our attitudes and behaviour - thus causing us to become aware of the kind of person (character) we are, or are becoming.

Such is the working partnership between soul and mind that was generally considered to represent the defining mark of *human* being. The combined psychological dynamic driving consciousness to participate intelligently and purposefully in the material world of time and space.... while also enabling us to feel and envision the pervading presence of a 'beyond time', spirit-like sphere of influence - the altered state that the poet Wordsworth describes as taking us beyond *'The touch of earthly years'*.

One wonders how the Greeks would have explained the concept of *psyche* and its working partner *menos* had they had known anything about the brain's 100 billion neurons and their trillions of interconnections. For possession of this neuroscientific information would have placed them in the same boat in which we find ourselves today - asking the question that the eminent British mathematician and philosopher, Jacob Bronowski put forward while giving the Silliman Memorial Lectures at Yale University in 1977:

> *'How (do) we get experience which is not directly physical through physical means?'*

---

* A distinction between what is meant by 'brain' and what is implied by 'mind' will be made later. For the moment – in order not to complicate matters – we will use them as synonymous terms.

One of the most vigorous thinkers of Greek antiquity, Parmenides (about 504-456 B.C.) initiated the distinction between a sensible world and a mental world, and in doing founded a metaphysical discipline based on observation and logic. He suggests that *menos* works in two ways. First, by assembling and analyzing the information provided by the five senses concerning the material, objective reality of the world - thus making it scientifically 'sensible'. Second, by pondering the meaning and relative significance of this knowledge to open up an abstract and imaginative 'mental' world in which persisted both the feeling and the persuasion that some sacrosanct, divine power - Aristotle's 'Prime Mover', for example - is lurking in the wings. To see *menos* as serving this dualistic function pervaded Greek philosophy, and has influenced Western thinking ever since. (But we should not forget that Eastern views of 'reality' have also embraced such a dualistic point of view.)

Yet we can address the enigma posed by Dr. Bronowsky in a way the Greeks never could. Neuroscience has brought us to the point where we can ask how the vast neural network of chemical and electrical circuitry alone - as a purely physiological mechanism - can 'dream up', and transmit to consciousness, highly abstract concepts, revelations, and sentiments *not* available to the neurons through normal sensory channels. It is a mystery leading one to legitimately ask: *'What the hell are the neurons up to... and why do we have such a plenitude?* Which, after all, is only another way of asking the question posed by Bronowski.

At this point it seems reasonable enough to wonder if the mental powers of the imagination - particularly at their most inspired and insightful levels - might not result from extrasensory energy forces operating beyond the laws of physics and biology as we know them? Certainly the Greeks suggested that those illuminating powers of mind that 'see' beyond the temporal and material facts of life; emanated from psychical resources that transcend physiological brain function, namely the soul. Many years ago in correspondence with the Greek scholar, poet, and novelist Robert Graves, I asked him about how he viewed the Greek concept of the psyche and the role played by the soul in originating our creative powers. His reply was humorously dismissive of such 'Platonic' speculation. He simply responded by saying that he thought 'psyche'.... *was a very strange way to spell 'fish'.*

In one way or another, every chapter in this book is concerned with the highly complex and profound proposition that soul and spirit, mind and brain, work synergistically in a psychical and physiological partnership to empower human consciousness. Consequently, distinctions between soul and spirit, and brain and mind, are discussed throughout the following pages. In Chapter 4, *The Compass Points of Consciousness*, these objective and subjective aspects of consciousness are presented diagrammatically in a way which helps to make the process more comprehensible. The illustration plots the courses we take in mentally navigating through life - in a way similar to the directional markings given on a compass card.

In 1908 the First International Congress of Psychoanalysis met in Vienna. It marked the medical profession's first coordinated attempt to seriously examine the nature of mental life - how to explain a healthy mentality as distinct from a dysfunctional one. The two most notable founding doctors were Sigmund Freud and Carl Gustav Jung. While both men worked in accord at the outset, their views on the nature of psychological types and the vital life forces responsible for our individuality soon diverged significantly, and in ways we shall later discuss. The term 'psychology' was applied to this newly structured and recognized branch of medicine dealing with the individual mind and its intellectual and emotional processes. It is a compound word combining the Greek *psycho* (breath, mind, spirit, soul) with the Greek *logy* (theory, or science of…). Thus, in a literal sense, the Greeks would understand psychologists to be 'soul-mind' scientists.

As the profession expanded throughout the 20th century there were those psychologists who followed the Greek *dualistic* approach. They set out to discover what was out of balance in the partnership between the inner workings of the patient's mind in terms of its spiritual and inspired imaginative energies…. and consciousness' sensory involvement with the outside world tempting the indulgence of physical appetites and ego. This approach to mental illness required lengthy analytical sessions 'on the couch'…. as the layman describes it. Whereas in the competing school of psychological thought practitioners left soul - or what may be described as spiritual health - out of it, and looked for 'mechanical' problems: neural malfunctions in the brain which could be treated by drugs that render a patient capable of dealing with life.

Now while the Greek philosophers were not privy to the remarkable discoveries of contemporary neuroscience, and could not have produced the systematic theories of a Jung, Freud, or Adler to account for the individualistic nature of human behavior.... the Greek playwrights had an innate and sophisticated understanding of human nature. They wrote comedies, dramas, and tragedies which disclose the potency and authoritativeness of the psyche's inner promptings - or lack thereof. They gave us heroes, moral weaklings, villains... and a variety of psychological types. The Greek hero was the one inwardly strengthened to overcome the harmful strategies of evildoers less well morally equipped; who found the courage to act and overcome impossible odds.... contest the 'slings and arrows of outrageous fortune', as Shakespeare had his hero Hamlet declaim. Yet those possessed of a creative mind, and the artistic gift of rendering its insights in one form or another, were also regarded as cast in the heroic mould.

Some years ago I wrote a book entitled *Art and the Creative Consciousness*, in which I attempt to reveal something of the inner life of painters and sculptors by describing the way they are led back into themselves after some memorable encounter with worldly things or events - mental experiences that are often called 'moments of vision'. It is difficult to explain such transforming, creative exploits in terms of the machine-like, chemical and electrical activity of the brain's neurons, despite their array of billions. Obviously, at times of heightened visual perception artists *are* automatically exercising 'sense' neurons to bring the factual presence of things to consciousness, before utilizing 'skill' neurons to coordinate hand and eye in making drawings, or models conveying the structural nature of what is observed. But the next stage - the process of creative transformation - being highly personal and subjective, is less neurologically intelligible. The deep feelings of empathy, and the sudden realization of the value of these unique encounters with the world that give significance and meaning to life - mental states of mind that must somehow find their way into the drawing or sculpture - suggests neural activity more creative than mechanical.

Back then to the quintessence of Dr. Bronowski's question. How do 'physical means' (neurons) manage to induce a moment of vision such as that experienced by Mozart when he 'heard' a complete symphony in his head while just walking around Vienna? And what kind of neurological activity is going on when the English novelist and essayist D. H. Lawrence, writes as follows about Vincent van Gogh's 'Sunflower' paintings?

> *When van Gogh paints sunflowers, he reveals, or achieves the vivid relation between himself, as man, and the sunflower as sunflower, at that quick moment of time. His painting does not represent the sunflower itself. We shall never know what the sunflower itself is. And the camera will visualize the sunflower far more perfectly than van Gogh can. The vision on the canvas is a third thing, utterly intangible and inexplicable, the offspring of the sunflower itself and van Gogh himself... It is a revelation of the perfected relation, at a certain moment, between a man and a sunflower...**

Just what - in neuroscientific terms - is a 'perfected relation.... between a man and a sunflower'? For obviously something is going on that surpasses straightforward sensory recognition of the sunflower simply as sunflower. For here we have the artist - visual perceptions heightened and passions aroused - transforming the flower's natural appearance by turning its corolla into a cosmic swirl of light and colour. And in so doing expressing some deeply-rooted, unconscious affinity he feels with the beauty and dynamic vibrancy of 'sunflowerness'. It is indeed a visionary moment for neuroscientists to get their teeth into.

So what *are* we to make of a consciousness that can utilize myriads of neurons to remove us temporarily from the present moment of time, and undermine the authority of the senses and reason? Because we all pull out of time's flow, and take to dream and reverie from time to time - an escape which I see as a significant feat on the part of the neurons. And when it comes to considering whether 100 billion neurons does not represent something of an 'overkill', it may be that such a number is necessary if every one of the planet's 6 billion plus people are to be endowed with a totally individualized psyche. Sometimes one wonders if it is not the purpose of this magnificent brain to render us alien beings on our own planet, enabling us to leave it at any time and wander beyond clock-time - a mental phenomenon perhaps occasioned when the latent powers of soul or spirit cause the brain-computer, to go 'walkabout' and transcend its more mechanistic functions. Neither should we forget that this phenomenon

---

* D. H. Lawrence, *Morality and the Novel: Selected Literary Criticism*, ed. Anthony Beal (New York: Viking, 1956), p. 108.

is responsible for the nature of our 'humanness' - our double-life embracing both the mundane and the spiritual. For on the one hand we can be compassionate and creative human beings, diminishing suffering and improving the quality of life; on the other wreaking havoc.... indulging a spiritless, dark and destructive side, mercilessly destroying our own kind and the civilizations we have created. It is a perplexing problem and one that will be discussed throughout the following chapters.

> *To make a prairie it takes a clover and one bee, -*
> *And revery.*
> *The revery alone will do*
> *If bees are few.*
> Emily Dickinson, *Poems*. Pt. ii, No. 97

There are those extraordinarily creative individuals in music, poetry, the visual arts... who need no sensory stimulus - no sunflowers or extraordinary happenings - to trigger the most profound levels of inspiration.

Mozart - a musical genius whose powers defy rational explanation - was frequently subject to the involuntary outbreak of complete symphonies which came unbidden to his 'mind's ear' as he went about his daily round. If we regard this phenomenon as purely a neurological issue then it would seem that veritable legions of neurons suddenly and autonomously 'ignite', firing untold salvos into consciousness - neurons specifically capable of generating the formed structures of sound we know as music. And, lo and behold, the 'Jupiter' symphony is born. Contemporary psychologists tend to see such unwilled musical intrusions as specifically a neurological phenomenon - pointing out that the brains of professional musicians can immediately be recognised as different from those of men and women in other creative fields: the neural bridge between left and right hemispheres for example - the corpus callosum - is enlarged, as is the auditory part of the cortex. Of course the brain and its neural paths must be held responsible for the conscious 'reception' of music - even if from unconscious sources. But one should not overlook the power of music itself. As Shakespeare put it in *Much Ado About Nothing...*'Is it not strange that sheep's guts should hale souls out of men's bodies'? And Mozart, replying to a patron who had asked how he set about composing his music, wrote that, *'All this...* (meaning the sudden takeover of original musical themes) *'fires my soul'*. Or perhaps he would have agreed to put it the other way round and say that it was some uncanny creative power of soul that fired the neurons (Mozart's letter is reproduced later in the chapter.)

But what then are we to say about those men and women who experience a different kind of inner illumination - no less transforming and imaginative than 'instant' music - that moulds character, inculcating an unconquerable spirit of idealism? A moral and aesthetic force no less, that inspires a high order of moral certainty: committing one to the principles of right and wrong; as well as inducing an awareness of the sublimely beautiful aspects of life, and the spiritual overtones that attend them.

Which brings me to the remarkable moral strength displayed by Hans and Sophie Scholl - brother and sister - and students at the University of Munich during the time of Hitler's Third Reich. These two young people displayed incredible courage in keeping faith with their moral convictions. And so steadfast was their commitment that one has to ask (as with the revelatory nature of Mozartian inspiration) how the chemical and electrical activity of neurons alone could be responsible for delivering such profound moral certitudes. They felt so deeply about the moral injustices, the lack of compassion and total disregard for the sanctity of life displayed by the Nazi regime... that they gave their lives in attempting to counter it. And in the process attained a level of wisdom and strength of spirit beyond their years, remaining faithful to the high ideals which illuminated their life to the bitter end. If one accepts the view of the Greek philosophers that the soul is the mainstay of the psyche, then the lives of Hans and Sophie Scholl exemplified the spiritual authority possessed by truly mature souls.

Yet Mozart and the Scholls shared something in common - the quickening of spirit brought about by music. For the greatest pleasure of the Scholl's was to play music *en famille:* it provided both a constant form of aesthetic expression and spiritual communion. And it is evident from reading Hans' and Sophie's letters that this deep feeling for music greatly influenced their humanistic outlook. Sophie in particular was the most innocent and gentle of women, easily transported by nature, music, and love of family into poetic reveries that verge on the mystical. Both she and her brother seem to have been possessed of a continuously growing spirit of wisdom and conscience from a very early age - innate sensibilities nurtured by a father and mother who were wonderfully sensitive human beings in their own right.

I find the tragic, yet heroic story of Hans and Sophie Scholl to be a moving testimony to a transcendent quality of humanness - one to which their Nazi judges were totally blind.

We get an amazing account from Mozart's letters of how a symphony dreams itself into existence. The letters written by Hans and Sophie Scholl likewise reveal an inner life rich in ideals and metaphysical speculation. After reading Mozart's explanation of how he composed, and the extracts from Sophie's letters with which we conclude this chapter, you might agree with the Roman poet Ovid when he penned the following sentence in his *Metamorphoses*: *'Those things that nature denied to human sight, she revealed to the eyes of the soul.'*

Let us go back for a moment to Gretel Ehrlich's informative and thought-provoking account of neural function. She writes: *'An electrical impulse travels down an axon - the transmitting arm of a cell - and when it arrives at the axon terminal, certain chemicals called neurotransmitters are released from a gated surface ... '* This description gives one some idea of how devastating a lightning bolt's surge of electrical energy can be on *'the body electrochemical'*. But what if we exchange the concept of a lightning bolt-strike for that of an 'inspiration-bolt' strike? We can identify the kind of energy responsible for a lightning strike that can overload the body's electrical circuits and 'fry' neurons and trailing dendrites. But do we know the source and the kind of energy responsible for an inspiration strike? Can it be thought of as a 'power' surge involving billions of neurons, called into action to stimulate consciousness to high levels of creative thought and action? In which case is the source of such a surge to be found within the neurons themselves, or does it result from the external radiation-like activity of some undetectable energy-intelligence - something like a discrete, quantum-like force operating below the line of traceable action? Or perhaps Ovid got it right two thousand years ago when he goes beyond physical experiences of the world to metaphysical insights that occupy the mind.

The extrasensory nature of all highly creative achievements is well conveyed by Plato in his dialogue *Ion*.

For the poet is a light and winged and holy thing, and there is no invention in him until he has been inspired and is out of his senses... when he has not attained to this state he is powerless and is unable to utter his oracles.

Let us assume that when Plato talks about *'the poet'* in the quotation above, he uses it as an all-embracing term that takes in all sorts of people who, like Mozart and the Scholls, move beyond their involvement with the world and wait for the promptings that come from within. Michelangelo writes in the same vein as Plato when he asserts that music is not a matter of the ear, but of the soul. And a famed 20th century American poet wrote:

Sweet sounds, oh, beautiful music, do not cease!
Reject me not into the world again.
With you alone is excellence and peace,
Mankind made plausible, his purpose plain.
Edna St. Vincent Millay: *On Hearing a Symphony of Beethoven*

'*Reject me not into the world again...*' - a plea which countless lovers of music will understand well enough. In my own case, whenever I listen to the *Organ Symphony* of Camille Saint-Saens, I lose awareness of time and place; even of myself as an entity. As the bass pipes of Chartres' grand organ (the solo instrument on my recording) begin to thunderously reiterate the musical leitmotif of the work, I find myself wrapped in a great calm: all mental activity ceased - no more questions; all bodily sensations closed down; gravity lost.

So what is happening to the neurons on such occasions? Are they responding to an inspiration-bolt strike, the poet being *'inspired'* and *'out of his (her) senses'*?

Mozart would seem to have undergone a similar kind of metamorphosis when composing, save for the fact that he was not quite so lost to time and place. One of the most extraordinary statements concerning the onset of an inspirational and completely unwilled state of mind – signifying its quite unconscious genesis – comes from Mozart's pen.

It is found in a letter written by the composer to an admirer – a local baron who had presented him with a case of wine and inquired about his method of composing:

When I am, as it were, completely myself, entirely alone, and of good cheer – say, travelling in a carriage, or walking after a good meal, or during the night when I cannot sleep; it is on such occasions that my ideas flow best and most abundantly. <u>Whence</u> and <u>how</u> they come, I know not; nor can I force them. Those ideas that please me I retain in memory, and am accustomed, as I have been told, to hum them to myself. If I continue in this way, it soon occurs to me how I may turn this or that morsel to account, so as to make a good dish of it, that is to say, agreeably to the rules of counterpoint, to the peculiarities of the various instruments, etc.

All this fires my soul, and, provided I am not disturbed, my subject enlarges itself, becomes methodized and defined, and the whole, though it be long, stands almost complete and finished in my mind, so that I can survey it, like a fine picture or a beautiful statue, at a glance. Nor do I hear in my imagination the parts <u>successively</u>, but I hear them, as it were, all at once. What a delight this is I cannot tell! All this inventing, this producing, takes place in a pleasing lively dream. Still the actual hearing of the <u>tout ensemble</u> is after all the best. What has been thus produced I do not easily forget, and this perhaps is the best gift I have my Divine Maker to thank for.

When I proceed to write down my ideas, I take out of the bag of my memory, if I may use that phrase, what has been previously collected into it in the way I have mentioned. For this reason, the committing to paper is done quickly enough, for everything is, as I said before, already finished; and it rarely differs on paper from what it was in my imagination. At this occupation I can therefore suffer myself to be disturbed; for whatever may be going on around me, I write, and even talk, but only of fowls and geese, or of Gretel or Barbel, or some such matters. But why my productions take from my hand that particular form and style that makes them <u>Mozartish</u>, and different from the works of other composers, is probably owing to the same cause which renders my nose so large or so aquiline, or, in short, makes it Mozart's, and different from those of other people. For I really do not study or aim at any originality.*

So there you have it: an amazing case of spontaneous mental combustion. And one which intrigues when it comes to wondering how the neurons were triggered, and what proportion of 100 billion were involved in order for the composer to hear – '*in my imagination,*' as he says – a complete symphony. In any event, such a prodigious creative feat certainly provides food for thought - testifies to the uncanny presence of an inspired

---

* Hans Mersmann, <u>Letters of Wolfgang Amadeus Mozart</u> (New York: Dover Publications, Inc., 1972), p. vii.

imaginative force capable of taking over consciousness, and without sound or sight of the outside world getting in its way.

It is important to note that Mozart did not hear the parts *'successively'*, but *'all at once'*. In other words, this inner hearing did not follow the linear time sequence in which sounds are usually structured by consciousness. Music which would take 45 minutes to be performed and heard when written down was experienced in its entirety in the composer's imagination without having to travel the normal highway of linear time. The great doctor and philosopher, C. G. Jung, would have no doubt suggested that when Mozart talks about all his *'inventing'* taking place in a *'pleasing lively dream'*, the composer was experiencing a 'constellation of the unconscious'. (Jung's terminology for a usurping of routine mental processes by a subliminal and autonomous part of the self: the unconscious, seen as an innate source of instant enlightenment delivered by the faithful neurons through the gate of intuition. Jung's thoughts on this phenomenon are outlined in Chapter 4, *The Chart of Consciousness*.)

The fact that Mozart writes, *'All this fires my soul...'* also supports Jung's theory that such creative 'dreams' radiating from the unconscious possess a numinous quality:* that is to say, are perceived as mental visitations which bring with them a certain mood of exhilaration, a feeling of satisfaction that one has come home to oneself in spirit, as it were; reached a level of awareness beyond the temporal and biological.

> *Whether angels play only Bach in praising God, I am not sure. I am sure, however, that en famille they play Mozart.*
> Karl Barth, quoted in *The New York Times* obituary, Dec. 11, 1968

After reading Mozart's letters telling of the up-rush of his muse, we can see what Plato was getting at when he writes *'there is no invention in him until he has been inspired and is out of his senses...'* Many personal statements by artist and scientist alike – although not describing such a total strike of inspiration as that experienced by Mozart – tell of the senses losing their hold on consciousness, of being but little aware of their

---

* Lest the reader think I am getting carried away here and ignoring the dark and destructive forces of human nature – compulsions which I do not consider to be forces of the true unconscious – I should mention that they are taken up in Chapter 12: CONSCIENCE and MORALITY - CHARACTER - FOR GOOD or ILL.

surroundings, or of the passing of time, when the creative imagination is at full flood.

For example: who amongst us has never stretched out in a hot bath, eyes closed, drifting, thinking of nothing in particular... and been beguiled by the unexpected arrival of words floating around in the mind - word-thoughts setting unfamiliar images and feelings in motion which sometimes provide the key to solving particularly difficult problems, or will set one thinking along new and significant lines. *'Surrender the World to the Bathtub'* would be my motto of choice - in Latin, of course - if one were to design a personal coat of arms. Even when taking a walk, the hypnotic rhythm of stride and pace can lull the senses into slacking off for a while, thus allowing the most intuitive and innovative of thoughts to slip through into consciousness. In fact, the best way sometimes to solve a persistent problem is not to try and rationally think it through, but forget it for a while, put it in storage; meanwhile take a bath or a walk and wait to be surprised. But it is during sleep, when the senses are pretty much out of it, that reason's guard is seriously off-duty, and compelling dream-images can break through which, on occasion, exercise a haunting influence. These are dreams powerful enough to persist after waking, linger on through the day continuing to occupy our thoughts, and challenging us to glean a hint of what we feel must be their meaning. An ongoing pattern of such dreams represents an important psychical experience, and good psychiatric help should be sought to interpret their overall symbolic significance. Sometimes their message and influence is sufficiently significant to demand drastic changes be made in one's way of life.

Mozart's *'pleasing, lively dream'* is obviously in a category by itself. It came about without apparent forethought, totally autonomously, overriding his involvement with whatever was going on existentially at the time. As a method of composing it remains something of a mystery – the kind of arcane phenomenon which would support Albert Einstein in the belief, expressed in *What I Believe* that:

> *The most beautiful thing we can experience is the mysterious. It is the source of all true art and science.*

There is a series of rock-cut temples at Ellora in west central India that is both beautiful and mysterious. Carved between the 6th and 11th centuries AD from a great stretch of basalt cliff, their high and low-relief sculptured figures are marvels of stone-cutting technique, and their expressive beauty - stirring enough nowadays even to Western eyes - must have been without rival in the period when the Hindu Gods were holding full sway. And the mystery arises when the master sculptor stepped down at the end of his life after thirty years of work, stood back, stared in amazement at the extent of his handiwork and is reputed to have exclaimed, *"Did I do that... when did I do all that?"* A long time to be inspired and *'out of one's senses'* wouldn't you say?

Mozart, with his innate and total passion for music, must have lived very much 'on the edge', so to speak - treading the narrow path between inner and outer worlds, so close to the inner source of his inspiration. A factor which left him little choice as to how he should live his life: at the beck and call of his own indwelling genius.

What a paradox! A flesh and bone creature, a biologically short-lived member of the animal kingdom who, nevertheless, hears from within himself a complete musical composition, delivered instantaneously and suspending the normal linear flow of time – living a life which was so frequently, as the mathematician André Weil put it, in a *state of lucid exaltation'*.

> *Every mathematician worthy of the name has experienced ... the state of lucid exaltation in which one thought succeeds another as if miraculously... this feeling may last for hours at a time, even for days. Once you have experienced it, you are eager to repeat it, but are unable to... unless perhaps by dogged work.*
> André Weil, *The Apprenticeship of a Mathematician.*

Lest one doubt that music is capable – as Michelangelo implied – of touching the deeper levels of one's being, then recent information concerning its power to alleviate stress and promote mental healing in those psychologically disturbed, should provide even the dourest of skeptics and materialists with food for thought. Hospitals which have introduced 'musical therapy' into the lives of children suffering from cancer have noted positive results. The distressing side effects of chemotherapy are mitigated – in part because the patients themselves seem to become more psychologically capable of dealing with them, more self-assured and upbeat.

There is a general rise of spirits all round, which is manifest not only in the children's demeanor, but in physical improvements in their condition. Saliva tests performed after the musical sessions show an increase in levels of the antibody immunoglobulin, thus indicating a boosting in a child's immune system. All of which confirms what many of us have experienced – that music, the result of rhythmically ordering sounds and tones in succession, has the ability to go beyond the purely sensory pleasure afforded by hearing itself, and induce responses deep within the psyche: reactions which can evoke a general sense of wellbeing, give rise to a positive attitude, recharge determination and willpower, and restore faith and peace of mind.

As Shelley put it in *Prometheus Unbound*: 'And music lifted up the listening spirit/Until it walked, exempt from mortal care/Godlike, o'er the clear billows of sweet sound.

And so we finally come to Hans and Sophie Scholl.

> *I've just been playing the Trout Quintet on the phonograph. Listening to the andantino makes me want to be a trout myself. You can't help rejoicing and laughing, however moved or sad at heart you feel, when you see the springtime clouds in the sky and the budding branches sway, stirred by the wind, in the bright young sunlight. I'm so much looking forward to the spring again. In that piece of Schubert's you can positively feel and smell the breezes and scents and hear the birds and the whole creation cry out for joy. And when the piano repeats the theme like cool, clear, sparkling water – oh, it's sheer enchantment....*
>
> *Let me hear from you soon.**
> A letter from Sophie Scholl (to Lisa Remppis), written at Munich, February 17, 1943.

Sophie Scholl, born May 9, 1921; was arrested by the Gestapo on February 18, one day after writing the above letter, and was sent – together with her brother Hans – to the guillotine at Stadelheim Prison, Munich, on February 22. The execution took place in the late afternoon on the very day they were sentenced to death by the so-called People's Court. Hans Scholl was born September 22, 1918. He died some seven months short of his 25th birthday.

---

* Inge Jens, <u>At the Heart of the White Rose: Letters and Diaries of Hans and Sophie Scholl</u> (New York: Harper & Row, 1987).

I have read many histories of martyrdom in my time, but none quite so heartbreaking as that of Sophie and Hans Scholl. They were members of a family which included two other daughters and another son, Werner, who was killed on the Russian Front. Their father was a most humane, learned, and gentle man, who as a pacifist had refused service in the German Army during World War I. Their mother emerges from Hans' and Sophie's letters as a totally unselfish and saintly character who, despite poor health, gave herself wholeheartedly to her children's welfare - always loving, always thinking of others. Together, they provided their children with the most cultured and stimulating of environments: a home in which literature, music, the visual arts, philosophy... and a naturally religious love of nature – particularly mountains – were matters of every day life. Sophie writes frequently of her love for music – how it would lift her up when spirits were low, and of the pleasure she derived from playing the piano and church organ. She was skilled in the graphic arts, and was asked to do the drawings for a projected new edition of J.M. Barrie's *Peter Pan*, some of which illustrations survived the war. Although less the scholar than Hans, Sophie was besieged by constant directives from her own intuitive resources, leading her to live a life intense in feeling and searching in thought - and possessed of a spirit constantly drawing her to see life as revelatory, evoking feelings of a naturally religious nature.

Inevitably, the clash of such sentiments with the totalitarian, atheistic, and racist philosophy of the Nazi regime under which they all lived, created a constant challenge to her faith - made it continuously necessary to question and reaffirm her attitudes and beliefs. Unlike Hans, she had little opportunity to find the support for such an internal struggle that can come through discussion with like-minded friends; or know the respite from soul-searching that an be found by participation in some active and ongoing undertaking such as training for a profession or vocation. Whereas Hans had his medical studies and, as a medical orderly in the German army occupying France found a great deal of support for his humanitarian views in his first encounter with the works of French philosophers and poets.

Dear Sophie was very much alone in her introspective dialogue: searching for a God who, despite her entreaties, was never there to provide either answers or comfort. Over and over in the letters one reads of her need to feel God's presence – for in the end, all she believed and stood for resolved itself into a search for spiritual truth. Yet even though her prayers were made to a silent and unapproachable God, she never lost faith in the spiritual authority of her ideals and aspirations – a faith which was tested so cruelly, and with such tragic results, by the barbaric and inhumane system under which she lived. One entry in her diary reads, *'I'd so much like to believe that I can acquire strength through prayer. I can't achieve anything by myself'*. And another, *'I've decided to pray in church every day, so God won't forsake me. Although I don't yet know God and feel sure my conception of him is utterly false, he'll forgive me if I ask him'*. And what humility is revealed in the following line: *'I pray for a compassionate heart, for how else could I love'*?

Her brother was possessed of a far more cosmopolitan character. A voracious reader of contemporary German humanist writing – Rilke was his favorite poet, as well as such classic authors as Schiller and Goethe – Hans Scholl was a young man whose intellect was finely tuned to literature, both as a means of learning and the most effective means of self-expression. During the time he spent in France he was greatly influenced by French writers and poets such as Claudel, Gilson, and Verlaine (who outdid Rilke, in his opinion) – earnestly learning the language so that he could read them in their mother tongue.

Yet having said this, it must be emphasized that his innermost character was almost a mirror image of Sophie's own: a strong feeling for the wonder of life, a belief in the beneficent influence of love, a commitment to human freedom, and a preoccupation with the spiritual battle constantly waged between what he saw as the forces of good and evil: moral and philosophical attitudes which could not stay as philosophical theories, but which had to be affirmed by action.

Hans Scholl was a man who, as his brief life reveals, led by example. He came to study medicine as a career not out of any strong sense of vocation, but because he considered it the most effective way to contribute to the well being of mankind. The impression received from reading his epistles to Sophie, his parents, and his girlfriend – all written between May 4, 1937 and February 16, 1943 – is that of a young man who never did anything by halves. In the first of his letters appearing in *At the Heart of the White Rose*, and written to his mother while doing his compulsory time in the State Labor Service in 1937, Hans states, *"I've changed a bit, I suppose. Inwardly and outwardly. It doesn't mean I've renounced my old principles and perceptions. I've taken another step up the ladder. This place is a mine of experience. I'm putting my heart and soul into my work, believe me – I never shirk...* His final letter to Rose Nagele, written two days before he and Sophie were arrested at Munich University after being seen by a university janitor dropping handbills against the Nazi regime into a lower courtyard, and just six days before their execution.... reveals the distance he has come on his life's journey:

> *Dear Rose,*
> *Your last letter saddened me. I saw your tears shine through the words as I read it, and I can't dry them. Why write that way? Although I live in a permanent transition from yesterday to today and tomorrow, the beauty of the past remains intact and no less beautiful. That bygone summer still reflects the light into the present. Must its radiance be extinguished by shades of melancholy?*
> *Nowadays I have to be the way I am. I'm remote from you both inwardly and outwardly, but never estranged. Never has my respect for your purity of heart been greater than it is now, when life has become an ever-present danger. But because the danger is of my own choosing, I must head for my chosen destination freely and without any ties. I've gone astray many times, I know. Chasms yawn and darkest night envelops my questing heart, but I press on regardless. As Claudel so splendidly puts it: <u>La vie, c'est une grande aventure vers la lumiere</u>.**
> *It might be better in the future if the sentiments conveyed in our letters owed less to emotion and more to reason. I'd welcome another letter from you soon.*
> *Affectionate regards,*
Hans

Taken from his cell in the Palais Wittelsbach on that black afternoon of February 22, Hans Scholl's last words before execution were: *Es lebe die Freiheit.* (Long live freedom). 'Whom the Gods love, dies young', wrote Menander the Athenian playwright - a famous line from *The Double Deceiver*. It is a view expressed by many other writers, none better in my view than Wordsworth in *The Excursion*.

> *The good die first,*
> *And they whose hearts are dry as summer dust*
> *Burn to the socket.*

---

* "Life is a great adventure toward the light."

Here, I would suggest, is the epitaph for the headstones of Hans' and Sophie Scholl's graves.

In the course of their brief lives, Hans and Sophie Scholl reached a level of being which went far beyond the compelling drives of ego, or the powerful urge to physically survive at all costs. When the danger was most apparent they seemed able to draw on great spiritual and moral strength. The power we think of as spirit. In the end, the truths they held dear, the sensibilities they had formed over the years, could not be ground under the jackboot of an informer-ridden and vicious regime.

Mozart would seem to have had no control over his musical muse; and the Scholls certainly seemed to be swept along by their moral one. And the fact that such *absolutely* creative and virtuous drives may be few and far between and are so incredibly compelling, overcoming even the tide of reason, renders them inscrutable and mysterious - certainly if one looks to neurological mechanics for explanation. Long before the days of neuroscience, Lucius Annaeus Seneca (4 B.C.? - 65 A.D.), Roman statesman and philosopher, makes short work of this mental enigma when he writes in *Epistulae ad Lucilium*: 'Why, then, is a wise man great? Because he has a great soul'. And much closer to our own time an American poet provides a similar answer by referring to the soul* as the metaphysical force responsible for ideas and feelings that evoke spiritual overtones:

> *Great truths are portions of the soul of man;*
> *Great souls are portions of eternity.*
> J.R. Lowell (1818-1891) Sonnets, No. vi.

It is now some 55 years since the Scholls were executed, and I cannot but wonder how many young people today would find themselves in accord with the sentiments expressed in their letters – would identify with the philosophical and spiritual beliefs of these two young Germans. How many would comprehend the selflessness, the compulsion to risk all for the sake of conscience and to keep faith with an ascendancy of spirit? And if it is suggested to postwar generations that developing intellectual and practical competence in order to compete in the economic marketplace, is only part of the battle on the road to fulfillment in life – that to cultivate

---

* See Chapter 19, *A Pyramid of Souls?*

the values and ideas which spring from discovering the nature of love, truth, beauty, goodness... constitutes the other side of the learning coin – will they smile a little pityingly at such romantic naiveté? Or would some perhaps crease their brows, wondering at the alchemy of neurons - or neurons *plus* soul (more difficult and old fashioned...) which determined the profoundly human character of a Hans and Sophie Scholl?

Some 80 years ago, the Irish painter, John Butler Yeats, wrote the following letter to his son, the poet W. B. Yeats.

> *My dear Willie, People have an idea that poets live in disorder, and feed upon it. If they feed upon it, it is only that the pain of disorder sets their intellects to work in bringing to it order, so that in the ensuing silence and quiet, they might have leave to sing. The poet is an orderly man, because he allows no single feeling to remain single, forcing it into harmonious relation with all the other feelings. He is a whole man, whereas others are only sectional, his anger that of a whole man, his love that of a whole man, his inner life crowded (and) complex.*

Mozart had *'leave to sing'*. Hans and Sophie Scholl knew the *'silence and quiet'* in which they found their wholeness.

There is now some recognition in Germany of the sacrifices made by members of the White Rose. A school and street have been named after Hans and Sophie Scholl. The six graves of those who were executed in Munich - four other members of the group followed Hans and Sophie to their deaths - are now places of pilgrimage. Many other arrests were made by the Gestapo of members belonging to other university cells, eleven of whom were killed - forced to commit suicide, or died in concentration camps.

3

THE NEURONS – ORDER & ENIGMA:
THE STRANGE AFFAIR ON EASTER ISLAND

*There are more things in heaven and earth, Horatio,
Than are dreamt of in your philosophy.*
Shakespeare, <u>Hamlet</u> I, v.cs

Fifty or sixty years ago any introductory course in psychology would drill into the student three magic words: *cognition, affection, conation* – signifying the three-part mental process accounting for human awareness and behavior. Cognition involves recognition of things in the world which are perceived through the senses, together with the recall of everything that is known about them. Affection is the type and quality of feeling which inevitably accompanies an act of cognition – the attraction, fear, curiosity, etc., aroused by the presence of the thing recognized. Conation denotes action – the behavior of the individual in response to the two preceding stages of knowing and feeling. And it was further emphasized that these three steps, working together as one, constitute what is called an 'act of perception.'

I well remember the illustration given. It went like this:

> *You are about to step off the sidewalk and cross the road as a car is bearing down. You recognize it as such, registering its color, size, make… estimate its speed and the distance between yourself and it: (the cognitive factor). As a result of this information you may feel confident, apprehensive, or even fearful about stepping into the road: (the affective factor). The way you feel, and the emotional intensity involved, will determine whether you stroll across the street, hurry across, or run like a scalded cat: (the conation factor).*

This is a very simple example of how the neuron-complex serves consciousness when our sensory systems are conveying the material facts of what is going on outside ourselves, and inciting ideational and feeling responses to it. But now I will recount a story which tells of knowledge gained without benefit of any of the objective, perceptual processes I have mentioned – where the only man who learned the true facts concerning a tragic situation did so many hours after they had happened, and while at home several miles away. Consequently, no normal act of objective perception was involved, and no explanation can be given which would satisfy a scientific criteria in attempting to account for it. Yet my reader will recognize a familiar aspect to this story – something akin to Mozart's brainstorms; those inspired moments when the neurons were performing their wizardry independently of any participation by the senses: a brain-phenomenon which, to say the least, represents something of an enigma – an anarchy of neurons where such unusual mental activity cannot be seen as attributable to an objective act of perception experienced at the time an event is taking place.

The dream of an Easter Island shaman – which changed the course of normal procedures on the island while I was there – bears out the brain's ability not only to present the evident impressions of external facts which occupy the waking day, but also to inform through the visionary images incurred during sleep. The history I am about to relate will likely be more difficult to accept or comprehend than the Mozart phenomenon – for the composer does at least inject a semblance of reality into his account of how inspiration strikes by telling us where he is, and what he is doing at the time. But in the case of the shaman's dream, we are not even dealing with a man who is awake, on his feet, going the rounds of his daily life. The dreamer is lost to time and events, perceptual processes more or less shut down.

It is generally considered that images received in the dream state can only be interpreted symbolically, not literally. Yet there is nothing symbolic about the imagery of the Easter Island dream: the events depicted turned out to have been as factual as if the shaman had personally witnessed the drama which had taken place on the island the day prior to the night of the dream. And so in contemplating both Mozart's creative frenzies, and the dream of the shaman elder, we face mental transactions where it can be said that the neurons are functioning lawlessly, 'doing their own thing' in the absence of any sensory stimulus. And in this dreamer's case, there could be no repressed memory of earlier witnessed events lurking subliminally, ready to break out at the appropriate moment when the reason-seeking processes of the workaday consciousness go off duty.

One might well ask, 'What the hell are the neurons up to? How do they manage to serve unknown prompters from beyond reach of the senses?'

The importance of the shaman's vision – (which is a 'big' dream rather than a 'little' dream; the significant differences between the two will be discussed in a later chapter) - is that it reinforces the proposition that such experiences which lie outside the boundaries of normal acts of perception deserve to be taken seriously; regarded as a form of extrasensory enlightenment which may be seen as complementary to the knowledge gained from a normal objective consciousness.

THE DREAM

Easter Island – sometimes referred to as the island at the center of the world, and certainly the most isolated spot of land in the Pacific – lies in the midst of truculent seas at the extreme easterly limits of the Eastern Polynesian Triangle. Steaming west from Puerto Montt on Chile's Pacific coast it took our little ship eight days to cross 2,300 miles of high-swelling ocean and reach the island. Had we continued westwards, the next landfall would have been Pitcairn Island, 1,400 miles away.

The seas around Easter Island lay constant siege to the place with the surging water continuously rising and falling from 10 to 20 feet at the base of the cliffs. There are no natural harbors, and few bays where small fishing craft and even the maneuverable Zodiac can find safe anchorage or be easily beached. Even small ocean-going vessels must stand offshore while passengers – 12 or so at a time – brave the formidable swell in a Zodiac or longboat in order to reach land. Approaching from the sea, fighting a quarter-mile of rising and falling white water, the island is not a reassuring sight (when one does manage to catch a glimpse of it between cresting waves, that is); particularly as the Zodiac is heading for what appears to be a scrap of sand about the size of a pocket handkerchief - a precarious landfall guarded by a small reef through which the driver must skillfully negotiate his way. High rock cliffs form the coastline: bluffs which, in the vicinity of the three volcanoes marking the three 'corners' of the island, are formed of a sinister looking black lava flow. The highest volcanic caldera has an elevation of 1,700 feet; the island itself is about 14 miles long by 7 wide.

Few trees have survived the ravages of Polynesian occupation. The profligate cutting to clear the land for agricultural purposes, in order to support a population constantly increasing since the early colonizing voyages of prehistoric times – as well as to provide fuel and material for boat building – has left the island a barren place of grass, rock, and rock-encrusting lava. In a word, it is a grim spot: and not solely from its bleak aspect and the overwhelming presence of a limitless surrounding ocean, but because there is a sadness to it – as if the constant fratricidal conflicts between one faction and another during the island's history has left a legacy of sorrow in the air.

The principal attractions of historic interest are the sites where the famous Easter Island heads are to be found: rows of giant heads on shortened torsos standing on altar-like platforms (known as <u>ahu</u>) which surround the coast. Recumbent individual heads dot the island, particularly in the immediate surrounds of the hillside where the stone was quarried – some blocks weighing as much as 60 tons: sculptures which apparently never reached their destination. Many theories exist among archaeologists as to how these huge loads were transported across the island, but none seem to have been accepted as the definitive explanation. Add to the grimness and the sadness a third and equally pervading atmosphere – that of mystery: for whether these long-nosed monoliths are standing upright on their platforms and staring sightlessly into the distance, or lying grounded in quarry or grassy hillside, they evoke the occult secrets of long gone rituals and beliefs. And then, intensifying the island's moods, there is always the weather: the suddenly rising wind bringing heavy, low-lying and dark squall clouds seemingly out of nowhere, through which shafts of light would switch on and off to illuminate patches of the newly dark landscape like searchlights going the wrong way. Sheets of water will suddenly drive against the face – mini-showers arriving from God knows where, or airborne spray from the sea – obscuring vision and making it all too easy to become disoriented in this macabre terrain. I have seen it like night at 3 o'clock in the afternoon, only the 'searchlights' providing hope for the return of the sun.

One incongruous intrusion breaks the spell cast by the island's total isolation, dramatic history, and stark geological aspect. An aircraft runway, close to being the longest in the world, cuts across the southwestern extremity of the land to actually reach the sea on the northern coast. It was built by the United States to serve as an emergency landing strip for the space shuttle, should weather conditions in the Northern Hemisphere demand a diversion. But Chile and its national airline, Lan Chile, have come to benefit most from its construction. Situated at a logistically strategic point in the wastes of the South Pacific, Easter Island and its runway provide a most advantageous staging post for Lan Chile on their long flights from South America across the ocean to Tahiti.

This brings us to my Easter Island tale, for it concerns a Lan Chile employee working at the airport, and the fate which befell him.

I was to leave the island on the Saturday morning flight to Santiago. On the Friday evening prior to departure, while sitting in the bar lounge of the Iorana Hotel, I was introduced by the charming proprietress to Benito Rapahango. Mr. Rapahango had served in the Chilean Navy and was now a government official on the island. An articulate and well-traveled man who put one at ease immediately, he had ancestral ties to Easter Island, and had an encyclopedic knowledge of the place and its history.

At one point I asked him if anything remained of traditional Polynesian mysticism and religious beliefs after so many foreign incursions, political and natural disasters, and the influence of Christian missions.

"I'll tell you a story that might answer your question," he said finally, after a long silence spent considering his response.

"Today is Friday. On Tuesday, just three days ago, there was an accident on the island. A man was killed." He paused for a moment, then turning his head sharply to face me, asked, *"Have you been onto the caldera of Rano Kau - walked out on the ridge to where there's a drop of 1000 feet into the sea on your right, and one on your left of about 300 down to the crater lake?"*

"Yes, I went up there yesterday. But I must tell you... I didn't go very far along. That ridge gets awfully narrow after a few yards – and it is a hell of a drop down to the sea; and not much better on the crater side. I doubt you'd survive either fall. And it's windy up there – difficult to keep your balance. I'm not much good on heights nowadays: get quite disoriented – lose sense of position, not sure where the feet are – then find it difficult to move forwards or backwards with any certainty. But I did get far enough out to see down both sides. Found it a pretty dangerous and scary place and backed-up pretty quickly."

"Well," Benito continued, "that's where it happened. The accident. Ira wasn't a close friend of mine, but everyone knows everyone else on the island. He worked for Lan Chile. Looked after their catering services here." He paused for a moment as if deciding how to organize the order of events he was about to relate.

"I'll tell you what happened. Now this was Tuesday – three days ago, don't forget. Ira had the day off and went out into the country with his girlfriend. He was married... but everyone knew about the affair. Anyway, his car was found at Rano Kau in the parking place. Apparently they set off up the trail to the caldera just before lunchtime – she said later she was carrying their food in a shoulder bag. According to a couple who saw them in the car park, it seems they were having a bit of an argument when they arrived – finally agreed to walk north along the cliff where those old below-ground dwellings have been excavated – rolling headland country there; the only terrain where you can sit and eat in any comfort and safety."

But then at about 4 o'clock in the afternoon the girl turned up alone in the township. She was in a very distressed state after walking the three miles back from Rano Kau – cross-country, it was thought. She was too hysterical to say. But she was badly scratched and banged up; also had likely cracked an ankle. At the police station she reported that Ira had lost his footing and fallen from the high caldera ridge into the sea."

But apparently it was common knowledge that Ira and his wife had had a fierce row that morning, and that he had promised to break off the relationship with the girl. It was also common knowledge that Ira had promised the girl that he would, one day soon, marry her. Well, the police constable put two and two together and arrested the girl on suspicion of murder: his case being that Ira had told her he could not marry her, and that she had then decided on revenge – suggested they go out on the caldera for the view, where she had pushed him off the edge. The overwrought girl completely broke down at this. She was clearly in a bad way – totally exhausted and inarticulate, out of her mind, you'd say – and was kept in police custody."

In the meantime they had been searching for Ira's body in the sea between Orongo and the rock island of Moto Nui – you know, where each year the eggs of the sooty tern would be found, and the swimmers would take part in that old ritual of racing back with one, hoping to be first in bringing the beneficent bird-spirit to the mainland for the coming season. But it's swift water out there and on Tuesday the searchers found no body. Yet they found the car where she said it would be, and brought it back to the township."

Here Benito paused – stared at the wall for a moment – swing round suddenly in his chair to look straight into my eyes.

"*Now I'll <u>really</u> answer your question by telling you what happened next. That night an old man in the village who has been highly respected for his shamanistic powers for as long as I can remember, had a dream. He dreamt that the spirit of the dead man came to him and told him that he, the shaman, must go to the magistrate at first light to say the girl was innocent. He must tell the magistrate that his, Ira's, death was the result of his own stupidity: that he was so upset at the thought of losing the girl that he had been drinking heavily before setting out to Rano Kau for the picnic; that he went out onto the caldera ridge alone, despite her pleas not to do so, and lost his footing – fell to the sea. And if the police would look under the driver's seat of the car they would find the empty whiskey bottle. Furthermore, said the spirit of the dead man, they should not be searching for his body in the waters off Orongo, but about two miles to the north in the bay beneath the <u>ahu</u> at Ana Kai Tangata. There they would find him – all but his head – for he had been decapitated in the fall. The head would be found at the base of the shallow cliffs at the most northern point of the bay.*"

At dawn the old shaman made his way to the house of the magistrate and insisted he be awakened. He then told him about the visitation in the night. And on this evidence the girl was released, then and there."

Benito lit a cigarette and inhaled deeply, blowing the smoke out through his nostrils as he gave me time to digest the import of the story. But just as I was about to speak, he moved his hand across the table and placed it firmly above my own.

"Three days ago this <u>happened</u> here. It is no tale from the past. And the girl was freed immediately. Do you know any other place in the world where the dream testimony of a spirit would be accepted without question? No. At rock bottom this is still Polynesia – the spirit world is still with us."

I nodded. There really wasn't anything to say.

"Oh... and I should tell you. The police found the empty whiskey bottle beneath the car seat. The body was recovered at Ana Kai Tangata. The head was where the dead man said it would be."

4

THE COMPASS POINTS OF CONSCIOUSNESS

Cogito ergo sum
I think, therefore I am.
Rene Descartes, *Discourses on Method*

CONSCIOUSNESS –
THE PEAK OF THE ICEBERG; THE SUBMERGED DEPTHS

The above statement by René Descartes (1591-1661), French mathematician and philosopher (of whom we will talk about later) - despite being much quoted - is not quite as self-explanatory as it seems. To *'think'* is a pretty complex operation of consciousness: one that goes far beyond the 'mechanical' workings of the five senses and the impressions they deliver concerning the appearance of oneself and everything else in the world. However keen such perceptions may be, the process of 'thinking' advances the rise of more abstract questions: thoughts concerning the personal value and meaning derived from each encounter with the world at large.... culminating in philosophical speculation as to *why* things should be the way they are - for what *purpose*, to what end?

To 'think', for Descartes - as scientist, philosopher and devout Catholic - was to cogitate at all these levels. As a mathematician he devised new systems of algebraic notation, and was the originator of analytic geometry. As a philosopher, he took a *dualistic* view of the Universe, believing in the existence of God as the supreme form of *spirit*; while the natural world revealed the existence of *matter*. He suggested that one proof for the existence of God lay in the fact that we possessed a mind capable of 'innately' - as opposed to 'sensorily' - envisioning spiritual levels of Being; while also possessing the ability through observation and reason to scientifically discover how the material world functions.

In his major psychological work, *Traite des Passions de l'ame* (Treatise on the Ardor of the soul), Descartes sees the soul as the spiritual nucleus responsible for those profound and heartfelt moments that spring intuitively to mind from the depths of the Unconscious.

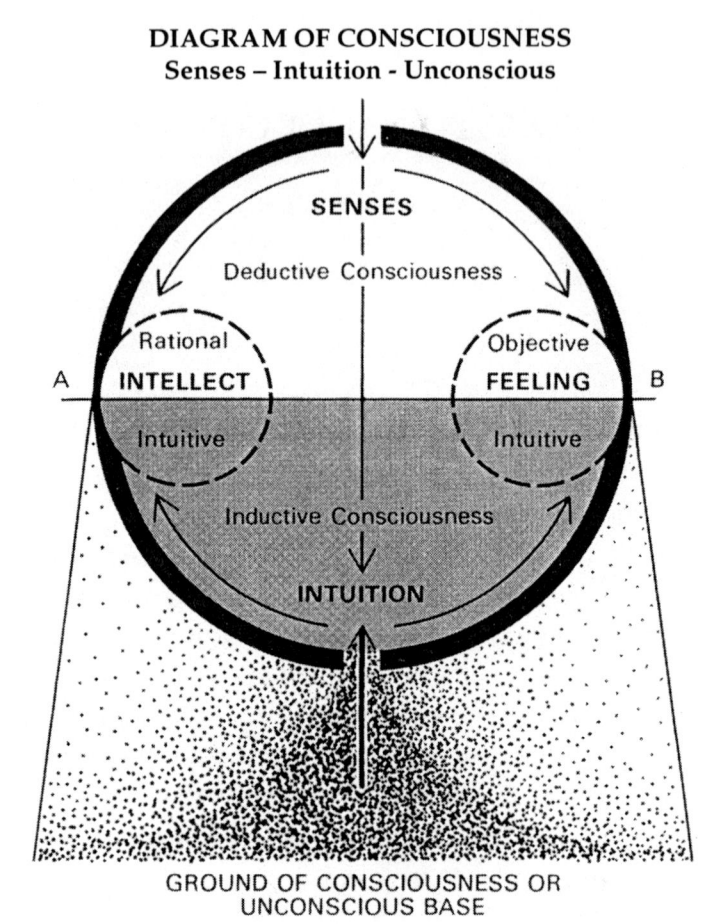

DIAGRAM OF CONSCIOUSNESS
Senses – Intuition - Unconscious

```
                    SENSES
              Deductive Consciousness
         Rational                Objective
    A    INTELLECT               FEELING    B
         Intuitive               Intuitive
              Inductive Consciousness
                   INTUITION
```

GROUND OF CONSCIOUSNESS OR
UNCONSCIOUS BASE

Many years ago I remember reading that Goethe had commented to a friend that when it came to dealing with abstract concepts he found diagrams more helpful than words. I hope that this will be the case here, inasmuch as our diagram introduces the difficult abstract theory of 'the Unconscious' - the term used by both Freud and Jung to signify a reservoir of mental energy operating below the threshold of normal consciousness, and thought to be the source of the most inspired, imaginative, and transcendent imagery that finds its way to consciousness through the gate of intuition.

It must be admitted that to suggest a diagram can convey the labyrinthine complexity of the brain's neural activity might well appear ludicrous. A gross over-simplification. Yet in the same way that a view of earth from a satellite gives little indication of either the details of surface configuration or the internal activity of the planet – yet, nevertheless, provides an image sufficiently informative for the viewer to gain an understanding of earth's basic characteristics – so may this drawing afford a similar introduction to the processes of consciousness. If nothing else, it should certainly help to pin down some of the terminology we have employed: make such words as 'intuitive' and 'unconscious' generally more intelligible within the context of these chapters. Also, there is much to be said for being able to visualize the basic 'departments' of consciousness – gain some idea of their interactive flow and sequence of operation.

Think 'iceberg' for the moment. When sailing past icebergs in the Southern Ocean one has always to remind oneself that however much ice appears above the surface, there is usually four times as much below - invisible, but there. The analogy of the iceberg to the overall structure of consciousness as laid out in the drawing is not as obscure as it may seem. The upper segment of the diagram, showing the intellectual and feeling centers that respond to the stimuli of the five senses, corresponds to the visible aspects of the iceberg. Whereas the lower level, beneath the A to B line, can be likened to the submerged and unseen mass of the berg. It indicates the presence of active 'underground' mental activity feeding into the rational and objective life of the 'sensory' consciousness - part intuitive; part unconscious. Both Sigmund Freud and Carl Jung recognized the importance of such intuitive and unconscious mental forces - regarding them as responsible, on the one hand, for conveying the influence of age-old archetypal themes and across the board human traits; on the other, as the source of our most imaginative creative insights and technical, practical skills.

Lancelot Law Whyte writes in his book, *The Unconscious Before Freud*:

> *The main purpose of this study ... is to trace in more detail the progressive recognition in the systematic thought which developed after Galileo, Kepler and Descartes; that it is necessary to infer that mental factors which are not directly available to our awareness influence both our behavior and the conscious aspects of our thought.*

Consequently, the heavy black circle of consciousness is shown as fitting snugly into its cradle-like support, the Unconscious (the really 'deep water' of the whole system of awareness about which we will talk later in the chapter.) So seen in its entirety, the diagram represents consciousness as a tripartate disposition involving: the objective Realm of the *Senses*; the subjective Authority of the *Intuition;* the subliminal Depths of the *Unconscious.*

The gateway of the Senses, and that of the Intuition, are at opposite poles and represent the two fundamental channels by means of which consciousness receives its impressions. That of the senses carries a constant and voluminous flow of information: we see, hear, touch, smell, and taste - each experience bringing its own special form of intelligence to consciousness by involving both the rational intellect and triggering many levels of feeling. Hence the two arrows indicating the initial mental targets of the senses.

For example, in watching a seagull as it hovers and glides in the air currents, the 'mechanics' of the visual sense come into play as the light reflected from the object travels via the retina of the eye and the optic nerve to the brain; thus bringing the observer to become aware of the bird's overall shape and behavior in its spatial context. Simply put, the existential *presence* of the gull becomes an event in consciousness, taking on a recognizable form when the rational, analytical processes of intellect are touched off. This in itself is a complex analytical and reasoning process - one involving the recall of past visual experiences of birds, thus enabling instantaneous identification of the gull as a bird in flight, as well as ascertaining whatever may be unique about this latest of bird-images. It is by such basic ratiocinative means that the intellect establishes the unique facts pertaining to every sensory experience... bringing itself constantly up-to-date, as it were, with every new act of perception.

At the same time - and equally importantly - the visual perception of the gull triggers the affective side of consciousness, evoking feelings of attraction, repulsion, pleasure, curiosity, wonder, sympathy, empathy... All of which, in their turn, serve as the means which enable us to assess the value and importance that we inevitably attach to every sensory experience. The range and quality of one's feelings - how shallow or deep run the currents of one's emotional life - plays a vital role in the development of personality and character. And as we go through these chapters it becomes increasingly obvious that it is the way we feel, rather than the way we think, that determines the nature of our actions.

> *Seeing is believing, but feeling's the naked truth.*
> John Ray, *English Proverbs*

Consequently, the upper half of the diagram is seen as the territory of the Deductive Consciousness - inasmuch as the reason and feeling brought into play is deduced from signals initiated by the Senses.

But then we have the third arrow, cutting straight from the portal of the Senses to directly activate impressions of a different sort arriving from the opposite pole - the Inductive half of this 'compass' of consciousness - the psychological territory home to the mental field of *intuition*. Intuitive impressions reflect spontaneous feeling-attitudes, or sudden breakthroughs of understanding - 'moments of vision' so-called, emanating involuntarily from within the self. They can be triggered by external things and events, providing likely answers as to 'how' things come to be the way they are, or 'why' they should be so. Also, such immediate insights can move of their own accord into consciousness, unrelated to any involvement with the outside world. In either case they reflect one's own most personal insights and values.... gleaned from some subliminal source of knowledge not readily available to the fact-oriented senses. Intuition has been described as 'instant wisdom' delivered by the 'inner eye': the means of apprehending the hidden dynamics, principles, truths.... that lie behind the surface of appearances in one's dealings with the world - as well as causing one to ponder the influence of the 'angels' and the 'demons' of one's own psyche.

The Greeks coined the word 'psyche' to signify the whole range of deductive and inductive mental activity we have been discussing, but they went further to lay great emphasis on that part of the diagram which I have not yet talked about: namely the Ground, or Unconscious Base of Consciousness - that mysterious psychical region wherein the metaphysical force of the *soul* works to charge consciousness with spiritual and humane intelligence. During the so-called Age of Reason in the 17th and 18th centuries the concept of 'soul' became secularized, and when the term was used it was to denote the 'mind as intellect'. However, in the school of

20th century psychiatry - established largely by Sigmund Freud and Carl Jung - the spiritual implications of the word were somewhat restored, particularly by those doctors who subscribed to Jung's view of the human psyche. For they were Greek-like in regarding every individual psyche as a psychological crucible where that famous triumvirate of powers - *body, mind, and soul* - operate at both conscious and unconscious levels in a grueling fight to achieve a purposive balance on life's journey: one in which sense, intuition, and soul (in the realm of the unconscious) play their part. Jung saw the aim of this process to be the creation of individuals who can equate their objective sensory life with the abstract wanderings of intuition, and the unconscious interjections of the soul's spiritual values. To learn how to allow the demands of a physical time-bound self to coexist symbiotically with those of both an intuitively-driven imaginative self *and* the metaphysical undercurrents of soul, is the mysterious life-process that Jung called Individuation.

In an essay entitled, *The Limitations of a Scientific Philosophy* (from his book *the Forms of Things Unknown*), Sir Herbert Read writes:

> *A distinction which runs through the whole development of human thought has become blurred during the past two hundred years. Implicit in all ancient philosophy, acknowledged by mediaeval scholastics and the natural philosophers of the Renaissance, and even by Locke and Newton, is a difference in kind, if not of value, between <u>wisdom</u> and <u>understanding</u>. By wisdom was meant an <u>intuitive</u> apprehension of truth, and the attitude involved was receptive or contemplative...*
>
> *Understanding, on the other hand, was always a practical or constructive activity, and <u>ratio</u> was its name - the power by means of which we perceive, know, remember... sensible phenomena...*

I suspect that most of my readers will appreciate the distinction Read emphasizes here between intuition and ratiocination. No doubt many of you can remember decisions made in the past based on unprompted thoughts and feelings which, having warned of difficulties or dangers ahead, have also 'advised' on the course of action most advantageous to take. In contemplating a move to a new job, for example - one which, on face value, would seem to be the absolutely right and obvious change to make - irrational doubts and negative feelings may persist which cause one to refuse the offer. A decision that often turns out to have been for the best. It is worthwhile to reflect on how many choices one makes in life that are not based on the facts alone, but on the involuntary intrusion of this 'sixth sense' of intuition. And to remember that when the way ahead is unclear, how important it is to sit tight - stop consciously thinking about the problem - and wait on intuition.

Later in this chapter I give an abbreviated account of Sir Ernest Shackleton's fateful Antarctic expedition with the ship *Endurance*. It is the saga of a man whose inner directives enable him to achieve the seemingly impossible: some decisions which bear the hallmark of the unconscious - that is far-seeing and transcend the kind of intelligence required to deal with immediate concerns - others being more intuitive responses enabling him to cope with the dangerous exigencies of the moment. For example, I can think of no more telling an example of the sudden intervention of intuition than that which occurred almost at the end of his famed, 800 mile small-boat voyage, in terrible weather, across the wastes of the South Atlantic. When some 90 miles away from the island of South Georgia - a mere speck in the waters of these sub-Antarctic regions - Shackleton suddenly told Captain Worsley, who was navigating by means of a few sextant shots at a mostly invisible sun, to change course to a slightly more northeasterly heading lest they missed the island. Reading Worsley's account of this epic small boat journey it is obvious that he was concerned that such a course change might have the opposite effect, and result in them becoming totally lost in the wilderness of the Southern Ocean. Yet he had tremendous faith in 'the boss', as Shackleton was called, and obeyed the order. As it transpired, without such a change of course, and given the fact that they were approaching South Georgia in the grip of a westerly hurricane, they would indeed have passed south of the island and been lost to history in the three-thousand miles of the South Atlantic that lay between South Georgia and South Africa.

A brief word here about intuition and instinct. They are often used as synonymous terms but are really quite distinct mental processes. As we have defined it, intuition is a form of insight providing knowledge that goes beyond the sensory stimulus and subsequent information provided by a rational intellect and objective feelings. Whereas instinct represents a reflex-like response - immediate and unpremeditated action - triggered by the mechanics of sense alone and bypassing the objective processes shown in our diagram of consciousness. It is the oldest of all responses, initiated by the oldest part of the brain: the primitive way by which we responded to signals from the senses by momentarily recognizing the dangerous or non-threatening nature of what was happening in the environment - and then reacting appropriately and practically instantaneously. Survival depended on this efficient mechanism which caused the adrenalin to flow and resulted in the 'fight or flight' syndrome. It was, mentally, the best we could do...until the brain developed its present physiological structure and sophisticated mental powers.

Given the above point of view concerning the semantic implications of 'intuition' and 'instinct', I would think that the author of the following quotation really meant 'Intuition' when he wrote 'Instinct'.

> *Instinct preceded wisdom*
> *Even in the wisest men, and may sometimes*
> *Be much the better guide.*
> George Lillo, *Fatal Curiosity*. Act I, sc. 3.

THE UNCONSCIOUS

There is no obvious straight dividing line in the diagram of consciousness between Intuition and the Unconscious. In fact it is extremely difficult to make a clear distinction between them, for unconscious mental content seems to pass through its own 'gate' into the intuitive zone, and consequently invade consciousness by utilizing intuition's own channels. But if the Intuition has been described as a 'sixth sense' in terms of dealing with the world, the Unconscious - given its most beneficent and positive aspect (for it has a dark and negative side) - has been seen as the 'power behind the throne' of intuition when described as 'Instant Wisdom'. The thoughts and compulsions it projects have a surreal aspect to them – 'out of time' and 'out of the blue' ... in that they are not under consciousness' control and generally have nothing to do with where one is, or what one is doing, at the time. Such intrusions may be inspiring in that they present solutions to deep-seated problems, or bring completely new scenarios to mind in searching for meaning in life generally, or purpose and guidance in one's own in particular. But there can also be a negative and destructive side to unconscious content. It can evoke vestigial memories of those long gone days in our evolution when consciousness was governed by the ancient pre-cortical brain of instinct - the days of kill or be killed - before soul or conscience brought the concepts of good and evil, love and hate... to mind.

There are those psychiatrists and psychologists who see the Unconscious as representing a more complex and profound form of intuition, rather than being a psychical power in its own right. Yet if one thinks of Mozart 'hearing' original musical structures in his head, at times when he was consciously employed going about the routine affairs of daily life, we are facing a kind of inspiration that represents *sheer* creation: sonata-sound and symphonic-sound independently generated - far removed from the auditory experiences offered by the environment, and so likely to be much less intuitively influenced. As Mozart describes it, such absolute creativity has all the hallmarks of what is described as 'an up-rush from the Unconscious'. It would seem that Mozart was one of those rare individuals who lived with one foot in the Unconscious much of the time. Yet history is full of famous scientists, artists, philosophers... who may not have been in Mozart's league when it comes to 'instantaneous discovery', but come pretty close to it.

There is no doubt that the proposition of the Unconscious is very difficult to deal with conceptually. The Flying Bishop who was my travelling companion on a flight to Dublin just after the end of World War II - and of whom you will read in Chapter 8 - described unconscious mental activity in a letter as: *'An undercover, high-powered clutch of neurons serving the uncanny and extraordinarily authoritative source of hidden knowledge possessed by the soul...* Yet... arcane and disembodied as such a psychological theory seems to be, there is one way in which it can be said we are all subject to its influence. We all dream. And when, in Chapter 6 - Time and the Dream - we discuss both 'Big Dreams' and 'Little Dreams', it will be seen how evident is the activity of this Unconscious when we sleep. There is a temporary suspension of sensory activity when we fall asleep, accompanied by an absence of chronological time. Consequently, with none of the barriers of normal consciousness in place, the incursions of such non-rational images as dreams moving in to affect the 'semi-sleeping' brain are facilitated. And so our 'night life' commences. A similar psychological phenomenon can occur without need of going to sleep in the normal sense. It is called 'daydreaming' - times when one goes 'off the air', so to speak: moving more deeply into oneself when the temporary suspension of sensory objectivity produces a slippage of time, and wonderfully imaginative ideas of all kinds, or inventive solutions to personal problems and concerns, present themselves.... unbidden.

Such 'absent-minded' moments cannot be produced at will - although one may know how to induce them: walking in quiet places without thought of time or commitments, in moments given over to meditation when time and the world are left behind.... or when very deeply moved by some compelling experience.

> *Somewhere, I feel, we inside ourselves, in the human spirit, know everything. We don't have to read. We learn to read what we already know. We read to re-prove what are truths for us. We have to listen to ourselves...*
> Teresa Stratas: quoted in *The New Yorker* magazine.

I remember discussing the writing of music with the composer Arnold Franchetti when he made a statement that might well apply as much to the pure scientist as to the artist and philosopher. *"Everything is out there,"* he said. *"Everything is preexistent. Every kind of sound, every possible tone-value, every possible combination of harmonies is out there – waves radiating through the ether: waves or particles of kinetic, electrical and chemical energy systems – a whole spectrum of energies. They come unexpectedly, and I can just wait pick up snatches – phrases of sound, you understand – write them down from memory as soon as possible... give them their shape and position on the staff; then edit and maneuver them into compositions."*

We do not know the nature of all the physical energy-forces at work in the universe, not to mention completely understanding the complexity of the electro-chemical interchanges in the micro-universe of our own small brain. Yet there are neuroscientists and biologists who assert that there is really nothing extraordinary, or even vaguely mysterious, about our ability to mentally wander beyond the given facts of time and place. They seem to be unaware of the paradox that a vast *physiological* network of neurons can work together in an immeasurable number of ways - (of which 'messages' from a hypothetical Unconscious are the strangest) - to introduce non-rational, and *abstract* images to consciousness: complex combinations of feelings, sensibilities, and thoughts.... psychical experiences that we name faith, hope, love, beauty, conscience, justice... to mention but a few;

courtesy of billions of chemically and electrically energized neurons, all operating according to their manner and inducing *subjective* and compelling states of consciousness.

There are also neuroscientists who view the whole phenomenon with a less rationally conditioned, more open mind. They see the brain's evolutionary, structural growth finally moving to fruition some three-hundred thousand years or so ago; after the stage was set some time between two to four million years previously when the left and right hemispheres of the cerebrum with their crowning cortex and communication link of the corpus callosum were in place - up and running. The result was to provide us with an amazing mental powerhouse: a brain capable of bringing to bear a remarkable range and complexity of objective, intuitive, and unconscious mental processes. And I know several 'brain men' who are certainly not prepared to say that sense-experience and the rational processes of deduction alone provide our only means of defining the nature of reality.

For with neuroscientists talking about one hundred billion neurons coming to nestle in the various divisions of the cortex (seen as the 'jewel in the crown', in the brain's evolutionary growth) - and discussing how labyrinthine neural interconnections and cooperative 'firings' create multiple electrical fields… it is surely an incredible phenomenon that can subject us to upsurges of sudden and inspired moments of discernment that reveal - as Sir Herbert Read puts it - the *Forms of Things Unknown;* or the unpredictable influence of the Unconscious as described by Freud and Jung.

After all, we live our lives against a background of radiation, and after half a century of unprecedented increases in knowledge concerning the forces at work in the universe, we now know that space is a seething field of energy. (It has just been reported that scientists have discovered immense 'rivers' of hot, electrically charged gas flowing beneath the surface near the polar regions of the sun… A discovery that could help them understand sunspots and other energy disturbances that cause electromagnetic storms and power failures on earth.)

> *A kind of metaphysical combustion seems smoldering in the fabric of things, a surge of incipient energy, breaking out of the bounds of its nuclear forms, and disappearing into the beyond.**

---

* William Everson, *The Excesses of God* (Stanford University Press, 1986), p. 13.

This evocative statement made by the poet William Everson would surely have received an affirmative nod from Aristotle. For it was he who, as I have mentioned previously, introduced the term *'meta*physical' to signify forces at work in the natural world beyond the reach of the physical sciences to uncover. And Everson's words reveal a visionary power of apperception that allows him - without benefit of reason - to talk of 'smoldering' cosmic energies in metaphysical terms that can both support and illuminate the discoveries of science. They also exemplify a poet's visionary ability to intuitively apperceive supramundane happenings - the *'metaphysical smoldering in the fabric of things...'* You may remember the lines from Plato's *Ion* which were quoted in Chapter 2, and which describe 'the poet' as follows: *For the poet is a light and winged and holy thing, and there is no invention in him until he has been inspired and is out of his senses... when he has not attained to this state he is powerless and is unable to utter his oracles.* These lines - poetic in themselves and written well over two thousand years ago - foresee the imaginative mental powers that many modern psychiatrists ascribe to the Unconscious and its near ally the Intuition.

Nowadays, the old adage that 'seeing is believing' has less relevance than at any time in the past, for science and its associated technologies are advancing beyond the frontiers of the observable. And I wonder if it is not possible to regard the mental mystery posed by the insights delivered through the medium we call the Unconscious, as having something in common with the phenomena associated with Quantum Physics. Or that of satellite communication through space... or with the transmission of human speech between cellular telephones having no visible connection, and hundreds of miles apart?

THE UNCONSCIOUS AS MAGMA CHAMBER

The reader may have noticed that our diagrammatic representation of consciousness underpinned by the Unconscious, bears some resemblance to a geological drawing – to an illustration of a magma dome protruding above the earth's surface, while beneath it lurks underground activity of a volcanic nature: thermodynamic and molten rock-pressures which can build to eruption levels.

The magma chamber - a large natural cavity with fissures through which molten rock ascends from deep within the earth, serves as a subterranean storehouse for this mass and the energy it generates. Here we have turbulent multiple forces comprising the kinetic energy and subsequent pressure of the upward-surging mass of lava, the thermal energy of the enormously high temperatures involved, and the convection currents generated by the tumultuousness of the molten magma and hot air – all of which must, at some point, be released through vents allowing the energy to erupt above ground. It is a geophysical phenomenon which serves as a useful analogy to help explain the sudden release of unconsciously-held notions and passions... via the vent of the Intuition. Think of such basal *'mental'* strata, overly replete with seminal imagery in such a fertile and turbulent state of readiness that it must, ultimately, escape into consciousness of its own force. Or, such a brainstorming might come about when one's psychological state is so demanding of resolution - vis-à-vis some pressing issue - that it produces an involuntary and intense excitation of the brain's neural, electrochemical resources.... thus activating the 'magma chamber' of the Unconscious. Such extreme passional urgency can cause reason and the objective processes of consciousness to be bypassed - thus leaving the neuro-channels free of normal 'traffic' and capable of receiving such primary and highly originative intelligence.

To the best of my knowledge, Mozart's description of the eruption of 'instant music' is the most illuminating account we have of the phenomenon of the Unconscious directly transmitting the 'music of the spheres' (as it has been called) to full consciousness. And although neither he or any of his contemporaries would be likely to call his sudden inspiration 'an eruption from 'the Unconscious'.... the idea that revelatory knowledge comes unbidden from within ourselves - and could be regarded as evidence of the soul's intervention - was present long before Freud and Jung talked about the psychical authority of the Unconscious.

Mozart notwithstanding, men and women in every creative field have testified to the unexpected arrival of mental 'directives' which provide stepping stones across unknown territory to bring inspired ideas directly to mind.

Writing in 1927 about Beethoven's reaction to the impending calamity of total deafness - (can there be anything more terrifying for a composer?) - J.W.N. Sullivan tells how this musical genius' response was one of defiance:

> He felt that he must assert his will in order not to be overcome. He would summon up all his strength in order to go on living and working in spite of his fate. 'I will take Fate by the throat.' He was, as it were, <u>defending</u> his creative power. But by the end of this summer of 1802 he found that his genius, that he had felt called upon to cherish and protect, was really a mighty force using him as a channel or servant. It is probable that every genius of the first order becomes aware of this curious relation towards his own genius. Even the more fully conscious type of genius, even such scientific geniuses as Clerk Maxwell and Einstein, reveals this feeling of being <u>possessed</u>. A power seizes them of which they are not normally aware except by obscure premonitions. With Beethoven, so extraordinarily creative, a state of more or less unconscious tumult must have been constant. But only when the consciously defiant Beethoven had succumbed to deafness, only when his pride and strength had been so reduced that he was willing, even eager to die and abandon the struggle, did he find that his creative power was indeed indestructible and that it was its deathless energy that made it impossible for him to die.

And so, with his hearing gone, Beethoven continued to compose: hearing the music although bereft of any auditory sense. A situation revealing the inner life of the Unconscious not unlike that experienced by Mozart. Albert Einstein in his essay *On Science*, wonders about creative revelation and concludes that, '*Such imagination is more important than knowledge.*' And Henri Poincaré, the French scientist who was one of the greatest mathematicians of his time, and who died in 1912, says in *The Foundations of Science*, that only "When the mind relaxes...' does the abstract world of mathematics take on a life of its own.'

SHACKLETON – MAN OF GREAT UNCONSCIOUS *and* INTUITIVE RESOURCES

> *In appreciation for whatever it is that makes men accomplish the impossible...*

So wrote Edward Lansing in dedicating his book *Endurance* – a tribute to all men and women who somehow manage to go beyond the normal bounds of human endurance in whatever testing situations confront them: trials which, if they are to be surmounted, demand far more than simply the exercise of sound objective perceptions and reasoning - something extra, difficult to define, and often referred to as an indomitable spirit. Anyone who is not sure what is implied by this attribute should read Shackleton's book *South* – an account of his 1914 expedition to the Antarctic in the vessel named *Endurance*. It is a masterpiece of understatement, totally devoid of self-aggrandizement: the chronicle of a leader drawing on a wellspring of fortitude within himself - a fount of wise counsel as emergencies develop; of calm resolution in dealing with near-disasters and excruciating physical hardship; and of an unshakeable faith in the playing out of his own destiny. All characteristics which tell of an extraordinary strength of spirit. Or which, put another way, signify the exceptional power of unconscious and intuitive psychological resources.

If Mozart is a prime example of the unconscious in action when it comes to the creation of great music, then I hold Shackleton to be a man similarly prompted - guided by an inner intelligence which, at times, endows him with seemingly superhuman visionary powers in his Antarctic expeditions. The Norwegian, Raoul Amundsen – the first man to reach the South Pole – said of him:

> *Sir Ernest Shackleton's name will for evermore be engraved with letters of fire in the history of Antarctic exploration. Courage and willpower can make miracles. I know of no better example than what that man has accomplished.*

When reading *South*, Shackleton's account of the long two years during which he finally brought the men of the *Endurance* to safety, one never feels that at any point did he seriously consider the likelihood of total failure and the death of himself or any of his expedition members. If so, it certainly does not show in these pages. Rather, the 'boss', as the men called him, seems to remain indifferent to the threat of such eventualities. It is a quality of the man which impresses the most: the absolute surety that this inner force we call spirit will come unconsciously, unwilled, to guide every action. Here is Shackleton, speaking in Sydney, Australia, in March 1917:

> *Death is a very little thing – the smallest thing in the world. I can tell you that, for I have been face to face with death during long months. I know that death scarcely weighs in the scale against a man's appointed task. Perhaps in the quiet hours of night, when you think over what I*

> have said, you will feel the little snakes of doubt twisting in your heart. I have known them. Put them aside...

My abbreviated account of Shackleton's Endurance expedition which follows cannot hope to convey the stark reality of the Weddell Sea drama. Go to Shackleton's book *South* for that. Even so, the events which befell both ship and men are so catastrophic that even the bare bones of the story will suffice to amaze the reader – have he or she wondering at the most sustained display of inner guidance, willpower, courage, resourcefulness, and faith they are ever likely to come across.

In the late autumn of 1914, Shackleton's expedition to the Antarctic set out for the Weddell Sea and the northern coast of the continent in the vessel <u>Endurance</u>. His plan was to penetrate the ice of the Weddell Sea on the east side of the Antarctic Peninsula (see map), and reach a point on the continental land mass from where he could embark on the first ever crossing of Antarctica - his route taking him to McMurdo Sound in the Ross Sea, involving a likely distance of 2100 miles. The journey was to be accomplished by a dog and sled party – five men and 54 dogs – and was calculated to take 120 days: an unrealistic estimate for such a distance over totally unknown terrain. No one knew what might be encountered in the form of mountain ranges, the crevassed condition of the icecap, or the violence of the weather.

As things turned out the land party never got under way. The summer pack ice that year was particularly extensive and dense - the weather, being unseasonably severe, did little to help in the breakup of the pack. Shackleton spent five weeks seeking leads of open water which would provide access to the coast known as Coats Land. Finally, after a succession of blizzards and gales, a survey of their situation on the 25th of January 1915, indicated that the ship was completely held by the ice: it was packed heavily and firmly all round Endurance in every direction as far as the eye could reach from the masthead. So reads the report.

There was no escape for the ship from this final stranglehold. Firmly locked in the ice she drifted about 6 miles a day with the pack for some 281 days until late October that year – covering a distance of 1500 miles to finish 570 miles north and west from where she was first trapped. During this time the men hunted seals and penguins for sustenance. On 27 October it was obvious that she was finally being crushed to death by powerful pressure ridges in the ice; that there was no alternative but to abandon the vessel to her fate.

Frank Worsley, the captain of Endurance, described her last agony in his diary as follows: 'Two massive floes, miles of ice, jammed her sides and held her fast, while the third floe tore across her stern, ripping off the rudder as though it had been made of matchwood. She quivered and groaned as rudder and sternpost were torn off, and part of her keel was driven upwards by the ice. The shock of the impact was indescribable. To us it was as if the whole world were in the throes of an earthquake.' Camp was pitched on the ice, but it was not until 21 November, almost a month later, that what was left of Endurance's stern lifted into the air and the ship sank below the waters of the Weddell Sea. It was when the floe beneath them began to break up with the approach of the Antarctic summer that Shackleton decided on 20 December to head northwest across the pack to Paulet Island 350 miles away (land which was closer to their position than Elephant Island which lay some 180 miles further north and west). But after taking four days to travel just 11 miles through deep snow and rotten ice, Shackleton abandoned the hike and decided to take the chance of drifting north/northwest on a large ice floe. Together with the ship's three small boats – which they had manhandled over the ice to this point – the party of 28 men retreated to a floe which, however, cracked beneath them during the night, necessitating a shift to a more substantial expanse of floating ice. This happened on 29 December 1915; they were now irrevocably committed to the winds and currents of the Weddell Sea.

The men had been camped on sea ice – on ice rafts, no less – from 28 October and, after the abortive attempt to cross the ice on foot, their leader's last hope was that the fast-drifting pack would carry them to open water where the boats could put to sea to reach Paulet Island. Unfortunately, the floe's drift was too westerly, causing them by 17 May to pass abreast of their desired destination by 60 miles – with no hope of crossing the decaying sea ice to reach it; and no open water that would permit launching the boats. It was now 8 April 1916 – the day on which misfortune struck again as the ice beneath them began to disintegrate, and their last floating home had to be abandoned. The expedition photographer, Frank Hurley, described what happened: 'Shortly after 6 p.m. the watchman raised the alarm that the floe was splitting. Our camp was reduced to an overcrowded rocking triangle, and it was evident that we must take the first opportunity to escape, no matter how desperate the chances might be. At 8.00 the next morning – after almost losing one man in the water when the ice separated – Shackleton ordered the launching of the three ship's boats: the 22 feet 6 inch long whaleboat James Caird was the sturdiest and most seaworthy; the Stancombe Wills and Dudley Docker were 21 feet 9 inch long cutters, designed for rowing, not sailing. But as open boats they were not the equal of the severe seas they were to face.

This is where Elephant Island comes into the story. For given their westerly position – and the fact that their only means of propulsion were oars and whatever small jury-rigged sails could be set up – it would be largely the ocean currents which would drive the boats forward. And Elephant Island was the only land ahead to which the sea might take them.

On 12 April, however, Shackleton realized that instead of running west, the little boats had drifted 30 miles east. Only Wild and Worsley were told of this. But after this setback tide and current were with them and the forbidding profile of Elephant Island was sighted to the north/northwest. On 14 April the men landed at the only place offering any sort of beach: a narrow strip of shingle – glimpsed through breaks in the foul weather – beneath the lowering 2000 foot high cliffs of Cape Valentine. The approach to Elephant had been fraught with dangers. A gale had separated Worsley commanding the Dudley Docker from the others during the night as they converged on the island, creating high and powerful seas which threatened to swamp them at every turn: frightening conditions to experience around an uncharted and rockbound coast in the dark – with the Stancombe Wills showing less than a foot of freeboard above the waterline. (In his book South, Shackleton acknowledges his debt to Frank Wild, the expedition's second-in-command, seeing him as a tower of quiet confidence and strength, who sat at the tiller of the James Caird seemingly untouched by fatigue, steel-grey eyes scrutinizing the ice-filled ocean as calmly as if he was on a day trip in the Thames Estuary.)

The near delirious joy of the men at being on solid land for the first time in 16 months was moving to watch. But Shackleton knew the place could not be a long-term haven: with the beach a mere 100 feet wide and 50 feet deep, gale-swept high tides would bring the ocean to the very base of the cliffs. On the other hand, he did not know whether the island offered any other secure ground for camping. At 11 the next morning he dispatched Frank Wild and four relatively fit men in the Stancombe Wills to search for a safer location. They returned in the dark having found a narrow spit of rock and sand 7 miles to the west which appeared the only alternative 'residential' site available.

The accounts left by both Shackleton and Frank Worsley of that 7-mile passage to Point Wild – as they were to name the constricted promontory he'd discovered – through raging seas makes riveting reading. Soaked to the skin, unsure of where they were making for around this grim and treacherous coast, and in constant danger of being swamped or pushed against a lee shore of cliff and rock, it speaks volumes for the seamanship – not to mention the high courage and leadership – displayed by Shackleton, Worsley, and Wild commanding the three small boats, who finally landed everybody on the narrow neck of land that was to be their home for 105 days. To the armchair reader there must appear something of the miraculous about their survival at this juncture. (In fact, Worsley's boat was almost lost when he decided it would be safer to leave the lee of the cliffs and steer out to sea to round a rocky point, lest there be shoals on the inside route. Beyond the protection of the high cliffs, wind and sea almost defeated him).

The rest of the story is well known. How Shackleton, knowing there was no hope of rescue from Elephant Island, set off with five men in the James Caird – on Easter Monday, 24 April, at the start of the Antarctic winter – to try and reach a speck in the vastness of the South Atlantic known as South Georgia lying 800 miles to the northeast. Frank Worsley went with Shackleton, and Frank Wild was left in charge of the 22 men marooned on the island. With the benefit of hindsight it seems likely that, without these two men, Shackleton's unyielding effort to save his company could have failed. It was Wild's gentle yet firm discipline, together with the example of his stoicism and serenity of spirit, that kept up the morale and *hopes of the slowly starving men in his charge. And it was Worsley's navigational skills, involving both dead reckoning and the handling of a sextant on a small boat buffeted by mountainous seas and gale-force winds – on those few occasions when a shot at the sun was possible – that finally got the James Caird close to South Georgia after 17 days at sea.*

Yet, in the final analysis, it was Shackleton's intuitive powers which saved them – the sudden awareness that they would miss the island if they remained on their present course. 'Can you be positive of your position?' he asked Worsley when they appeared to be about 90 miles from the western reaches of South Georgia.' Not to 10 miles, but can easily allow for that,' the navigator replied. But the 'Boss' was not so sure – feeling that they might be passing south of the island's western tip, and insisted on a change of course that would give them a little more easterly heading. He was right. Without the eastward change of course they would have missed the tip of South Georgia and disappeared into the wastes of the Atlantic.

The account of the James Caird voyage – generally regarded as the greatest small boat voyage in maritime history – is a spine-chiller in itself. It seems nothing short of a miracle that icing conditions, continuous gales producing mountainous seas (and one giant tsunami wave), infrequent appearance of the sun for sextant navigation (on a small and heaving deck), and their constantly wet bodies hovering on the edge of hypothermia... should not have combined to defeat them. And without the care and steadfastness of Shackleton they probably would have done so. Worsley, writing in his diaries of Shackleton, says: 'He always, as it were, has his fingers on our pulse and, at the psychological moment orders a hot feed. This saves any bad effects and possibly our lives.' Even when South Georgia's precipitous and ice-girt cliffs loomed for a brief moment in a break between squalls, their troubles were far from over. The triumph of crossing 800 miles of a wintry Southern Ocean – the most violent and elemental sea in the world – must have been short-lived as the weather worsened and hurricane force winds drove the James Caird onto the island's wild and dangerous southwest coast – a coastline which ships today still prefer to avoid. Desperately, and against all the odds, they managed to pry the small boat away from the shore and the surf-pounded shoals and hidden rocks – aided at the very last moment by a sudden and fortuitous (to say the least) change in the wind – and standoff in the heavy swell.

After trying to beat to the northeast in order to round the northern tip of South Georgia and make for one of the two whaling stations on the east coast – an impossible task against the relentless force of the western swell – they fought their way back to the mouth of a bay which Worsley identified as King Haakon Sound. Only after hours of trying to make headway through the narrow and dangerous reefs – against an offshore wind and retreating tide – were they at last able to get into the fiord, find a small cove, and beach the boat: 17 days after leaving Elephant Island.

Now the island had to be traversed: which meant that Shackleton had to attempt a 17 mile crossing of the island's unmapped and unknown alpine peaks, glaciers and frozen lakes, in order to reach help at Stromness whaling station on the east side – a feat never before attempted. No one had ever penetrated more than one mile inland – and this only on the relatively benign east coast. He set out on 15 May 1916 with Worsley and Tom Crean (second officer on Endurance), leaving the sick men of the boat journey behind to await rescue. Thirty-six hours later Shackleton was in Stromness. The achievement in making this first crossing of South Georgia in an Antarctic winter – immediately after the harrowing (and at times seemingly miraculous) voyage of the James Caird – reveals the truly impressive power of Shackleton's guiding spirit: of the commanding role of his intuition at those vital moments when all could have been lost, the strength of his will, and faith in his own resources when it came to surmounting adversity. It should be noted that 40 years later, in 1965, South Georgia was crossed again by the British Combined Services Expedition – hardened men, properly clothed for the severity of the elements and using the latest and best equipment. They were under no pressure of time, could select their route carefully (survey maps were now to hand), yet found it treacherous and difficult going.

Duncan Carse, leader of the team, said that Shackleton travelling at night had, perforce, taken the most difficult route. 'I do not know how they did it,' he wrote, 'except that they had to – three men of the heroic age of Antarctic exploration with 50 feet of rope between them... and a carpenter's adze.' The day after descending the mountains of the Allardyce Range into Stromness, Worsley was on his way back in a whaling steamer to pick up the three sick men left on South Georgia's southwest coast. And the day following Worsley's return to Stromness with the remainder of the James Caird's crew, Shackleton, Worsley and Crean were heading for Elephant Island in the Norwegian whaler Southern Sky. This rescue attempt failed – as did two subsequent efforts - ice conditions and bad weather prevented any approach to the island. But on his fourth try – in a small steamer loaned by the Chilean government named the Yelcho, under the command of Captain Luis Pardo – a brief 'window' in the weather allowed them to find a channel through the ice, and the ship was able to maneuver between icebergs and reefs, put a boat ashore, and take off the 22 men who had lived for 105 days beneath the upturned Dudley Docker and Stancombe Wills. It was 30 August 1916. Thus Shackleton became known as the leader who, when in personal command of an expedition, never lost a man.

Frank Worsley recounts how Shackleton, scanning the Elephant Island beach, counted the black figures emerging from beneath the upturned boats. On reaching 22 he yelled to Worsley, They're all there, skipper! And when he lowered his binoculars it seemed to Worsley that the 'Boss' had shed years.*

Well, there you have it - fearlessness, resourcefulness, courage, stamina, confidence... all the necessary characteristics of the great explorer. Yet

---

* Graham Collier and Patricia Collier, <u>Antarctic Odyssey</u> (New York: Carroll & Graf, and London, Constable-Robinson, 1999), pgs 25-33.

the ability to withstand such prolonged physical hardship, and maintain faith in one's purpose over so many months and in the face of impossible odds, requires a special kind of inner direction. In Shackleton's case I think the clue to this can be found in the fact that from time to time he isolated himself from the rest - sitting at the edge of the ice to distance himself from them as much as possible. On these occasions there were those who thought that the 'Boss' was despondent, had given up, whereas the exact opposite was the case. He was, in Plato's words, 'out of his senses', a man waiting - waiting patiently for instructions from within, musing you may say, no effort of will seemingly involved. The uncanny sense of timing he then displayed in taking the appropriate action, sometimes flying in the face of sound logical advice proposed by others, has the mark of an active Unconscious 'magma chamber' to which Shackleton (in his musings) had access. I feel sure it was Shackleton's deep insights which the veteran Antarctic geologist Sir Raymond Priestley had in mind when asked, in 1956, to sum up the capabilities of the men who led the three major South Pole Expeditions, he is said to have recommended Scott for his zeal in gathering scientific data, and Amundsen for swiftness and professional efficiency in attaining his goal. But in the face of adversity, with no salvation in sight, one should…. *'get down on one's knees and pray for Shackleton.'*

I would add a footnote to this brief account of the *Endurance* expedition. Both Shackleton and Worsley - writing independently and at different times - about the hazardous winter crossing of South Georgia's Allardyce Range with its totally unknown 14,000 feet high peaks and glaciers - tell of a strange phenomenon each experienced independently on the journey. You will remember that there were three men in the mountain party, yet both Worsley and Shackleton write of a *fourth* member - of sensing the presence from time to time of an invisible companion lending them guidance and support on the last crucial leg of the crossing.

> *Personality is the supreme realization of the innate idiosyncrasy of a living being. It is an act of high courage flung in the face of life, the absolute affirmation of all that constitutes the individual, the most successful adaptation to the universal conditions of existence coupled with the greatest possible freedom for self-determination.*
>
> Carl Gustav Jung: *Collected Works* (1954)

THE UNCONSCIOUS and the SUBCONSCIOUS.

The terms 'unconscious' and 'subconscious' tend to be used synonymously nowadays. Yet they refer to different forms of awareness operating below the level of normal consciousness which, admittedly, do at times appear to overlay each other. Both Freud and Jung pointed out the distinction that must be made between them, and I find myself constantly surprised by how frequently the terms are mistakenly assumed to be interchangeable.

Freud regarded the subconscious as a neural storehouse of *engram complexes*: the imprints (*engrams*) left in consciousness of particularly significant events that have happened in a person's life - be they traumatic and distressful or benign and pleasurable. As such they represent deep-seated memories… either repressed or simply lost to time and relegated to the subconscious: a 'memory-bank' that can be regarded as a 'storehouse' below the threshold of consciousness. If bitten by a dog, say, when just a toddler, the persisting memory carried in the subconscious may return years later when, as an adult, an unexpected dog suddenly appears on the scene - causing one to spontaneously adopt a wary attitude.

However, the content of the mental images emanating from the *true* Unconscious is - as we have seen - original and inventive; neither dependent on, nor related to, leftover impressions of things that have taken place during one's lifetime. And their manifestation has far more profound and long-term results: creative breakthroughs in both art and science, molding character…. even involving the course of one's destiny:

> *The uttered part of a man's life, let us always repeat, bears to the unuttered, unconscious part a small unknown proportion. He himself never knows it, much less do others.*
> Thomas Carlyle: Sir Walter Scott.*

---

* I use this quotation from Thomas Carlyle's *Sir Walter Scott* again in Chapter 16 where it is particularly apt.

5

SOLITUDE and the SYMBOLIC ENCOUNTER

A day spent without the sight or sound of beauty, the contemplation of mystery, or the search for truth and perfection, is a poverty-stricken day; and a succession of such days is fatal to human life.
Lewis Mumford: *The Condition of Man* (1944)

More people than you might think dislike being alone for any length of time. They find the days drag, and become easily bored with their own company. Likely as not they would hardly understand what the 17th century English physician and writer, Sir Thomas Browne, was getting at when he wrote: *Be able to be alone. Lose not the advantage of solitude, and the society of thyself.*

As a species we generally see ourselves as social creatures who have consistently lived as members of large or small tribes or groups, subscribing to a relatively collective view of life. Which is still the case to a greater or lesser degree: family, neighborhood, church... being examples of the most readily available affiliations available for people in a world becoming increasingly homogeneous and socially 'neutered' through the hegemony of international commerce, and constantly accessible, time-consuming, stay-at-home entertainment systems. Yet however these and other demographic factors work to change the social fabric of communities, three distinct - and I suspect, age-old - personality types exist who differ in their attitudes to, as they say, 'getting together'. There are those people who constantly need the company of others, comfortable only when in the presence of 'the group', whatever the nature of the situation; those who are gregarious within limits and mix with others of similar interests when political, business, sporting... or family occasions demand; and those who would be considered unsociable because they do not seek - or respond to - the company of others, preferring to spend as much of their time alone as possible. And yet all these individual preferences can go by the board when a common danger threatens, drawing everyone together to work for their mutual well-being. All three types have their part to play in the functioning of any society. Those who feel comfortable in large gatherings learn to foster respect for

71

the opinions of others, and this induces the kind of goodwill that promotes tolerance and compassion - the two attributes which make any community truly human. While the less gregarious prefer their own company because it is only when detached from the crowd that they experience their more reflective moments, and develop independence of mind and a personal way of seeing things - mental attributes which, when called upon, bring perceptive and intelligent insights to bear if communal difficulties arise. But when Sir Raymond Priestley suggested that if caught in the face of adversity, with no salvation in sight, one should get down on one's knees and pray for Shackleton... he is advocating sending up a prayer for the type of person who is, as they say, 'his own man': the so-called unsociable individual who, as Sir Thomas Browne puts it, will *'lose not the advantages of solitude...'*

I can vouch for the sagacity of Priestley's observation. My own experiences in the Second World War were revealing in this respect. As aircrews in Royal Air Force Bomber Command came and went - the rate of loss on operational squadrons was about seventy percent - I noticed that many of the pilots who survived (bringing their crews back with them) were *quiet,* self-effacing men: not 'loners' in the conventional sense but, while not likely to stand out in a crowd, preferred to converse in small groups and possessed what I can only describe as a certain quiescent steadiness to them - a quality which was communicated through their gaze and relaxed posture, the slow, deliberate movements of hands and arms.

It was only many years later - after reading extensively of Shackleton's exploits and visiting the hut on McMurdo Sound from which he had set out on his near-successful effort to reach the South Pole in 1908... that I came to see how many of my former aircrew colleagues were cast in the contemplative mould of a Shackleton, and benefited from, *'... the advantages of solitude.'* I would think that few of my readers will be in doubt as to what these advantages are. They can be deduced from all that has been said in these pages about the profound levels of self-discernment in the form of courage and resourcefulness which can be uncovered when one is content to spend much of the time in one's own company.

Shackleton was such a man. Early in his professional life as a deck officer in the Merchant Marine he discovered the pleasures of mind that came with being alone. During long night watches on the bridge of liners plying between Cape Town and Southampton he would read to himself the poems of

Keats and other favorites, for he found poetry evoked the essential nature and significance of human experience - gave form to the philosophical values and sense of '...*a man's appointed task*' which had been with him since childhood. The isolation of ocean and the solitude of night worked wonders in promoting Shackleton's musings with himself. But, as I have recounted, his need of solitude was not always understood by those under his command. During those long drawn-out months on the ice – both before and after the *Endurance* was crushed and sank – when he would sit apart from the men for hours at a time, incommunicado, waiting on himself for answers to the problems they faced, the men would surely have cried out, *'Give the Boss his space...'* if they had known how their hopes of survival depended on their leader's periods of solitariness and quietude. One cannot imagine Shackleton experiencing the sort of 'poverty stricken day' envisaged by Lewis Mumford.

On more than one occasion I have mentioned how important is the Shackletonian ability to shift attention from what is going on outside and join what Sir Thomas Browne called the *'society of thyself'* - that is if one is willing to encounter the deepest levels of the psyche, and be ready to accept both the light and dark aspects of its auguries. On these occasions I think of the neurons readily availing themselves of the opportunity to take time off from dealing with the world, concentrating themselves on serving the magma-like upsurges of the unconscious and its ally the intuition. And although we obviously cannot all have the innate ability of a Marcus Aurelius, a Shackleton or a Mozart... in this regard, it is possible for the rest of us to experience moments of such self-revelation when least expecting them - while lying solitary and somnolent in a steaming bathtub, or meandering alone in garden or countryside, for example. For it is usually on these latter occasions when, far from the madding crowd, that whatever is brewing in the ground of the unconscious breaks through and usurps the normal routine of consciousness. Some of the earliest records of western civilizations reveal the need felt by urban citizens to escape the town for the country in order to bring about some degree of solitariness, and experience for a short time the life of the contemplative which is induced by quiet communion with nature.

> *And this our life, exempt from public haunt,*
> *Finds tongues in trees, books, in the running brooks,*
> *Sermons in stones, and good in every-thing.*
> Shakespeare: *As You Like It* II. i.

However, I should point out that while one usually thinks about solitude and quietness in a physical, existential sense – in bodily separation from the din of the city and the multitude – it is, nevertheless, possible, as Marcus Aurelius, Roman Emperor and Stoic philosopher (121-180 A.D), points out in the *Meditations*, to be alone in a psychological sense wherever one happens to be.

> *Men seek out retreats for themselves in the country, by the seaside, on the mountains... But all this is unphilosophical to the last degree... when thou canst at a moment's notice retire into thyself.*

Marcus Aurelius was a remarkable man. Here was the most powerful man in the known world at the time, governing the Roman Empire, spending about two-thirds of his reign engaged in military campaigns preserving the empire's borders, and writing a book (*Meditations*) on ethical, moral and religious values - on a philosophy of life which is still in print. He seemed able to retire into himself as and when he needed to restore the mental and emotional link to the source of imaginative and abstract thought - to that contemplative inner world where he knew the seeds of wisdom to lie; and where the hovering presence of spirit may be encountered. It is not easy to do simply by willing it to be so, but requires the ability to let the world and its affairs drop away, put the senses on hold for a while and drift into a meditative state. Marcus Aurelius was obviously able to do this while on campaign, in his tent, surrounded by an army. Would he be completely alone? I doubt it. Members of the Roman imperial guard would surely be hanging around fairly close to hand.

THE SYMBOLIC ENCOUNTER –
WHEN THE SENSES LEAD TO NON-SENSE

> *Solitude in the presence of natural beauty and grandeur is the cradle of thoughts and aspirations which are not only good for the individual, but which society could ill do without.*
> John Stuart Mill, *Principles of Political Economy*, 1848.

Now we run into a paradox. For when John Stuart Mill (1806-1873), English philosopher and economist, introduces the words *'natural beauty and grandeur'* into the idea of wandering alone in the peace and quiet of nature, he is talking about a countryside that transcends the norm in terms of simple pastoral qualities - one where the topography, quality of light, flora and fauna... are particularly eye-catching: where it is not just a matter of responding osmotically, as it were, to a rural ambience, but of having the visual sense consciously and actively engaged. For to recognize and be moved by a panorama of beauty and grandeur requires an act of heightened visual perception: one's eye is caught and looks with intent, with eager attention, dwelling on the natural features which capture the imagination to imprint the scene in memory, and so retain the intensity of the unaccustomed levels of thought and feeling they inspire - the *'cradle of thoughts and aspirations'*, as Mill describes it.

And solitude is the key. For such moments of vision are far more likely to occur when one is quietly alone, not distracted by others, ensuring that the vital nature of the perceptions responsible will proceed directly to stimulate a strong intuitive response - one which is all the more evocative if it also brings with it soundings from the unconscious. Frank Wild, Shackleton's second-in-command - a man enticed by the beauty and elemental grandeur of the Antarctic wilderness for most of his life - found his responses to the place brought out the deep strengths and qualities of the 'real' Frank Wild: realizations of his own incredible stoic capabilities of endurance, his quiet and confident faith in the positive outcome of events, which brought meaning and purpose to his existence. He owed this he said to '...the little voices'. Talking to himself? Talking to the wilderness? The ice-clad continent talking to him? Whatever the nature of the dialogue, it was one he seemed only to experience when in Antarctica, the place where the most absolute solitude is to be found.

Consequently, I would say that consciousness functions at its highest and most meaningful levels when one is in a state of splendid isolation - a condition which brings eye and ear (in particular) to full alert. And the more discerning and absorbing the visual and auditory revelations which ensue, then the more intense and gripping will be the deep-seated thoughts and feelings evoked. There is really something quite remarkable about this process. For it is reasonable enough to conclude that with the eye locked onto a tree, a flower, or a face... the viewer will come to thoroughly 'know' the structure and overall form of the object. But it is altogether less obvious that such an objective involvement can cause consciousness to turn inward and release thoughts and sentiments, indwelling aspects of one's own nature hitherto dormant. In other words, we depend on the senses not only to introduce *sensible* knowledge of tangible realities, but also to elicit *non-sensible* levels of awareness - of immaterial states of one's being which are imperceptible to normal sensory scrutiny. Quite a feat, and one which we assume - for we have no reliable way of knowing otherwise - is peculiar to human consciousness.

When Vincent van Gogh was in his 'sunflower period' he produced spectacular paintings of sunflower-structure and color. Yet they went far beyond simply yielding such factual information to the keen eye of the painter. For the sunflower, caught in all its baroque magnificence as it swayed in the golden Provencal light, acted as a catalyst - the agent which released a deep well of emotion within the artist - betokening the powerful empathy the flower evoked in him, and which was expressed in the painted image no less commandingly than were the tangible realities of its organic form. Man and sunflower joined forces, creating a union of great expressive and symbolic impact which communicates itself immediately to most viewers of the paintings. Michelangelo, in talking about the awakening and creative expression of such inner sensibilities and the transformation of nature into the work of art that ensues, attributes this complex mental transaction to the workings of the 'inner eye'.

(Yet we have seen that spontaneous mental breakthroughs of inspired creative intensity can occur *without* the presence of some 'special' outside object to excite the senses, and work as a catalyst - a symbol - to stir the imagination. The brainstorms experienced by Mozart did not occur in conditions of solitude, nor did the intrusion of any suddenly perceived outside thing or event inspire his musical discoveries).

As the Second World War came to an end, a series of 'thorn' paintings was created by the English artist, Graham Sutherland. They very quickly achieved public recognition as evocative images symbolizing the sufferings of war. In the aftermath of a conflict which had seen cities badly bombed and thousands of civilians killed, with hundreds of thousands of men and women lost while serving in the armed services, the drawings and paintings of thorns touched a collective national 'nerve' - evoking the memory of hardship and the pain of human separation and loss. It is significant that the impact of the thorn images was not limited to only those interested and versed in the appeal of the visual arts. Certain works in the first London exhibition were reproduced extensively in the popular press, and their expressive and symbolic power was immediately picked up by the general public. There was no need of connoisseur or critic to explain them to the man in the street. Such was the graphic forcefulness of Sutherland's bleached and bony spikes - embodying years of wartime danger and suffering in their forms - that people were brought to reflect on their own and the nation's losses - to recall the years of war, ponder human folly and our natural mortality, and feel for those who suffered a premature death in such fearsome circumstances; wonder at their courage.

Yet there is a universality and timelessness to Sutherland's images - their impact is not limited to a wartime generation. The series can bring any viewer up short, whatever their generation, wherever they come from. For when confronted by his gateway-like arcing and thrusting spines of thorn, most observers feel discomforted, are made sharply aware of the hazards which must be faced in every human life - and braved - if safe passage through to the green and mysterious inner depths of the design, the goal at the end of the journey, is to be accomplished.

The story of how Sutherland came to create these graphic constructions of 'thorniness' – his outstanding achievement as an artist in my view - bears telling. It attests to the way, when one is alone, how profound feelings and attitudes can be awakened by the sudden appearance of some external catalyst - inner awarenesses which demand to be expressed and shaped through language, drawing, music, dance... In Sutherland's case the catalyst was the thorn bush: a surprising confrontation which resulted in the making of over 40 drawings and paintings. 'Surprising' is the key word here, for truly significant encounters of this kind are almost always unpremeditated: the thorn bush - which had never previously invited the artist's attention although he passed it regularly on his walks -appeared 'out of the blue', as it were. For the day came when there it stood before him, inescapable. The artist said that it seemed to detach itself from the hedgerow which grew alongside, and exist apart, demanding attention. And for the first time he perceived, with startling clarity – as if blinkers were suddenly removed from his eyes – the structural complexity of points and scimitar-like thrusts of thorn spines; the interlocking weave of branches

Thorn Bush. Graham Sutherland

shaping thorn-tree-space in extraordinary ways. Sutherland was fascinated by the organic tenacity and menacing appearance of nature's design. And following quickly on the heels of these visual perceptions came a rush of sympathy for – to use the poet Dylan Thomas' words - *the wire-dangled and suffering human race:* the millions who suffer extremes of physical and psychological pain through the vicissitudes of human life. He felt a thorn's sharpness without having to touch it; thought of the shrike bird which impales its victim on a thorn; of a crown of thorns lacerating the scalp; of the frailty of all flesh when pierced by the lance of death.

So he began to make drawings – sketches of details, surgically precise; and studies of the complete bush from several sides – all from a need to capture and understand the structure; yet, at the same time, to re-form nature in order to create the forms which would convey the strength of his feelings and the run of his thoughts.

In this way the Thorn Tree series came about. And from then on his art became more and more profound - always celebrating and expressing the intriguing complexity of natural things, and always managing to suggest the impermanence of everything in nature, the constant process of moving from being to becoming: images which evoked in myself a certain sad contemplation at the finiteness of all living things. An artist's creative and philosophical life.

Courtesy of an affair with a thorn bush.

Special experiences of this sort are not the prerogative of poets or painters. Many of us have surely been brought more intimately into contact with ourselves, surprised by the direction thoughts have taken, the strength of feeling aroused by the sudden and surprising sight of some object or scene - a particularly spectacular sunset, a venerable and gnarled old tree, or the marvel of a radiant hummingbird. Or by any of the other senses abruptly jumping into high gear: an unforgettable sound - a person's voice, music, the song of a bird, the strike of surf; or an acutely evocative smell - the aroma of the kitchen, the fragrance of perfume or flower… the lingering odor of violent destruction, decay and death; even the sensation, well remembered, of a most unusual taste or quality of touch - *Of all smells, bread; of all tastes, salt,* wrote George Herbert, (17th Century English divine and poet.) In the poem *Tintern Abbey,* Wordsworth writes, *Sensations sweet/Felt in the blood, and felt along the heart.*

Why should this be so? Why should a particular object or scene, a distinctive sound or smell, strike through to heart and mind? What is William Blake going on about with his sand and flower quatrain? *To see a world in a grain of sand/And a heaven in a wild flower/Hold infinity in the palm of your hand/And eternity in an hour.*

Well, there is nothing puzzling about the fact that shapes, colors, sounds… which are out of the ordinary will attract interest and sharpen perceptions. It is when involvement persists beyond an appreciation of physical qualities alone, and arouses sensibilities within the individual which were hitherto dormant, or at least unfocused, that 'the plot thickens,' as the mystery writers say. For this switch from the objectivity of an act of perception, to the subjective experience of self-revelation, is one of the wonders of consciousness' labyrinthine ways.

But why should such symbolic confrontations not be commonplace, considering the thousands of impressions of the world we receive every day? In taking Graham Sutherland's thorn images as an example of an induced and altered state of consciousness, I would say that the thorn bush was only able to make its impact because the artist himself had reached the necessary degree of readiness - the development of a high degree of analytical perception, and the onset of strong intuitive feelings and reflections.... all sufficiently charged with a sense of mystery and a sympathetic response to all forms of life. And the thorn bush, taking him by surprise, acted as the catalyst which served to release these, by now potent, sensitivities and awarenesses. It might also be said that on that particular day, at that particular time, Sutherland had arrived at the crucial point when this now finely tuned circle of consciousness was *also* invaded by the spirited contents of an unconscious.... just waiting to breakthrough.

This is how catalysts work as symbols. They strike when the iron is hot, so to speak. Had the thorn bush not moved Sutherland so profoundly because of its symbolic significance – merely interested him solely as a visual phenomenon due to, say, the intricacy of its natural design - the artist in him might still have been attracted to make some fine drawings resulting in technically good 'portraits' of a thorn bush. But they would be representations, however well done, of only its physical appearance – not images showing the bush transformed, re-designed, to tell you also a great deal of the man Sutherland.

Let me talk a little more about the quite extraordinary performance of a consciousness which brings about this interplay between aspects of an external world, and the internal notions which make up the life of the human psyche. It is a process which may be described as one of *symbolic awareness*. Suppose, for example, you tell me in conversation that you have a great deal of hope for mankind, and I respond with a cynical smile and ask what you mean. *'What is hope?'* say I, *'Let me see it, touch it...* You would certainly not take my request seriously, knowing full well that hope represents an abstract state of mind; is not a material entity capable of being described in physical terms. The nearest you could get to conveying what is meant by hope would be to allude to something which, through its appearance, behavior, and associations, corresponds to the emotive experience we call hope. Point to a soaring bird for example and declare, '<u>There is hope...</u>' leaving one to pick up the symbolic significance of an ascending bird. Then the question might be asked, *'When is an upsoaring bird more than a bird in flight?'* And the answer would be, *'When it becomes the embodiment of hope.'* For the freedom of feathered creatures to ascend skywards and escape the earth, the pull of gravity, has for thousands of years served symbolically to convey wishful thoughts of escaping the futile, the trials and tribulations of the world - usually dreams of attaining a life of more promise and purpose. Hope.

> *Hope is the thing with feathers*
> *That perches in the soul,*
> *And sings the tune without the words,*
> *And never stops at all.*
> Emily Dickinson (1830-1886)

Our old friend William Wordsworth, contemporary of Coleridge and living in the mountainous country of the English Lake District, would take long solitary walks over the fells of that wild domain. One of his best-known Lake Country poems, *The Daffodils,* describes a sudden and unexpected encounter with a 'host' of lakeside daffodils, and commences by vividly setting the scene for the reader. He starts by describing the setting, *Beside the lake/Beneath the trees...* ; and goes on to convey the scale and liveliness of the spectacle, *Ten thousand saw I at a glance/Tossing their heads in sprightly dance...* Yet after several more verses presenting all the facts, so to speak, it is the final verse which carries the ultimate impact of the panorama on and within the poet's consciousness.

> *For oft when on my couch I lie*
> *In vacant or in pensive mood,*
> *They flash upon that inward eye*
> *Which is the bliss of solitude.*
> *And then my heart with pleasure fills*
> *And dances with the daffodils.*

Thus Wordsworth reveals how he was led into himself – how daffodils, 'ten thousand' of them, set in a scene of elemental grandeur, first triggered a high level of perception before acting symbolically to quicken 'that inward eye', and evoke an exalted state of complete happiness.

And here is his neighbor Coleridge on the same theme:

> *In looking at objects of nature... as at yonder moon dim glimmering through the dewy window pane, I seem rather to be seeking, as it were asking, a symbolical language for something within me that forever and already exists.*

THE TEST OF ULTIMATE SOLITUDE

There is a remote part of the world where Coleridge's *'something within'* - that shadow-self normally kept at bay by the overbearing pressures of time and the complicated logistics of life in western civilization - will make itself known to consciousness. It is a place where any lengthy stay will put fortitude to the test in a way rarely experienced when living in the relative physical and psychological security afforded by contemporary urban cultures. Hidden aspects of personality and character can come to the fore which lower morale - fears, phobias, pathologies, weaknesses of will, failures of nerve, faintness of heart or spirit.... Or, conversely, hitherto undiscovered resources which inspire courage, confidence, intelligence, strength of will... can take over and provide one with the volition to adapt and be resourceful in coping with the harshness and isolation of a hostile environment. Such is the nature of the internal dialogue which automatically becomes the order of the day in these climes. Thoughts and feelings concerning one's adequacy flow constantly, welling like the waters from some hidden spring: self-revelatory apprehensions - occasioned by the surreal topography of the place; ferreted out by the frissons of fear which accompany the onset of bad weather with its attendant dangers; and driven to surface by the symbolic power of ice-rock vistas illuminated by the most subtle and evocative light on the planet - all of which has the imagination running wild in a free fall of mental and emotional associations. Even when in the company of a few others, it always seems that one is alone. And if one has built up a comfortable, self-important image over the years, it is likely to be substantially diminished when faced with this raw, primordial world.

Well, we have been here before with Shackleton when recounting the saga of his small boat journey to South Georgia following the sinking of his expedition ship, the *Endurance*. So, you will no doubt have gathered that I am talking about Antarctica - that landmass larger than the United States, isolated in the unpredictable waters of the Southern Ocean. A place of vast emptiness and stark purity - the ice cap on the high plateau extending downwards onto the continental bedrock some 10,000 feet or more, running down to form huge glaciers feeding extensive floating ice-shelves like the Ross, which is the size of France. For millions of years the continent has known no human presence - sailors only began to venture purposely into the Southern Ocean towards the end of the eighteenth century. The first man to winter over with his expedition on the land itself was Carsten Borchgrevink, the Norwegian, in 1899. And despite the scientific bases around its shores nowadays, and the tourist traffic that plies to the relatively accessible finger of the northeast Peninsula, the Antarctic Continent - its ocean and its ice - manages to remain the most pristine elemental force of nature in the world: all stillness and silence until the great autumnal and winter storms turn the sea into 70 foot swells and roar over the land creating vast whiteouts. It is a place where a man's spirit will rise or crumple; where I believe that most of us average types - caught in seemingly unsurvivable circumstances - would need a leader of Shackletonian qualities in order to help us persevere and keep ourselves alive.

As an example of what Antarctica can do to men who would seem to lack any inner reserves of courage and will, there is the legendary case of the doctor at Almirante Brown - a former Argentinean scientific base in Paradise Bay (in the Antarctic Peninsula) - who, when faced by the darkness and fearsome weather of the winter months, suffered such a total loss of identity and absence of inner direction that he set fire to the base and burned it down.

Obviously, such utter isolation is not for everybody. There are those who possess qualities of character which take over during situations involving solitude and danger, and which militate against accepting the possibility of defeat; others will at least try and rise to the occasion; still others will be overpowered by the place and give up easily or go berserk. Isolation can sometimes introduce a very real sense of a 'shadow' self - a mysterious presence bolstering the psyche and activating the intuition when the situation appears to be hopeless. In my abbreviated account of the crossing of the Allardyce Range in South Georgia by Shackleton, Worsley and Crean after that incredible small boat journey in the *James Caird*, I mentioned that both Worsley and Shackleton, independently of each other, reported feeling the presence of an invisible fourth member of the party guiding their progress. Shackleton took this ghostly visitant to be part of a genuine mystical experience - and I sometimes wonder if his 'shadow' self, so powerfully energized as to subtly transform his appearance, was ever able to be 'perceived' by others. The poet T.S. Eliot acknowledged that when he wrote the following lines of *The Waste Land*, he was perhaps subconsciously influenced by Shackleton's account of his other-worldly visitant when crossing the Allardyce Range on South Georgia.

> *Who is the third who walks always beside you?*
> *When I count, there are only you and I together*
> *But when I look ahead up the white road*
> *There is always another one walking beside you*
> *Gliding wrapt in a brown mantle, hooded*
> *I do not know whether a man or a woman*
> *- But who is that on the other side of you?*

Frank Wild, Sir Ernest Shackleton's second-in-command – and a great but unsung hero of Antarctic exploration in his own right who accompanied Scott, Shackleton and Mawson on five South Polar Expeditions between 1901 and 1922 – said he could not keep away from this land at the bottom of the world because once you have been to *'the white unknown'* you could never escape *'the call of the little voices'*. It would seem that only at the bottom of the world did Frank Wild experience the degree of solitude, together with the harshest of all fights for survival, which imposed the ultimate test on his great spirit. His achievements testify to the calm resolve, courage and stamina he found within himself, leading him to lead others - the internal, dialogue between body and spirit speaking with authority: the psychical power commanding the physical. It was the only place where this much travelled heroic figure found the elemental grandeur and untouched natural purity which possessed the symbolic power to evoke *'the call of the little voices'*.

Sir Ernest Shackleton's greatest heroic accomplishments took place on and around the Antarctic continent. In common with many others he found it difficult to stay away, and in the end, in 1922, he died at sea, embarked on yet one more expedition south. He was buried on South Georgia, the island which was the scene of his most consummate triumph – the first crossing of the Allardyce Range at the end of the small boat journey from Elephant Island. And after his death it was Frank Wild who took command of the expedition and attempted to complete the 'Boss's' plans for the voyage. Shackleton and Wild were a great team in the history of the Heroic Age of Antarctic exploration, for nowhere else in the world seemed to afford the experience of hearing the 'voices' which never failed to guide them through all manner of perils.

I can attest to the fact – after many varied experiences on and around the Antarctic Continent – that many who take the long voyages attempting to traverse West or East sides find themselves subtly changed. Those holding inflexible and traditional religious beliefs, or given to self-importance with a need to measure personal success in life in terms of wealth and possessions, seem to be particularly vulnerable to sabotage by the 'white unknown.' During the 12 long days sailing the Ross Sea from the Antarctic Peninsula to McMurdo Sound, one hears individuals talking of the increasingly dreamlike quality of the days; of the overall feeling of having drifted into a dimension of time and space which takes on a surreal quality bearing no relationship to calendar or clock. I have heard the odd reference to 'life back there' - of how distant, even petty, it seemed with its check-book-materialism and social conventionalities. Even subjects normally taboo - not 'philosophically correct' in social gatherings - such as the brevity of human life and how to face the prospect of death, were aired as if these were the most normal things in the world to talk about.

The place makes philosophers of the most unlikely candidates. Yet unending ice, infinite reaches of space, a crystalline translucency of light, stillness and absolute silence... can do it. I'm at a loss to explain the phenomenon of the Antarctic silence. There were times spent sitting on the lower slopes of Mount Erebus overlooking the Ross Sea when, in the total absence of any external sound in that brittle air, it seemed I was actually *hearing* silence – hearing the cosmos going about its affairs in the absence of all local, extraneous noise. Silence as a transcendent sensation in its own right. For myself, I can only say that weeks spent in nature's last glacial stronghold force a confrontation with the undeniable fact of one's own physical frailty and brief life span. And yet 'the little voices' even manage to find an answer to that. Spirit makes itself known and comes to the rescue.

> *Where does the strange attraction of the polar regions lie, so powerful, so gripping that on one's return from them one forgets all weariness of body and soul and dreams only of going back? Where does the extraordinary charm of these deserted and terrifying places lie?*

So wrote the French explorer, Jean-Baptiste Charcot, who made his last Antarctic voyage in 1908. (Today he would not find it quite so deserted. Yet the scientific bases - most sprinkling the coastline - are really lost in the immensity of the continent and create, as yet, minimum pollution. But I suspect that Charcot would turn about and flee like hell if he found himself standing off the United States Base in McMurdo Sound with its attendant garbage, where well over a thousand personnel – together with their equipment – arrive to invade the solitude every summer season.

6

TIME and the DREAM

*I believe it to be true that dreams are the true
interpreters of our inclinations; but there is art
required to sort and understand them.*
Montaigne: *Essays III.xiii.*

We dream at night while asleep. And even during the hours when supposedly awake we can find ourselves involved in reveries or daydreams – involved in thoughts and creating imaginative scenarios which have nothing to do with the present moment or the job in hand. But it is the night dream which, for most of us, can provide the most extraordinary and surreal of mental experiences: particularly those dreams which, by virtue of their bizarre content and out-of-present-time setting, question the authority and absolute credibility of the 'reality' presented by consciousness when we are awake.

When it comes to being whisked off into dreamland, the senses and their neurons which take stock of the outside world - activating the rational and emotive processes of awareness - have basically gone off duty, leaving but a skeleton crew of neurons to break through and inform of anything happening beyond the confines of the bed. Now, with the sensory gate to the outside world largely shut down…. those neuron-complexes which serve the *opposite* gate of intuition are free to create the dream-life of the sleeper. When one is 'dead to the world', the subliminal content of intuitive and unconscious intelligence - manifesting in dream-form - is far less likely to be challenged by the neurons serving a wide-awake consciousness. Daydreaming can induce a somewhat similar situation, but it is during night sleep that the most symbolic and perplexing dream-messages from deep within the psyche can break through, and present to a mind free of worldly involvement the sort of 'night life' that can be either positive and enthralling or negative and nightmarish.

There are BIG dreams and LITTLE dreams. Such is the distinction made by many earlier cultures (including the Australian Aborigines) in order to separate dreams of consequence from those of little account. The general view has been that little dreams are relatively topical: they possess the required element of 'dreamy' gravitationless in moving through bizarre repeats of life's familiar routines.... in relatively familiar places; they may be pleasant and benign, or reveal hidden menaces which one must try and avoid. But in the final analysis they bear an obvious relationship to things and happenings in 'real' life. One important clue as to whether a dream falls into the 'big' or 'little' category is the degree to which its pictorial sequences and general content fade on waking-up. If it fades rapidly when awake, and any recall proves to be impossible, it is a dream of little consequence: the result, perhaps, of an overactive brain that continues to dwell on the affairs of the day - or of one over stimulated by the events of the day, or by too much good food and drink!

On the other hand, if a dream unfolds as a bizarre, surreal series of events which bear no relationship to anything experienced in the normal course of your life - be it either in time or place - and in which you play a leading role, vaguely aware that what is taking place has a direct bearing on the course of your life, then a big dream is likely underway. Time, as experienced in such dreams, can be extraordinarily fluid and non-chronological: you can be back in the historic or prehistoric past, in the present, or be transported abruptly into what is seemingly the future. And when you find that dreams of this nature persist strongly throughout the days and weeks which follow - indeed, a big dream can remain in memory for months or even years - it is important that you recognize and seek to understand its significance. Another big dream characteristic is that on waking, the events dreamed seem to be more 'real' than the familiar world of daily life; so much so that you find yourself incapable of simply shrugging the dream off... and continue to dwell on its possible meaning.

Graham Sutherland's moments of heightened thorn-bush perceptions can be seen as a 'big' daydream.... releasing thoughts and feelings which changed the course of his life as an artist, turning exceptional talent into creative genius. Big nightdreams work in a similar kind of way.

On several occasions throughout these pages we have noted that clock-time has no absolute hold on consciousness – that one minute can seem like ten, or ten minutes like one, depending on one's emotional state and general frame of mind. Many of us are able to recall how time drags while waiting for any important telephone call - say from a lover, wife or husband - which is supposed to arrive at seven o'clock in the evening: how incredibly slowly the last five minutes drag themselves out *before* the hour is reached - while every single minute *after* the hour, when the phone does not ring, becomes a veritable eternity. Yet when finally the call comes, and he or she is there, the reverse is true – time speeds up: you talk together for half an hour, yet would swear the conversation had lasted but a few minutes.

It is a matter of experiencing time in *duration*, or time in *intensity*. During the normal course of events when one is engaged in life's predictable routines, time is experienced very much in duration; as long as there are no surprises clock-time rules, being automatically and unselfconsciously assessed. But when urgent and strong feelings are invested in the situation – such as those present while waiting for important news, when confronted by the unexpected, or while absorbed in truly absorbing, creative activity... then the clock-measure of time goes by the board, disrupted by the intrusive presence of any strong passional intensity affecting one's state of mind. And there are no occasions when clock-time is thwarted so startlingly as when we trustingly embark on the venturesome night passage known as sleep, when the big dreams impose the most dramatic transpositions of time and place. They are obviously the ones to wait for. Throughout history they have been regarded as auguries of personal destiny and thereby as guides to the outcome of events, demanding interpreters (now psychoanalysts) to make their implications known. And certainly, since the medical skills and insights of modern psychiatry were introduced by Freud and Jung early in the 20th century, the importance of understanding such dreams in order to help determine the nature of a patient's psychological inadequacies is acknowledged by many psychiatrists. Big dreams were seen - especially by Jung - as the means by which the unconscious communicated its most pressing concerns to consciousness: acting as a kind of 'seventh' sense - a phenomenon similar to the so-called 'sixth' sense of the intuition.

There is no standard way to reveal the symbolic significance of big dreams. Sometimes continuous and patient reflection - usually an involuntary mental preoccupation after such powerful dreams - produces a level of understanding which can lead one to make the necessary changes in attitude and behavior subtly projected by the dream experience. And sometimes it is possible for the dreamer to know while actually dreaming - or immediately on waking - just what the dream is getting at.

However, the most surprising prophetic and revelatory examples of our dreaming life occur when scientists, writers, musicians, philosophers.... receive information in dreams which provide the solutions to the final, seemingly intractable problems they encounter in their work. While those who find themselves in dire and urgent need in their day-to-day lives can 'sleep on it', as is said, and find answers.... There are also those psychically sensitive men and women in everyday life who receive prophetic images of future events in their dreaming - revelations concerning happenings in which they themselves would not necessarily be involved: knowledge which has enabled preemptive action to be taken, disaster averted and lives saved.

Let me provide some illustrations as to how big dreams work. Suppose for example that you, as a psychiatrist, have a young woman patient who has an absolutely obsessive attachment to her father, a good and kind man - a love which she describes as the most important thing in her life. And that one day she tells you of a disturbing series of dreams she is experiencing – dreams in which her father appears in one guise or another as a violent, even evil person. Being overcome by feelings of guilt in dreaming of him in such a derogatory way, she asks your advice on how to deal with these troublesome nightmares. As a doctor versed in dream lore, I expect you might say something like, 'Well, it's about time...' and then proceed to explain what you mean by this somewhat unsympathetic statement. You would no doubt point out that for the child who has become an adult to still regard her father so idealistically as a man who can do no wrong, is pretty unrealistic, given the frailty and unpredictability of human nature. More to the point, you would tell her that to idealize a parent to such an extent can work adversely in restricting the development of her own self-image as an adult - militate against her going her own way in life, for the struggle to live up to the unrealistic standard of perfection she has imposed on her father will more than likely result in a sense of failure, unworthiness, and a low level of self-esteem on her part. Living with such a fixation will severely inhibit the growth of her own unique character and make it difficult for her to realize her own potential.

Consequently, by projecting a series of unfavorable father-dreams, an unconscious 'regulator' seeks to correct the imbalance in the daughter's everyday consciousness which can result from such an intense paternal relationship. For in bringing her to face the dark and potentially destructive side of human nature - present in her father as in most human beings unless they are truly saints - the dream's repeated insinuations that the father, like most men, is far from perfect, can work subliminally to remove him from the ranks of the Gods and free her from subservience to an unrealistic and unattainable ideal.

It is really quite remarkable how often – without benefit of an analyst – a pattern of 'big' dreams of this sort becomes intelligible: how its symbolic allusions are recognized by the intuition and subtly make their way into consciousness as a convincing series of thoughts.

One night, late in March 1960, I did not get to bed until well after midnight. I had spent weeks rewriting pages of a manuscript that were covered with more editorial blue pencil than I care to remember. And I was getting nowhere. The concepts still refused to shape themselves verbally and the new writing was without form or style. Mentally, a grand confusion reigned; physically I was tired out. I remember falling asleep very quickly when I retired, and the dream seemed to take over almost immediately. Everything was taking place at sea – water a calm azure blue, and flat as the proverbial pancake to the distant horizon – myself placed at the stern of a naval ship, holding a rope which ran out over the rail to a small floating raft-like platform occupied by a single figure, and situated about a quarter-mile out on the port side. The wing-bridge of the ship was just forward of midships and to my right. The captain standing perched up there was clearly visible. And stationed at the prow was a seaman – vaguely familiar – also holding a rope running through the sea to the raft. The captain gave instructions through a megaphone. We would commence steaming straight ahead at 10 knots and, on his command, I and my partner on the prow were to start hauling in the raft to reach the gangway situated just above the waterline at dead center of our vessel's hull. Off we went, a white wake streaming out behind us straight as a die. On the bridge, the captain raised his hand and I, together with my colleague up forward, began to pull. All went well for the first 200 yards or so and then, thinking that if I did not exert myself the raft and its occupant would miss the ship completely, I started to pull wildly on the rope losing all rhythm and synchronization with my colleague on the prow. As a result the raft came hurtling towards the stern, finishing up in the wake and dangerously close to the propellers.

The ship slowed. The captain came down from the bridge and gave me a lecture, the gist of which was that I was trying too hard, was too impatient and ignoring his directions from the center; that I appeared oblivious of the need to coordinate my efforts with those of my fellow hauler at the prow. It must be a *balanced* pull. Try again. We resumed speed. Up went the captain's hand. I began to haul in. All went well. The two of us with the ropes pulled in unison, each with an eye on the bridge. But with only 50 yards to go I could no longer contain myself. The vessel's speed seemed to have increased, the bow wave was causing the raft to bounce and lose headway, and I saw victory slipping away if more effort was not exerted to maintain the raft's momentum. Again, the same result. My frenzied attack on the rope had man and raft almost swamped in the wake at the stern. Another visit from the captain. We were to make a third and final attempt. *"You cannot go it alone,"* he said sternly. *"There are four members to this team – you, your colleague on the bow, your good friend on the raft, and myself. You'll get nowhere if you don't understand we must act as one – and especially if you're blind to directions from the bridge. You must listen to me; have patience, and trust the rest of us. Then all will be accomplished in due time."*

Out went the raft yet again; once more we commenced the exercise. Long measured hauls on the rope now – hearing the captain and keeping one eye on him as he orchestrated rhythm and timing; and synchronizing my movements with my colleague up front. Slowly, steadily, the raft approached. A hundred yards to go and the captain indicated we should ease off for a moment... recommenced slowly... slowly... until the raft slid gently alongside and the occupant – giving me a grin and a wave – clambered aboard at the midships gangway.

Immediately I was no longer viewing things from the deck of the ship, but from directly overhead – way above mast height. And as I watched from this new vantage point the ship suddenly changed course, turned hard-a-port and made a complete 360-degree rotation. The sparkling white water of our wake – vivid against the blue of the sea – closed on itself to form a whole and perfect circle.

I cannot say with any certainty when I actually awoke from the dream, for the reflected light of the jewel-like ring of foam persisted – a great halo illuminating the ocean to some depth – even though at this point my eyes were open and I was dimly aware of the familiar layout of the bedroom. I lingered in this fashion for some time, weightless on the bed and possessed of an exceptional tranquility and sense of well being, before falling into a dreamless sleep which lasted the rest of the night. And on waking, sequences from the dream were still present in the mind's eye – vivid and arresting after-images which seemed no less tangible than the physical realities presented by the 'real' world as one went about the work of the day. The extraordinary calmness and serenity which followed in the dream's wake has always been subject to recall, simply by visualizing the blue ocean and the iridescent white circle formed by the ship's wake. Even forty years later, when beset by doubt and apprehension occasioned by the direction life is taking, I can conjure the white-on-blue immaculacy of the circle to appear; hear the captain's instructions as I struggled with the rope.

The mysterious influence of big and important dreams is usually felt fairly quickly and tends to leave a lifetime impression on consciousness. I recollected, even as my dream of the sea-born circle faded, that the *circle* as such has long been used to symbolize an age-old metaphysical concept. For as a geometrical figure described by a line moving continuously around itself, having no discernible beginning or end, it represents the idea of a cosmological continuum: a coherent whole embracing the progression of space, time, and the dynamic forces by which everything is held in a state of existence, flowing seamlessly in a universe which has been mythically referred to as the Great Round. The circle formed the basis of Hindu and Buddhist mandalas, and lent itself to the intricate construction in stone and stained glass of the 'wheel' or 'rose' windows found in Europe's Romanesque and Gothic churches: wonderful circular designs symbolizing the unity of all cosmogenic forces and events; as well as the complementarity of opposites such as life and death, body and soul, finite and infinite, light and dark, good and evil... In the great wheel-windows of mediaeval cathedrals - usually found on the west facade, the east wall of the sanctuary, or high on the walls of north and south transepts - intricate designs of stone tracery spring like the spokes of a wheel from the circle's intrinsic center or nucleus to interconnect at points around its perimeter: visual reminders that everything and everywhere is equidistant from, and connected to, the structural authority of the hub from which they spring. And when filled with a pictorial record of biblical events created from pieces of colored glass, the windows acted as powerful religious icons, particularly for the illiterate members of mediaeval congregations.

But to return to the dream. To have been involved in the forming of such a perfect circular figure - however ineffectively at the outset - had the effect of infusing self-confidence and imparted a sense of purpose at a time when these qualities were badly needed. The 'transformation' was immediate. Throughout the following day the normal passage of clock-time was suspended, in the sense that I never gave a thought as to what time of day it actually was. And, stranger still, was the feeling that I was not alone: that this physical self, the face in the mirror, was merely the aspect of myself that could be seen and touched… whereas the truth of the matter was that the visible man played host to invisible selves going by the name of 'mind', 'spirit', or 'the other'… whose presence, once recognized, opened up new avenues of comprehension down which to wander on the journey we call life.

But the exercise at sea had been a close call. It would have failed if, on the last attempt, I had been unable to curb impatience and self-centeredness and recognize the superior skill and intelligence of the other members of the crew. Over the next few days I came to realize what a self-absorbed and self-important little scribbler I had become. Sitting aloof in my ivory tower, having little or no time for the love and companionship of family, friends, dog… I was living life as a 'straight line': a narrow, blinkered, and selfish existence in which all the benefits of solitariness were lost - as they are always likely to be lost - if an overweening self-absorption in wearing the mantle of 'the writer' renders one insensitive to what is going on in the lives of those who share one's world. A hermit-like severance from those who love and are loved - undertaken in the mistaken belief that it is the sure way to be in touch with the muse - can have exactly the opposite effect. For after several weeks of misanthropic seclusion the words I was expecting to flow, simply stopped coming altogether, resulting in a growing sense of frustration and increasing levels of anxiety. And no amount of effort on my part could force the relevant thoughts and insights to come to mind. Obviously, my particular muse looked askance at such self-conscious efforts; would not be pressured to speak however earnestly I interned myself like a broody hen, and whatever the degree of urgency I felt.

The dream put paid to this mannered and pretentious attitude, demonstrating the negativity and ineffectiveness of such psychological insularity - witness the impatient and egocentric way I pulled on the rope, incapable of cooperating with the others in bringing the raft and its occupant safely home to the central position of our ship. It was only after the lecture by the captain that I was able to act as a member of a team and, in so doing, contribute to a balanced and collective effort. At which point the man at sea was brought aboard, and from above I saw the white wake form the great circle. The day after the dream, the words came, and the book was finished within two months.

Dreams such as this cause one to ponder how their imagery works to expose the psychological problems which are putting a spanner in the works, so to speak - and then goes on to set in motion the changes in attitude and behavior which can put one on the right path. In my own case I found it impossible the next day to relegate the dream to the back of my mind, and it was while walking the dog in the afternoon that, quite suddenly, the final pieces of the puzzle fell into place: there were not four independent members of the ship's company involved in this labored rescue attempt. Just myself.

There stood the 'I' in the stern, all existential ego. The one out on the water in the raft embodied the Unconscious - the seat of mystery and the soul - drifting at the full extent of the lines barely securing its raft to our collective 'ship of state', and urgently needing to be restored to its central position at the heart of the vessel. The captain, supervising the operation from the bridge, was a manifestation of the surpassing and guiding force of spirit - the soul's dynamic ambassador to consciousness. While the crewman in the bow represented my intuitive self, constantly trying to moderate ego's impetuosity. These three psychical powers had a hard time in teaching ego to recognize its limitations and cooperate in making the rescue operation a success: overcome the danger of losing contact with a very lonely and lost soul....

Even the daydream can work effectively to lessen the dominance of the senses and reason; draw attention to the other sides of consciousness' coin - soul, spirit and imagination. We need the balance if we are to have a window onto the world and into the psychical landscape of oneself. Remember Alice? Lewis Carroll sets the scene for the daydream of all time in his evocative story of *Alice in Wonderland*.... when Alice, feeling sleepy on that hot afternoon - mind off and free, open to whatever unconscious needs will opportunistically launch into the symbolic world of the big dream - sees the White Rabbit running past and, pulling herself up from the bank where she is sitting, follows him...

> *Alice was beginning to get very tired of sitting by her sister on the bank, and of having nothing to do: once or twice she had peeped into the book her sister was reading but it had no pictures or conversations in it, 'and what is the use of a book', thought Alice, 'without pictures or conversations?'*
>
> *So she was considering in her own mind (as well as she could, for the hot day made her feel very sleepy and stupid) whether the pleasure of making a daisy-chain would be worth the trouble of getting up and picking the daisies when suddenly a White Rabbit with pink eyes ran close by her.*
>
> *There was nothing so <u>very</u> remarkable in that; nor did Alice think it so <u>very</u> much out of the way to hear the Rabbit say to itself, 'Oh dear! Oh dear! I shall be too late!' (When she thought it over afterwards, it occurred to her that she ought to have wondered at this, but at the time it all seemed quite natural); but when the Rabbit actually <u>took a watch out of its waistcoat-pocket,</u> and looked at it, and then hurried on, Alice started to her feet, for it flashed across her mind that she had never before seen a rabbit with either a waistcoat-pocket, or a watch to take out of it, and burning with curiosity, she ran across the field after it, and was just in time to see it pop down a large rabbit-hole under the hedge.*
>
> *In another moment down went Alice after it, never once considering how in the world she was to get out again...*

We may not all have had such surreal adventures during our own afternoon periods of drowsiness - yet we can recognize the aptness of the rabbit-hole metaphor, knowing how easy it is to fall into the dark cavity of the imagination as languor overtakes one around siesta time, and how only a terrific effort of will can arrest the descent. But Alice was well and truly gone, lost deep in the underworld of a remarkably intriguing and whimsical fantasy. Lewis Carroll's story is a joyous tribute to the ease with which the young person's attention can slip away from the here-and-now and be engaged by the deep levels of an uninhibited imagination - by pantomime-like tableaux that run the gamut from the credible to the incredible, the farcical to the tragic; telling of the presence of both the powers of goodness and those of malevolence in the world: a series of events that often parody 'real life' situations. In so doing, they sabotage the belief that the objective information provided by the five senses when we are wide awake is sacrosanct - unquestionably true as the only criteria by which we judge something to be 'real'. The topsy-turvy nature of life in 'Wonderland' leads Alice, naturally enough, to consider the 'why's' and 'wherefore's' of the daydream's bizarre encounters, and to contrast them with the way things happen in 'real' life: a process which, as the reader notes, induces her to cast an ironically appraising eye on the unpredictable and often illogical behavior of people in life proper. And the fact that animals play many roles in her daydreaming is significant, for generally children experience an easy empathy with other creatures and intuitively find them vehicles to which they attribute human personality traits. I suspect it is one way in which they can deal with the vagaries of character exhibited by the adults in their world. In addition, and I would think that certainly in Alice's case, her wanderings in never-never land helped her develop a certain breadth of vision when it came to reconciling fact with fancy - accepting the relative ease with which the mind can move from indulging in a reverie, to reporting on the 'real' world of time and material events offered by the senses. She seems to have had little difficulty in reconciling the two experiences - allowing each their particular level of credibility.

Alice is the archetypal young person caught in the very spring of youth - increasingly sensible concerning affairs of the external world, yet still equally responsive to the reality and significance of her own inner realm of thought and feeling. She is representative of all those youngsters standing at the crossroads of childhood and adolescence who find it difficult to let their imaginations succumb to the growing onslaught of adult reasoning and logic - and seem to know intuitively that their individuality, their wholeness, depends on reconciling the world of imagination and dream with the material facts of life. Alice was obviously a very vital and curious child before she followed the White Rabbit down his hole. A less alert and inquiring child might well have remained indifferent to his appearance and just dozed vacuously away. Certainly, it seems to me, she was even more interesting and dynamically alive *after* her sojourn in Wonderland - 'curioser and curioser' to quote her own words.

Curiosity is one of the most permanent and certain characteristics of a vigorous intellect, wrote Samuel Johnson (1709-1784), the English lexicographer and author, in *The Rambler* (No. 103). And we would do well to remember that memorable dreams work not only to bring one to know oneself, but also to stimulate intellectual curiosity concerning life's paradoxes and mysteries. In Alice's case they obviously did both.

Yet it is also worth noting that despite the dream's overall whimsicality, the macabre nature of some events is but thinly veiled. I have always thought that the coming-and-going apparitions of the Cheshire cat gave a somewhat sinister air to the dream; while the use of the poor old flamingoes as croquet mallets reflected a bizarre and sadistic humor. Constantly quarreling with her guests the Queen of Hearts was constantly crying, 'Off with his head...' - executions which, however, were not carried out; much to Alice's relief, her innate sense of justice aroused by such arbitrary and unfeeling sentences. And then consider the Mock Turtle's obvious unhappiness as he dwells on the culinary fate which befalls many of his kind. Funny, amusing, and fanciful - the dream is all of these things; yet there is also more than a hint of the dark side of nature, and of human nature in particular - the unpleasant, the cruel and the harmful... sober realities which Alice, as the dreamer, must be prepared to face in her life. And I wonder if today the shriek of the Gryphon in Alice's Wonderland, the Mock Turtle's heavy sobs, and the squeak of the Lizard's slate-pencil, may be too symbolically and imaginatively subtle for young readers brought up on a diet of television fare: science fiction and special effects, criminal dramas and gratuitous violence.... and the all too frequent inane advertisement.

Albert Einstein is said to have disclosed that the inspiration for his theory of relativity came to him in a dream. The German chemist, F.A. Kerkule describes a dream which resulted in the discovery of the formula for benzene: *"I turned the chair to the fireplace and sank into a half sleep. The atoms flitted before my eyes...wriggling and turning like snakes. And see, what was that? One of the snakes seized its own tail and the image whirled scornfully before my eyes. As though from a flash of lightning I awoke. I occupied the rest of the night in working out the consequences of the hypothesis."* ('Consequences' which led to the discovery of the benzene formula.) Some years later Kerkule, speaking to a gathering of scientists in 1890 advised them as follows: *"You must learn to dream, gentlemen,"* he said. George Cabanis, French physician and philosopher, writing in 1802, asserts that Benjamin Franklin - *'the most wise and enlightened of men'* - considered that there were times when it was through his dreams that he learned how matters which were presently absorbing him were going to turn out. And as a prime example of revelation by dream I can do no better than refer my reader to the account in Chapter 3 of the dream-intelligence – recounted firsthand to myself – which came to one of the Polynesian elders on Easter Island; through which he learned of the facts concerning a mysterious death that had occurred earlier – facts which were shown to be true – and which served to clear an innocent person held under arrest in the local gaol on suspicion of murder.

Many more people than might be expected can recount instances of receiving vital information by 'dream telegraph', even in the course of routine living. Students of history and anthropology discover that creative dreaming is widespread in human history: the prophetic clairvoyance of the African Pygmy during sleep, for example, is well documented. The acclaimed writer, Graham Greene, kept a diary of his dreams, published after his death in 1991 as *A World of My Own*. Yvonne Cloetta, who contributed the Foreword, writes:

> *It is well known that Graham was always very interested in dreams, and that he relied a great deal on the role played by the subconscious* in writing. He would sit down to work straightway after breakfast… He was in the habit of then reading, every evening before going to bed, the section of the novel or story he had written in the morning, leaving his subconscious to work during the night. Some dreams enabled him to overcome a 'blockage'; others provided him on occasion with material for short stories or even an idea for a new novel (as with It's a Battlefield, and The Honorary Consul). Sometimes, as he wrote, "identification with a character goes so far that one may dream his dream and not one's own"- as happened during the writing of A Burnt Out Case, so that he was able to attribute his own dream to his character Querry and so extricate himself from an impasse in the narrative.***

Even more curious are those dreams which prophetically tell of the future - serve as premonitions of things to come which are far beyond the purview of the dreamer's normal range of perceptions. And, stranger still,

---

* Both the unconscious and the subconscious would seem, on separate occasions, to have been involved here.

** "Foreword" by Yvonne Cloetta, from A WORLD OF MY OWN by Graham Greene, copyright © 1992 by Verdant S.A. Used by permission of Viking Penguin Group (USA) Inc.

such visionary knowledge does not necessarily have to wait for sleep to divulge its information. On such occasions the unconscious seems able to roam freely in time and break through with images relating to the past, present, or future - even when one is fully awake, and particularly at moments when such intervention may avert or minimize disaster. All of which causes me to imagine the presence of a few élite regiments of neurons, specially placed under the command of the unconscious, ready at the appropriate times to fulfill their undercover missions and, at the most crucial of moments, bring prophetic intelligence to cerebral headquarters.

During the Second World War, aircrew in both Fighter and Bomber Commands of the Royal Air Force experienced dreams which foretold the danger which lay ahead – warnings which, survivors say, served to prepare them for trouble. Air Chief Marshall Lord Dowding, Air Officer Commanding in Chief, R.A.F. Fighter Command during the Battle of Britain – and the man whose pre-war preparations were largely responsible for the defeat of the German Luftwaffe – tells in his book, *Many Mansions,* of Hurricane and Spitfire pilots within Fighter Command who encountered such dream premonitions; and goes on to tell of paranormal phenomena which intervened directly, without benefit of a dream - of injured pilots having their severely damaged aircraft taken over and kept in the air; even landed by (or so they were convinced) some invisible presence.

The following account of a 'time switch' which occurred in flight, suddenly presenting a new field of visual perceptions to the pilot which overcame those by which he was flying at the present moment - and allowed him to 'see' into the future - is an example of what I would say was the active intervention of inexplicable unconscious forces in a perilous situation. Victor Goddard, a Royal Air Force pilot flying a Hawker Hunter biplane over Scotland in 1934, lost his way during a bad storm. At a critical moment he had the thought that he was in the vicinity of Drem, an airfield he knew to be abandoned. Descending through the clouds to a dangerously low altitude there was the disused aerodrome, a quarter-mile or so ahead. Then something quite amazing happened. Goddard (who was to become Sir Victor Goddard) was to write later: "*Suddenly the area was bathed in ethereal light as though the sun were shining on a midsummer day.*" And he goes on to say that Drem was not deserted and falling into ruin, but displayed restored hangars and outbuildings and was a hive of activity:

mechanics in blue overalls were working on a variety of yellow-painted planes. He flew over the field at about 50 feet, yet no one looked up as he roared overhead and back into the clouds, now sure of his position. In 1938, with war looming on the horizon, Drem was reopened by the R.A.F. as a flying training school and, following Air Ministry instructions, the color of all training aircraft was changed from silver to yellow.

It seems that Victor Goddard had descended from the storm clouds to fly, for a few moments, four years into the future.

"*There was never any doubt to me that something was going to happen,*" said David Booth, a 23-year-old office manager in Cincinnati, Ohio, after undergoing the same nightmare for ten nights in a row. "*It wasn't like a dream. It was like I was standing there watching the whole thing – like watching television.*" In the nightmare he heard the sound of falling aircraft engines, then saw a large American Airlines passenger jet swerve and roll in the air before plunging to earth in a fiery crash. Booth phoned the Federal Aviation Administration at the Greater Cincinnati International Airport on 22 May, 1979. On 25 May an American Airlines DC10 jetliner crashed at Chicago's O'Hare International Airport killing 273 people. The FAA had taken David Booth's nightmare seriously enough to try and check it with all available facts and likely regional scenarios but, after all, they did not have very much to go on in order to speculate on time and place.

"*It was uncanny,*" commented Jack Barker, public affairs officer for the FAA's southern region. "*There were differences, but there were many similarities. The greatest similarity was his calling (identifying) the airline and the airplane... and that the plane went in inverted.*" With hindsight, it is possible to see that the crash site as depicted in the dream resembled O'Hare International, but not positively enough to provide anyone with an 'educated' guess beforehand that the accident would occur there.

> *It was the wise Zeno who said, he could tell a man by his dreams. For then the soul, stated in a deep repose, betrayed her true affections: which, in the busy day, she would rather not show, or not note... The best use we can make of dreams, is observation: and by that, our own correction or encouragement. For 'tis not doubtable, but that the mind is working, in the dullest depths of sleep.*
>
> Owen Felltham: *Of Dreams* (c.1620) - referring to Zeno, Stoic philosopher (c.300 B.C.) - who talked of dreams as influences shaping feelings and states of mind.

7

TIME and the WHIRLPOOL

Lewis Carroll made it very clear in *Alice in Wonderland* that from the moment the White Rabbit went rushing by, glancing at his watch, time had begun to play some very strange tricks on Alice. For in the dream world where she found herself, things did not happen with the kind of understandable, sequential continuity that is experienced in normal consciousness. Events came and went without rhyme or reason. I remember reading the book on my 10th birthday and realizing that this is the way things are in dreams, and accepting Alice's topsy-turvy world without hesitation as being real enough.

Yet this is not just a story for children. The older one becomes, the more one is aware that life is very like the wonderland experienced by Alice. For in discussing consciousness and its billions of neurons, we have seen that our sense of the passing of time is by no means mechanical, uninfluenced by our state of mind. Every significant event in life develops its own psychological context – a particular state of thought and feeling in which the tidy division of time into past, present, and future, can go by the board to greater or lesser degree. It is possible to experience present time so intensely that past and future cannot break in – there is no such thing as 'five minutes ago;' and 'tomorrow' does just not exist. On the other hand, the sense of 'now' – by which the present is defined – can desert one completely, and prospects of only what the future might hold would dominate consciousness. Similarly, unforgettable memories of the past may hold complete sway and constitute the sole temporal reality. Time's divisions of past, present, and future are not linearly inflexible, and indeed can come and go arbitrarily, certainly in the dream, and also in the light of day depending on how we are mentally affected by the nature of what is taking place.

> *Time, to the nation as to the individual, is nothing absolute; its duration depends on the rate of thought and feeling.*
> John William Draper, *History of the Intellectual Development of Europe.* Vol.1, ch.1

Does the universe itself have any built-in agenda concerning time? Would our practice of conceiving time as an ongoing 'flow' – a stream coming from the past, arriving in the present, and continuing into the future – mean anything to the 'powers that be,' so to speak, of the cosmos? It is, after all, a construct of *our* consciousness – and a very necessary one at that. Without it we would have no 'measure' of life – no way to regulate and lay out our span of existence; no ability to recognize the sequential nature of events and so learn by comparing the impact of the most recent happening with its antecedents; no chance of estimating when and where anything might be expected to happen, therefore having no ability to comprehend cause and effect.

And yet, vital as such a construct of time is for us, can it really have much relevance for a cosmos in which the flow of energy – and therefore time – is perhaps *not* linear but vortex-like, or is even discontinuous in that energy flows in parallel systems, or as a series of operations, simultaneously rather than consecutively.

Who knows when the first humans began to mark the passing of day and night on some rock or cave wall, and so gain the means of first comprehending, then regulating and organizing, their brief span of existence through anticipating and planning events: a practice which would lead to an elementary concept of time as such, and develop into the sophisticated understanding which has made the early science-fiction dream of space travel a reality. From scratchings on rocks, to an apportioning of time based on the earth's orbit and rotation, we have finally arrived at the atomic clock which measures the oscillations at the atomic level of a rare metal known as *cesium*. We have come a long way in seeking to comprehend the nature of time.

Augustine, saint, church father, and bishop of Hippo at the time when the Roman Empire in the west was disintegrating, lived from 354 to 430 A.D. Two of the books he wrote had a profound effect on Christian thought. The work now entitled *The Confessions of St. Augustine*, and a second book, *The City of God*, both go to some lengths in discussing the question of time: time which Augustine distinguishes first as that comprehended by human consciousness which is 'secular' time; and then that which is beyond our understanding because it lies beyond nature in the realm of spirit, and to which he refers as 'sacred' time.

The gist of his thoughts is that 'secular time' reigns within the material universe and is perceived by consciousness to manifest itself in terms of the linear and sequential progress of events as we experience them. A rudimentary awareness of what Augustine meant by 'secular' time would likely have been an early attribute of human consciousness. The rising of the sun each new day, for example, would likely have been the most provocative and significant event in the life of early man. And with the disappearance of the sun and the onset of darkness, the experience would have surely been equally as powerful. The regularity of this phenomenon would ultimately have led the thinkers in a group to define a 'day' in terms of the *duration* of sunlight (the time factor), and associate it with the *distance* travelled by the sun from rising to setting (the space factor) - thus introducing the belief at some point in our evolution that we live in *space-time*. (And if the concept of matter and *energy* had also been a factor in this early stage in recognizing the significance of the sun's journey... the science of physics might have got off to a flying start.)

So I am suggesting that once the chronology of a single day was perceived, the principle of *linear time* as the yardstick defining the whole of existence - us and the universe together - became generally accepted. And as the basic principle of linearism recognizes movement from a beginning to an end, so we came to see a life span as one embracing *past, present,* and *future* time. Augustine's 'secular time'.

But then he goes on to postulate the presence of an immaterial, *meta*physical realm of spirit in which time does not flow in a linear way as past, present, and future - is time*less* in the sense that it has no beginning and no end: a concentricity of 'being'. Augustine calls it 'sacred time' - but I wonder what terminology he would have employed had he known about the theories of twentieth century quantum physics? For quantum theory - as formulated by Max Planck in 1900 - holds that energy radiation and changes of material state do not occur in a linear continuity, but discontinually as *quanta* - thus rendering the traditional concept of time open to question. The time*lessness* to which Augustine refers in the following lines has a quantum-like ring to them. He writes:

'...forget what is behind, not achieved wasted and scattered on things which are to come and things which will pass away... and contemplate Thy delight which is neither coming or passing.' Such, for Augustine, is the way to escape the world of 'secular' time and approach the 'sacred' time-zone of spirit within one - an experience with which the Roman poet Ovid (who lived some 300 years before Augustine) would seem to be familiar when he wrote: *'There is a God within us'*: (A statement which I feel would have received an affirmative nod from Mozart.)

We have seen throughout these pages that under conditions of great stress, or when subjected to compelling moments of insight or revelation, it is possible to become lost to the world's time. Step inside for a while... and enter a realm where - as Aldous Huxley implied by the title of his 1944 novel: *Time Must Have A Stop*.

THE ANALOGY of RIVER and WHIRLPOOL

Imagine walking alongside a river running straight as an arrow to the sea. It has reached its mean width, and is flowing at a steady rate. As such, it may be regarded as a linear phenomenon: a moving length of water coming from somewhere behind one – out of the *past* as it were; then reaching the position where one is actually standing on the bank, thus arriving in the *present*; before continuing on beyond oneself into (theoretically) the *future*... It is this association between movement (the river's flow) and time... that has likely led us to talk, metaphorically, of 'the river of time.'

However, let us say that as the observer, totally detached from 'river-time' as such, one proceeds a mile downstream to arrive at a position where this life-flow of the river is interrupted by a large and powerful whirlpool - a place where confusion reigns as water from upstream (the *past*) is engulfed, its forward momentum suspended as it now swirls in circular fashion, and its continuance (into the *future*) temporarily held up until it can escape the centripetal force of the whirlpool. And if the conditions are such that the whirlpool creates a large and powerful vortex, water will be sucked back into the circular maelstrom, even if it has managed to escape and briefly regain a downstream or *future* course. Such an extensive and powerful whirlpool facing one's position represents an abnormal volume of water - going nowhere for a moment or two - caught in time, as it were, and so may be thought of as *timeless* water.

In the larger river estuaries of the Kimberley coast in northwest Australia, where the tides can run as high as thirty feet, and huge whirlpools are encountered, a launch heading upstream can lose headway near the center of the vortex and appear to 'stand still' for a time, lost to the past while still not in the future. At some point the whirling mass of water will release is hold, and the boat will start forward once again, back into forward moving time.

Whether one talks about a whirlpool or a whirlwind, it is the 'whirl' which denotes the energy-force involved. If you look up the word in *Webster's Dictionary*, you will find it defined as, '*a magnetic or impelling force by which something may be engulfed.*' And if one is going to liken the flow of river-time to the passage of time in human experience, then the river used as the analogy must contain a strong whirlpool. Without it, the river and time would flow equably and predictably. It is something akin to the 'whirl effect' in a river's continuity that I believe happens in consciousness when the regular flow of everyday linear time is suspended - when past and future cease to exist, leaving one immersed in a seemingly timeless present. These are usually occasions when one is swept up into a whirlpool of powerful feelings, or suddenly carried off by a brainstorm of swirling and compelling thoughts, or taken over by a condition of inner serenity and quietude when time seems to stop: episodes in life when one experiences the 'eternal present' of the poets, composers, visionaries... or the 'moments of truth' when inspired research is attaining its goal. What Augustine describes as 'sacred time'. I would think that it was on such occasions when time stood still that Christ knew his destiny; Mozart heard his music; Aristotle, Plato, Galileo, Jung, Einstein... had their greatest ideas; Michelangelo 'saw' the form waiting to be released from the rough quadrangular block of marble before the stone was even quarried; Shackleton sat alone at the edge of the ice floe and harnessed his will; and Hans and Sophie Scholl found the moral and spiritual power to deny evil and oppression.

If such a timeless condition results from the constellation in consciousness of intuitive and unconscious forces - as we have previously held - then could we not liken their effect on the 'flow' of normal chemical-electrical brain activity to that of the whirlpool's interruption of normal river movement? Groundbreaking research in 1998 revealed that the brain's electrical circuitry is changed when a person is subjected to frightening experiences occasioned by actual outside events; that fear can engulf and overwhelm normal brain function even when internal and subliminal forces are not involved. This investigation revealed that some connections between the nerve cells within the amygdala (a small almond-shaped structure within the brain) become intensified when a rise in the level of fear raises the rate at which nervous signals flow through the brain's 'fear-center'. And if this is so when daunting situations in the outside world are threatening one, then similar intensity changes in 'brain electrics' - in terms of higher voltages and higher speeds of conductivity - are likely to occur when unexpected insights, and compelling passions and sentiments, break through from the psyche's inner world. I have this image of the unconscious and the intuition mustering untold hosts of neurons, and trillions of synaptic and dendrite connections... creating an electrical and chemical storm of circuitry engulfing normal consciousness: a whirl of neurons scrambling the brain's 'time center'.

> *Time present and time past*
> *Are both perhaps present in time future*
> *And time future contained in time past.*
> T. S, Eliot. *Burnt Norton*, in *Four Quartets*

The neurological scientist tells us that rates of travel for neuron impulses vary from about one mile an hour to speeds of well over a hundred and fifty, depending on the size and physical characteristics of a particular axon – the neuron's stem-like, impulse conductor.

Consider, for a moment, that if during normal exercising of the neurons – while scanning this sentence, for example – your brain is taking but a few milliseconds to bring together the precise and complex arrangement of nerve impulses required to convey the meaning of the words... then how much more extraordinary is its ability to bring to bear, practically instantaneously, the increasingly large numbers of neuron signals, and the great speeding up of existing ones, by which you experience the brain storm that delivers the brilliant idea - the brain storm that may be described as a whirlpool of neurons. It is hard enough to picture the standard operation of the millions of split second neuron 'firings' which get us through the normal day – without having to go further and strive to imagine a 'full house' of one hundred billion neurons in action, utilizing the one hundred trillion synaptic connections available to them. And all called into play to bring highly imaginative inner stirrings into the light of everyday conscious.

Does it ever happen? – The 'full house,' I mean. For awesome as the numbers are, it might be that the highest levels of inspired achievement in any area of human activity could not be attained without such a mustering of neuron power - that it requires an incredibly high count to produce the electronic circuitry responsible for Mozart-like feats of creativity and the suspension of chronological time.

The chemical interchanges which produce significant electrical voltages are described by Robert Ornstein and Richard F. Thompson writing in their illuminating book, *The Amazing Brain*. And I feel I can do no better than quote them in full – not only to advance the plausibility of my own hypothesis, but also to impress upon my reader the sheer physiological complexity of the organ which is responsible for the marvelous phenomenon we call consciousness - an information system operating at both objective and subjective levels.

> *The nerve impulse... is a process that involves primarily sodium ions, which have a positive electric charge. When a nerve cell is at rest, almost all the sodium ions are outside the cell. But because of its protein molecules, the inside of the cell is electrically negative relative to the outside, so there is a very strong electrical force trying to pull sodium ions into the cell. In the nerve cell membrane are sodium channels, holes that allow sodium to pass through. At rest, these sodium channels have gates that are closed so the sodium ions cannot rush into the cell. But when the nerve impulse develops at a particular place on the axon, the sodium channel gates in that region pop open very briefly and sodium ions rush in. The voltage just inside the membrane at that place shifts from negative to positive because the sodium ions have positive charges that they carry inside. This is the nerve impulse – a brief local rush of positively charged ions.*
>
> *When the nerve impulse occurs at one place on the axon membrane (and it occurs all the way round the axon membrane at that place), the closed gates on sodium channels, which are voltage-controlled just like electrical switches, pop open briefly and the nerve impulse moves to this next place, and so on all along the axon. This is how the nerve impulse travels along the axon.*
>
> *Although the basic operative mechanism of the nerve (neuron) impulse involves movements of ions across the cell membrane, and changes in the electrical potential of the cell membrane are critically involved, the... impulse itself is not an electric current. (Yet) because*

> the cell membranes of neurons do have changes in their electrical potentials, however, they generate electrical fields that can be quite large and easily recorded.*

In one context or another throughout these pages I have touched on the circumstances which may give rise to such a whirling field of electronic energy: namely, those occasions when the intensity of 'the knowing' - the intelligence and understanding - generated within the unconscious and its borderline state the intuition, can no longer be subliminally contained, and 'brainstorms' its revelations into consciousness. And we have seen that such a jolt to the even tenor of consciousness appears to come about in one of two basic ways. It can manifest of its own volition as a spontaneous breakout from within; or it can be set off by the presence of an outside catalyst. Mozart is the most obvious example of the former; Graham Sutherland, with his thorn bush, of the latter.

However, I should point out that such electrical superactivity resulting from powerful unconscious influences does not always result in beneficent and creative actions and insights. There are dark and destructive aspects of the human psyche also lurking at a subliminal level, equally urgent and forceful in their ability to control consciousness; to the extent that a sense of time and place can be lost and personality changed - and without having to imbibe the kind of chemical cocktail described by Robert Louis Stevenson in his story of Dr. Jekyll and Mr. Hyde. But I will wait to discuss the potential influence of the psyche's dark and malign powers in a later chapter.

SPEAKING OF THE ELECTRICAL CONNECTION...

You may recall that my stepfather sent me off to bed one night with the departing observation, 'We're electro-chemical creatures, my lad... and without the electro we'd be dead meat... we've got to keep these millions of plus and minus charges in balance, otherwise we could blow up.' And how this last possibility really had me worried as I lay in bed in the early morning, juvenile water tap suddenly and unexplainably transformed by explosive charges of this electrical stuff over which I had no control. After

---

* Robert Ornstein and Richard R. Thompson, *The Amazing Brain* (Boston: Houghton Mifflin Company, 1984), p. 77.

reading Dr. Robert Becker's and Gary Selden's book *The Body Electric** I have to think that my old man was not too far off the mark – and that there were some grounds for my apprehensive acceptance of his theories concerning physiologic electricity.

Dr. Becker spent 30 years working experimentally to determine the vital role played by electricity in the physical existence of animals and humans, and his part in a 'little known research effort' (his words), has led to significant advances in our understanding of how body chemistry and electricity govern the course of life. It is a book dense in information about their life-giving role; particularly when the author discusses the overseeing authority of the brain in regulating bodily functions. In his Introduction he states:

> *My research began with experiments on regeneration, the ability of some animals, notably the salamander, to grow perfect replacements for parts of the body that have been destroyed. These studies... led to the discovery of a hitherto unknown aspect of animal life – the existence of electrical currents in parts of the nervous system. This breakthrough in turn led to a better understanding of bone fracture healing, new possibilities for cancer research, and the hope of human regeneration – even of the heart and spinal cord – in the not too distant future... Finally, a knowledge of life's electrical dimension has yielded fundamental insights... into pain, healing, growth, consciousness, <u>the nature of life itself</u> (my underlining), and the dangers of our electromagnetic technology.*

Among the many provocative and truly illuminating accounts of physio-electrical research which crowd the pages of his absorbing book, are many which impinge directly on our speculations here, especially those concerning brainstorms and the powers of inspiration - leading me to think that perhaps the unconscious can receive its extrasensory knowledge by tapping into the electromagnetic and bioelectromagnetic fields surrounding us. For it has been demonstrated that thoughts can manifest as radio waves, and possibly the converse is true - that radio waves can manifest as thoughts. Here is Dr. Becker again:

> *Over and over again biology has found that the whole is more than the sum of its parts. We should expect that the same is true of*

---

* Robert Becker and Gary Selden, *The Body Electric* (William Morrow, New York) 1985

> bioelectromagnetic fields. All life on earth can be considered a unit, a glaze of sentience spread thinly over the crust. <u>In toto</u>, its field would be a hollow, invisible sphere inscribed with a tracery of all the thoughts and emotions of all creatures... Given a biological communications channel that can circle the whole earth in an instant, possibly based on life's very mode of origin, it would be a wonder if each creature had <u>not</u> retained a link with some such aggregate mind. If so, the perineurial* DC system could lead us to the great reservoir of image and dream variously called the collective unconscious**, intuition, the pool of archetypes, higher intelligences deific or satanic, the Muse herself.

I must resist the temptation to overload our own circuits by quoting too freely from Dr. Becker's arresting story, but here are a few last lines I simply must borrow. They summarize most succinctly all the references hitherto made to an 'electrical connection.'

> Variations in the current from one place to another in the perineurial system apparently form part of every decision, every interpretation, every command, every vacillation, every feeling, and every word of interior monologue, conscious or unconscious, that we conduct in our heads.

A headline in this week's London Sunday Times – ELECTRONS INJECT LIFE INTO WILTING FLOWERS – caught my eye today. Here is the gist of a report which provides an interesting sidelight on the theme of this chapter:

> A new electron generator could bring benefits to horticulture. Florists could use it to extend the life of cut flowers, garden centres to make plants resistant to disease, and warehouses to protect grain.
> The discovery was made at Radiation Control Systems... while it was developing a machine that can clear houses of 'alpha' radiation caused by radon gas... An electron generator was left running in an office next to a vase of wilting flowers. The follow morning the flowers had been restored to their original freshness even though the heating had been on all night.
> A series of experiments were made with cut flowers, soft plants

---

* <u>Perineuria</u> are the sheaths of connective tissue which surround bundles of nerve fibers.

** C. G. Jung's term to connote the continuous presence in the human mind of commonly held ideas, irrespective of time, race, or place.

and seedlings. It was found that all the plants exposed to the generator showed more vigour, longevity, and resistance to attack by fungus and parasites...

The machine emits a stream of electrons, which are thought to increase a plant's metabolic rate. Keith Foster, the firm's owner, says: 'We used the theory that, as all life evolves against a background of radiation of some kind, our device would work better if streams of electrons were created similar to those found in nature.

At this point I find myself dwelling on George Steiner's cogent remarks concerning the workings of the human brain - from which I quoted earlier - for as a statement regarding the potential range of human awareness, it serves as the perfect epilogue to this chapter.

There is too much of our cortex. We could do with far fewer cells and synapses and still have an excellent information system. Something much deeper is going on. Man has a marvellous excess of invention. He can say 'No' to reality.
                    Professor George Steiner, Churchill
                    College, Cambridge University.

8

TWO SIDES OF LIFE'S COIN: THE FLYING BISHOP & HIS 'SERMON' ON OPPOSITES

To perceive the necessary interdependence of spirit and matter, creation and destruction, life and death, 'light' and 'darkness,' yet never to fail to be involved in the eternally necessary struggle against the forces of brutality and evil, at every level of life.
Dorothy Norman, *The Heroic Encounter*

The Second World War had been over for six years, and for the first time since the end of hostilities I was flying again – as a civilian this time – and about to become airborne en route to Dublin in a DeHavilland Rapide, a pre-war twin-engined biplane with a dubious safety record.

Sitting next to me, occupying the window seat, was a Church of England Bishop – dark suit, crimson stock, small silver crucifix suspended over its dark red ground by means of a delicate silver chain. His secretary sat behind, clutching the ecclesiastical briefcase.

The aircraft's port engine had fired; the pilot was running it up, checking magneto performance, boost and oil pressure gauges. But the starboard engine was proving recalcitrant, ignition only half-catching, engine turning for a couple of exhaust-cracking seconds before shutting down leaving the prop idly windmilling. Trestles were wheeled into position; mechanics began to remove engine cowlings. Not an auspicious start.

The Bishop was clearly nervous, drumming fingers on his knees, pursing his lips, running his eyes around the cabin. Then he spotted my R.A.F. tie.

"Ah," he said, obviously glad to start a conversation. "*I see you must have been in the Service. You don't mind flying, I take it?*"

"Well," I responded, "*I've never felt very comfortable flying; told myself I'd never get into an aeroplane once the war was over – assuming of course I was around to make the decision.*"

"*Mm... well I suppose it is a little different in wartime.*" He paused, obviously wanting me to go on.

"It's simply that you're dealing with a machine," I continued, "with any number of moving parts in every engine and some very crucial control linkages in the airframe: statistically there's a fair chance that any one of them may pack up. And if that happens, the air is not the place you want to be. On the ground or on the sea you at least have the chance of walking or swimming away from trouble. I'm afraid a few thousand feet of airspace don't give you those options."

I thought I had perhaps been too unfeelingly realistic. The Bishop was looking very pensive, his brow furrowed. Yet his eyes were calm and the agitated drumming on the knees had stopped. By now the mechanics had finished working on the engine and were pulling the trestles away. The ground crew chief had boarded the aircraft and was conferring with the pilot in the cockpit.

"I suppose you're wondering why I should be worried," said my companion – "as a 'man of the cloth,' I mean, having an unquestioning trust in God..." He waited for my reply; yet as I had never spoken confidentially to a Bishop before I was somewhat diffident about broaching the subject of faith. The starboard prop began to turn – slowly and with much hissing and spluttering from the engine's exhaust ports. I pretended to give it my undivided attention until once again the pilot cut the ignition switch. Out came the trestles again. Off came the cowlings.

He was watching me quizzically. "You think that for a priest every potentially risky situation in life becomes simply a matter of sitting back and handing the whole thing over to God – exercising complete faith, mm...? Well" – without waiting for my response – "that's a typical lay point of view and, if you don't mind me saying so, a pretty simplistic one. By the way, are we going to get off the ground, do you think?"

"If the pilot has any doubts about that engine, we won't."

"Good." The Bishop was smiling and for the first time I wondered how old he was. Sixty-five? Seventyish? Difficult to tell. He could have been younger. His hair was gray and his face elegantly lined – but the blue eyes were ice-keen and penetrating; the mouth generously proportioned and flexible, nothing pinched or illiberal about it.

"Good," he repeated. "You've just implied that we should have faith in the pilot, in his judgment – rather than in God's. No?" Now he was looking impish and a little pleased with himself.

"Yes: obviously his judgment is vital."

"Right – so you've hit the nail on the head."

Suddenly the starboard engine burst into life; the pilot opened the throttle gradually and pushed it through the full range of running-up tests. Not once, but twice. It was impossible to talk above the racket. And then down came the revs and the decibel level. We began to taxi slowly onto the perimeter track.

"How do you mean, I've 'hit the nail on the head?'"

The Bishop was wrestling with his safety belt. When it was finally adjusted to his liking, he looked up and prodded me in the chest with his forefinger – not aggressively, more a 'put you at your ease,' companionable sort of gesture.

"Well, you see, neither of us are in charge of this thing, and we can't expect God to assume responsibility for us as we are both here entirely of our own free will, prepared to abandon our natural element. So we must put some degree of faith in our own decision to fly in the first place; then in getting on board this contraption in the second; and finally our highest level of faith must reside in the pilot who is going to do the flying. His judgment is vital, you said so yourself."

"Highest level," said I, "but surely..."

The Bishop cut me off: *"No – not God. The pilot."*

Silence for a moment while I marshaled my thoughts. At the same time the aircraft's brakes came on suddenly and we jolted to a stop. I could see a couple of light planes moving onto the perimeter track ahead of us. My companion had turned away and was looking out of the window, shoulders hunched, left hand gently pulling down on the lobe of his ear; I thought how physically frail he suddenly seemed to be. Then he straightened up, regarded me in a most kindly fashion and took hold of my wrist.

"No – not God. We all make terrible mistakes in attributing everything to the will of God – success and failure, good fortune or tragedy. Remember Christ saying, 'Give unto Caesar the things that are Caesar's. And unto God the things that are God's?' Now there's a reminder of the two distinct sides of life's coin with which we, as human beings, must contend. There is that of our physical and temporal nature which demands that to survive in a material world we live governed by practical and pragmatic concerns. And that of our spiritual side where the soul, utilizing the authority of spirit and the power of mind, allows us to escape from the world of toil and time into a space of calm detachment, where peace - the peace of God, if you like - enfolds one. Christ uses the name 'Caesar' to denote all earthly

happenings and struggles – *from both the good and the terrible things men do, to the life-threatening vicissitudes of nature herself. One must do one's best in practical ways to improve the human lot while in 'Caesar's world', says Jesus - both in terms of man's inhumanity to man, and in the face of Nature's predations. But – He urges – don't, in the meantime, overlook the other reality: the life of soul and spirit urged upon us by some ineffable mental comprehension which draws us away from time and the world. So we must see reality as two-sided. A duality. Ideally, living with a metaphorical foot in both camps.'"*

He let go of my wrist and smiled self-consciously. "But good grief man, I'm giving you a sermon..."

Turning away, he gazed again out of the window, muttering to himself, "Why on earth can't we get going? What's holding us up?" Then, swinging around to give me a worried look, "Is there something else wrong, do you think?"

"No, I wouldn't think so. We'd taxi back to the hangars if there was. You know, this stop-and-go taxiing reminds me of a good friend in the war – was lost in '42 – who hated hanging around once the engines were started. Said the plugs got oiled up – engines running too long at low revs and overheating, which could result in a loss of power just when it's needed to get you airborne. It was an experience to see him hustling a Lancaster round the perimeter track..."

The Bishop was not happy with this. Unthinking of me to have mentioned old Matthews and his theory, which was really quite sound. However, the story obviously encouraged my travelling companion to continue his 'sermon.'

"But that's exactly what I'm talking about: your friend was in charge – totally responsible – making all the decisions. Because he was a 'Caesar' pilot flying a 'Caesar' aeroplane in a 'Caesar' war: it was a situation all of our own making. You know, the politicians are on shaky ground when they invoke the 'God' side of the coin and claim Him for their warlike causes. Which is not to say that prayer, honest and true, might not invoke divine intervention in any struggle which is clearly between right and wrong, good and evil. Perhaps like the war that's just over... wouldn't you say?"

We were moving again. Bumping along on a poor surface.

"Yes, I think some of us thought along those lines. Hitler was generally considered to be an evil so-and-so, and there was a kind of crusading spirit among us. But it was tempered by the knowledge that the politicians should have seen the Fuehrer as a pathological liar, and nipped him in the bud in the mid-thirties. Then there'd have been no war. But they chose to do nothing. And without Churchill – who we in the forces did regard as a genuine crusader – the war might not have been seen as such a just cause."

"You're right: before Churchill the British government was gutless – failed to take either the moral or political initiative. Now listen," continued my own personal ecclesiastic, "before we get up into the 'bright blue yonder,' let me run over the gist of what I have been trying to say. Simply this: at some point in our evolution we became possessed of an imagination, found ourselves free to conceive of things quite beyond the limited requirements of mere subsistence living – one such very recent fancy being the idea of flying. So we went into the science of it – remember Leonardo's notebooks? – and finally constructed the machines in which to do it. And now here we are, you and I, aboard one such flying machine, choosing to take our destiny into our own hands."

"Now... hang on a minute, and then I'll shut up. If we believe in the presence of some omniscient power behind this ongoing universe – whether we call it God or not – then it follows that to have been endowed with the talent to dream, imagine, and invent... for at least the last 100,000 years, represents a unique feature of evolution's Divine Blueprint for us. And with the vista of mental possibilities that are now brought to mind comes freedom of choice: one can say 'yes' or 'no 'to following this or that fancy - decisions which ultimately shape our existential and moral destinies; determine the kind of individuals we become: each one of us a mystery in his or her own right. But such a complex mental life inevitably carries a price: that of having to accept responsibility for the outcome of one's choices and actions - for success or failure in both materialistic and psychologic terms. The privilege of free will in choosing between alternatives makes little sense without such accountability, wouldn't you say? So if we want to conquer the element of air, and learn how to fly - risking our necks in the process - we may succeed or fail, that's the chance we take. We can't expect God to be hanging around with a safety net when something goes wrong. It would sabotage the scheme, wouldn't it - that which has seen the growth of a consciousness presenting us with personal choices and inclinations; thus setting us up to learn by our own efforts. To shield us from the consequences of our decisions would negate the very purpose of the human journey. One can't expect to have one's cake <u>and</u> eat it. For mind and spirit can only be put to the test when there is no expectation of a safety net. Only in accepting the degree of autonomy which attends our choices and actions, and recognizing our personal responsibility for the good or ill which ensues, can we come to know the reaches of our mind, the extent of our courage, and the moral standards we embrace."

"No: give unto Caesar when in Caesar's world. That's why I hope our chap's a bloody good pilot, and that the people who put this aircraft together knew what they were doing. That's where my faith lies at this moment - in the level of human competence attending our present situation. And I've done my bit in this regard. I've checked with the airline and our pilot flew with your lot during the war - and obviously survived, so he's got to be O.K. Also, I've found out that this aeroplane is just out of a major service. And I've checked with the meteorological service, and nature and the elements have no surprises in store. God helps those who help themselves, as the saying goes. And if something does go wrong and we crash... then I will transfer my faith to God because I have already made my peace with Him. And it will be appropriate to do so at that juncture because I shall, after all, be leaving the physical world."

The Bishop lay back and closed his eyes as if, having resolved the worries of flying for both of us, he could now relax knowing that he at least *he* was covered on all fronts. The Rapide had reached the departure end of the runway; our pilot was running both engines up hard, holding her trembling, bucking against the brakes. Satisfied with his instrument readings he released the brakes and let her go. My neighbor opened his eyes then shut them again. At about 2000 feet, still climbing, we encountered strong crosswind turbulence – a factor which no Rapide handles very well – requiring constant kicking of the rudder pedals to bring the aircraft's nose into the wind before edging her gradually back on course... which makes for a crab-like motion across the sky. Five thousand feet and we were clear of the south-flowing air stream. The Thames River valley lay stretched out beneath us.

Now in calm air my companion sat up, reached into his pocket, pulled out a handkerchief and blew his nose heartily. "*Ah, that's better*" – turning to regard me quizzically. "*I hope I haven't bored you stiff with all this 'Caesar' stuff, but I should tell you that I'm only able to proselytize so unashamedly – out of the pulpit, that is – when I feel I have a sympathetic ear. Anyway, now the plane's finally in the air I can do a better job of shaping my thoughts. Don't often have such an amenable captive audience. But having talked about what we can do for, and to, ourselves, I haven't touched on the problems of life's uncertainties in the face of what Mother Nature can do to us."* He paused, eyeing me as if a little uncertain of my response.

"Mm..." I murmured, encouraging him to go on.

"Well, what about earthquakes, floods, plagues, epidemics...? We have no freedom of choice in these matters – we can't opt out, say 'no thank you' and keep ourselves out of harm's way. These hazards are not of our choosing. Unless, of course, one elects to fly in bad weather, live under a volcano, or go to dangerous places in the role of explorer or missionary, say. Yet a new strain of 'flu can sweep the world and take all before it, however much you try to 'will' it away, or even attempt to outrun it."

"Now – I don't mean to lecture – but here is a most important point which I never seem to put across very well. We should understand that everything that goes on in nature – volcanic eruptions, hurricanes, lightning storms, 'flu virus outbreaks, cancer virulence – also belong to the 'Caesar' or existential side of life's coin. They are part and parcel of the same physical phenomena which shape our physical bodies. The laws of physics, biology, geology... apply to us no less than they do to the flora, fauna, rocks, and all other material forms which go to make up this planet. Therefore it is inevitable, is it not, that our own physical structure will be subject to nature's dynamic goings-on."

He paused, as if to give me time to sort this out. I took a surreptitious glance out of the window. The Severn estuary lay immediately ahead. Headwinds were slowing us down; the note of the engines deepened as the pilot demanded more power. I estimated we'd be on the ground in about 45 minutes. My fellow traveler seemed suddenly withdrawn, looking down at his feet, head slightly to one side. When he turned in my direction again his expression was pensive, his voice lower in tone; eyes looking beyond me into the distance.

"Often, the most difficult part of my job is getting people to accept the fact that because we are physical organisms, we are therefore mortal beings. Sickness, suffering, death... so often provoke the cry, 'Why has God done this to me?' – or 'to my child?' – or 'to my wife?' And when I say that in this world our bodies, however magnificently designed, are inevitably vulnerable to genetic flaws or runaway cell growth leading to tumors and cancers, or just old age... it is, of course, small comfort. For as the God I talk about is by definition omniscient, the bereaved naturally ask why then can't He intervene to save a loved one. Which, as I point out, does happen from time to time as the result of prayer – which again

isn't much help as they likely have been praying themselves silly without result. And one can't get into the selective workings of faith and miracles when trying to console a mourner."

"I sometimes paraphrase Jesus: 'Give unto Nature the things that are physical. And unto God the things that are spiritual – making the point that we *are* a dual creation. That what has died – in the manner of all things organic – is the creature of flesh and bone. But that what continues to live when its residence in the body is over, is the essential self, the soul-spirit which, not being of the natural world, has gone on questing to discover its source."

"Do you know those lines which Shakespeare gives to the Queen of Denmark, Hamlet's mother, as she tries to persuade her son to accept his father's death?"

"No, I don't recall."

"Well, as usual, the old Bard manages to say in 15 seconds what I have been trying to say for half an hour:

> Do not for ever with thy vailed lids
> Seek for thy noble father in the dust.
> Thou know'st 'tis common; all that lives must die,
> Passing through nature to eternity.

"There you have it. Old fashioned Christianity. The duality of being. The temporal and the eternal. But, you know, the mystery of matter and spirit conjoined is difficult to 'sell' nowadays. 'Ah, given the secular and hedonistic culture of our time,' say many in the Church, 'it is easier, more relevant, and certainly more gratifying, to minister to the social ills of society, or to bring religion into politics'. Well, I'm afraid that says more about the spiritual poverty of the Church than it does about the spiritual needs of society: 'O Ye of little faith...' I'm tempted to mutter at diocesan meetings – not arrogantly, I hope -' it is into the mystery of being in general, and human being in particular, that we should be taking our congregations... for that is the question which myth and religion have striven to answer for thousands of years.

The nose of the aircraft dropped: we surely couldn't be on the approach to Dublin already – from somewhere in the middle of the Irish Sea? Weather was clear. No sign of the coast. Yet we continued to lose height gradually; leveled out at about 6000 feet. We were both preoccupied – the Bishop looking down at the white-capped water; I struggling to grasp the implications of something he'd said a short time before. Finally I asked the question.

You said earlier that when we exercise free will and things go wrong, we shouldn't expect God to be hanging around with a safety net to help us out. But I wonder, following this train of thought, if when it comes to death from natural causes - a matter over which we can exercise no choice - does the fact that we have not <u>chosen</u> to die indicate that some sort of supernatural safety net will be in place to help us out and save us from the nothingness of the abyss? That God will be in attendance, so to speak? Whereas if we choose to die, and do so by our own hand, that Divine assistance will be withheld - simply because we have exercised free will?

He took his time to reply, looking first down at his feet and then out of the window before turning in my direction.

"*Well, I don't see God 'hanging around', as you put it, in either event - whether death is the result of suicide, or is natural and unwilled. Death is death - the supreme metamorphosis from the physical state to the spiritual: a process which, once underway, is so awesomely sacrosanct unto itself that the causes whether natural, or self- willed and self-initiated, no longer have any bearing on the nature of the metamorphic change. I would say that the end - being the end of the physical body - will follow the same course, whatever the circumstances of the demise. When the coin flips - from whatever cause - the soul-spirit is on its way from 'Caesar's' world to its supernal home. I tend to think that the 'mechanics' of departure from this realm are automatic: a cosmic transference of radiant energy - wave rather than particle - carrying one's uniquely individual essence. We may all receive some form of supernatural guidance at our passing - a 'safety net', if you like - but as we cannot define or personify God, better leave it at that. But whether the emergent soul-spirit of a suicide has to answer in some way - once passed over - for the choice and action to which it acquiesced when incarnate, endowed with a human body, can only be a matter for theological speculation.*

"*No: it's not a matter of God being personified and 'hanging around'. Rather that some small part of the force and intelligence we think of as God is active within the process of metamorphosis, the means by which the transformation of matter to soul-spirit-energy is accomplished.*

The coast of Ireland was now clearly visible; the Rapide stabilized on a shallow angle of descent, engines cut back, cabin suddenly quiet. I realized we had been carrying on this conversation despite the punishing racket of the two close-in engines – a feat demanding that we practically had to shout at each other, and that listening required absolute concentration. I felt whacked out. But the Bishop was clearly not fatigued. "*Been shouting all my life from the pulpit in big churches, my boy,"* was his response to my comment on the welcome quiet of the cabin.

We walked into the Arrivals building together, the Bishop's secretary – who hadn't uttered a word in my presence – ambling a few paces behind. With the formalities of arrival completed we stood together for a few minutes. Neither of us had any baggage to claim; neither had we any urgent need to rush out and find our respective hotels. Standing together I was surprised – now that he was not buried in an aeroplane seat – to see that he matched my own six feet in height, and that he possessed a presence which could only be described as saintly. Those incisive eyes of his gave me the impression that I was facing the embodiment of all goodness... truth, charity, compassion.

"Here's my card," he said. *"Please, if you're ever in the vicinity, come and see me. This was my first flight, you know, and with you and the pilot being both from the R.A.F. the omens were good, no? Here: write your name and address on the back of this for me."*

And so we shook hands and went our separate ways.

Three weeks later I received the following letter:

My dear Graham,

I have been thinking about our conversation in the plane and realize I was very remiss in not going beyond the rather simplistic 'translation' of the 'Give unto Caesar...' and 'unto God...' declaration.

I should have pointed out that our thoughts concerning the dual nature of things are entirely constructs of our own consciousness, and only convey the reality of how things are <u>within the capability (limitations) of human comprehension</u>. And as consciousness perceives the world and life, it <u>appears</u> that 'opposites' rule in both the workings of the physical world and in the psychological states manifested by human beings.

The physicists talk of matter and antimatter, positive and negative electricity, magnetic attraction and repulsion, gravity and weightlessness, light as wave and particle... While those dealing in the various branches of the human condition use terms such as matter and spirit, conscious and unconscious, rational and irrational, love and hate, pleasure and pain, good and evil, right and wrong, life and death, beauty and ugliness, cruelty and kindness, the mundane and the transcendent, etc., etc.

But these are <u>words</u>... resulting from the evolution of language, and it was language which enabled us to give a particular 'mental shape' to those perceptions (or constructs) by which our increasingly analytical consciousness came to explain the workings of a world beyond self and, ultimately, of self. With the coming of the words we were free - because then we ceased to be in an undifferentiated state with nature. By which I mean that the unselfconscious operation of instinct was overruled, set aside by virtue of a higher mental authority, and 'I' and 'It' came into being. And then only by having a word handy... to catch and identify a thought, can that thought have a mental form, be said to exist, take on meaning. And because it is 'mine,' then I also am confirmed in my own personal state of existence. Before man had a vocabulary there could be little profound awareness of 'I' and 'It,' 'I' and 'You,' or 'I' and 'God.'

I don't know whether such self-consciousness existed for the first walking-erect creature, Homo erectus, so-called. But with the coming of Homo sapiens, we have a thinking hominid, an advance I would attribute to the onset of speech (In the Beginning was the Word) – a human being possessed of a consciousness capable of talking to itself saying, 'Hey, I've got a great idea...' or, 'Have you noticed that the thing I call a <u>horse</u> goes down to where the <u>water</u> is and we can catch him when it puts its <u>head</u> in?' Without the 'naming' of things and events – providing us with the means of identifying their characteristic features, recognizing the sequence of events, the operation of cause and effect – happenings would likely register, at best, as strong visual impressions which would trigger an instinctual, as opposed to an intelligent response.

Here's the point I am trying to make.

Ages before language brought an awareness of self and the world to <u>Homo sapiens</u>, the planet and its forms of life were out there, established already for millions of years and, it is reasonable to assume, with everything held together and vitalized by the same types of forces which operate today. Therefore if we see this universe as God's creation, what we perceive – courtesy of the late arrival of the 'sapiens-language' factor – to be contrary forces at work (involving such perceived phenomena as, say, matter and space, or body and soul) - can have only one implication: that such contrariness is the Creator's way of setting things up. Or, put another way, all the apparently opposing factors and principles – the dualities I mentioned at the beginning of this letter – can only be seen as the means by which an overall working unity is achieved throughout the cosmos.

So diversity is really unity. As a philosophical proposition this statement says that wholeness – when thought of as <u>stability</u> of being – comes

> about through the equilibrium achieved between 'parts' in a necessary and dynamic opposition. And that includes matter and anti-matter, positrons and neutrons, good and evil, love and hate...
>
> Here you have it; from the horse's mouth.
>
> My Dean preached a sermon the other day to which he gave the title, 'God is my Pilot!' Hadn't the heart to disabuse him.
>
> Don't forget to visit. As always... S^t

More than 40 years have elapsed since I received this letter. During this time I have visited the Bishop fairly frequently – the last occasion when he was in retirement as a member of a small Anglican monastic community. His 'sermon,' delivered during our flight to Ireland has stayed in my mind as a philosophical milestone – a marker directing thinking along the only metaphysical lines which, I believe, make sense.

I once asked him whether or not he thought that animals possessed a spirit. His response was that the spiritual spectrum comprised a wide band of ethereal levels, and that all living things capable of overriding the autonomy of instinct to act intelligently, as well as being able to express feelings, have a spiritual 'home' on one level or another. And that the highest levels of spirit-attainment for all such animals – including the human one – comes with the capacity to give and experience love, and feel compassion. There are obviously degrees of caring in the animal world which do not carry the intense, personal devotion we humans associate with love. Yet we really have no way of knowing. One can spend hours observing the straightforward maternal, protective and instructional care given, say, by a lioness to her cubs - which surely has a strong instinctual drive for survival of the species behind it - without discerning particular marks of what we think of as affection; yet, on occasion, the mother seems to display a level of tenderness to one or other of her offspring which goes beyond the requirements of formal nurturing.

A short time ago I watched a television documentary which was filmed over seven years and entitled, *The People of the Forest*. Made in Africa by Hugo van Lawick who was working with his former wife, Jane Goodall, it is a remarkable record of life in a chimpanzee society which, at one point, reveals the depth of a chimpanzee mother's grief at the death of her baby. Her emotional suffering was intense; her eyes bleak with the pain of her loss, and she was driven - it would seem by some forlorn hope - to cradle and carry the small dead body around with her for days, as if sheer devotion could bring him to life again. Everything about her displayed a maternal sensitivity I would not hesitate to call love - an awareness of her bereavement which surely surpassed the kind of detached response associated with behavior triggered solely by instinct.

Finally the film focuses on a particularly moving relationship between a chimpanzee mother and her recalcitrant son - a youngster who was a social misfit in local chimp society, but who could always rely on his mother's protection and care when other members of the group showed their displeasure at his aggressive, and sometimes mean, conduct. He always turned to her when he was in trouble and she never failed him. And when she died he was probably two or three years old. They were together at a watering hole when she suddenly collapsed. Seemingly inconsolable, he never left her body unattended. He sought no food and was clearly not concerned for his own survival. Finally, he climbed onto the branch of a tree some six feet above her corpse, where he remained until he himself died.

I think the Bishop would say that chimpanzees – who have also been observed 'burying' their dead beneath piles of leaves and vegetation – have high claims to a spirit 'home'; as indeed, I'm sure he would assert, do other creatures that exhibit such seemingly strong levels of devotion, the one to the other. It has been observed that many a wandering albatross - the magnificent, twelve-foot wingspanned, pelagic bird of the Antarctic that endlessly soars on the winds and updrafts of the Southern Ocean latitudes known as the Roaring Forties and Filthy Fifties - will not usually survive for long after its partner's death.

We meet all kinds of people in life, many of them as the result of chance encounters. Most such random meetings fade from memory fairly quickly. But from time to time one comes across an individual whose personal magnetism is compelling, or who has something particularly memorable to say. And then one forgets neither the face nor the quality of mind. The Bishop scores heavily here on both counts. Forty-five years after our flight in that old sky-weary Rapide he still occupies a special place in my thoughts.

The question I ask myself is why such meaningful confrontations are so rare: can the fact that I found myself seated next to the Bishop be attributed simply to luck, chance... or, more fortuitously, to the workings of what might be called 'intended coincidences' - what, in the days of Classical Greece would be seen as destined events, happenings that significantly affect one's *moira*, one's fate or destiny. Certainly, I had little to offer the Bishop in this regard, but his airborne 'sermon' profoundly influenced my thinking and the way I thereafter conducted my life. He took me from the shallow water where shoals of minnow-like thoughts concerning life's 'why's' and 'wherefores' had been swimming around aimlessly in my head for years, and released me to sink or swim in the deepest oceans of the mind where the songs of whales make of spirit a truth; of mystery a revelation.

Shortly before my last visit I had been reading about the effects of what is sometimes called the 'Little Ice Age' - an incredibly cold period affecting Europe and the West that lasted throughout the 14th, 15th, and 16th centuries. Millions died either from starvation or illnesses resulting from weakened immune systems when plagues such as the bubonic - otherwise known as the Black Death - which, in 1349, eliminated one-third of the population in England. I asked him how his theory of the two 'realities' by which we lived - the *dualistic* system of the *material* self and the *spiritual* self.... could be reconciled with such terrible disasters and suffering. He simply said that as hybrid creatures it was the price we paid for being biologically born into the world.... but that once our biological life was over - whatever the natural circumstances - the soul went on its way into the immaterial realm of spirit. Once again, so many years after our encounter on that flight to Dublin, he saw my doubt; pressed my hand and said, *'If you keep searching, you will come to know.'*

9

LEADING A DOUBLE LIFE: MATERIAL NATURE: ENLIVENING SPIRIT

*Outside of a dog,
a man's best friend is a book;
inside of a dog it's very dark.*
Groucho Marx

EXISTENCE AS MANIFEST BY THE BALANCE OF OPPOSING FORCES...

This was the sentence which headed my notes of the lecture given by Sir James Jeans, the Astronomer Royal, in October 1944 - the occasion to which I referred in the opening pages when Sir James talked about electrical energy to a group of Royal Air Force officers: an address which ranged impressively and informatively over the general dynamics of the cosmos, leaving one in no doubt that he was talking not only as an astronomer and physicist, but also as a philosopher; and that, like Aristotle, he came at the end of the day as a physicist, to face the metaphysical questions with which science had ultimately presented him. And the reader may remember that although opinions from the audience were invited, I was not able to ascertain his thoughts concerning my 'theory' of *hyper*electromagnetic energy - of radio waves operating beyond known frequencies, and capable of being received by the brain's own galaxy of neurons: signals emanating from some ultramundane source which took the human mind into areas of imagination lying beyond the ability of the five senses to engage. Or to ask his views on the possibility that a controlling intelligence was at work in the universe - one linked to the vital life-enhancing drive we think of as the human spirit? Such would have been the nature of my aborted inquiries.

I might well have used Groucho's whimsical yet thought-provoking verse to stand alone as an aphoristic Introduction to the overall theme of this book. For it will be obvious by now that my intention throughout these pages is to show how much more there is to 'reality' than meets the eye: that the physical aspects of the outside world are there to be read like a book; but that the 'inside' story of the human mind, and the complexity of a man's mental drives is darker even than the internal machinations of his dog.

Essentially, this chapter is written to affirm - insofar as human ingenuity can make accurate suppositions concerning how things work in the universal scheme of things - that for anything to persist in a state of being, a balance between opposing elements and forces must be achieved; that, as the saying goes, it 'takes two to tango'. Take electricity, for example. Early in the eighteenth century, Carles Francois du Fay demonstrated that there are *two* kinds of electricity which Benjamin Franklin later designated as *positive* and *negative*. And it was discovered that an electric current only occurs when a source of negatively charged electrons is connected - by a conductor through which electrons can flow - to a material which, having fewer free electrons, is positively charged. Which is what happens when one connects the negative terminal of a car battery to its opposite pole with wire of a suitable strength. At that point a circuit between negative and positive has been completed.

Physicists tell us of the interaction between the strong nuclear force and the weak nuclear force; of the challenge to unify gravity with electromagnetism; and of matter and antimatter – the latter being a form of matter in which the electrical charge (or additional property) of every component particle is the reverse or opposite of that in the normal matter of the universe: an atom of antimatter, for example, has a nucleus of antiprotons and antineutrons encircled by positrons. Very contrary. But 'contrariness', as the Bishop suggested in his letter, can be seen as 'God's way of doing things.'

It is a contrariness that theoretical physics today sees as a challenge – sparking a quest to find a unifying theory which will bring beneath its umbrella all universal energy forces.

Yet unifying theory or no, the fact is that up to now the cosmos has not come 'unglued' – in the macrocosmic sense, that is. Meteoroids and their breakaway meteorites may enter the earth's atmosphere and fall to the planet' surface and, if large enough, make large craters and destroy flora and fauna. But the stars and planets stay on their courses, their gravitational force countering the centrifugal force caused by their spin. And we conclude that the counteractive play of the major interplanetary forces has resulted – over the long haul of billions of years – in a measure of equilibrium sufficient to allow our own planet at least, with its sun and moon, to continue in existence: a balance that would be lost were any change in force reciprocity to occur - any breakdown in the fine-tuning of these opposing forces.

> *The gravitational force keeps the earth and the planets in their orbits and binds the galaxy. Without the gravitational force of the earth we would be flung into space like rag dolls by the spin of the earth. The air we breathe would be quickly diffused into space, causing us to asphyxiate and making life on earth impossible. Without the gravitational force of the sun, all the planets, including the earth, would be flung from the solar system into the cold reaches of deep space, where sunlight is too dim to support life. In fact, without the gravitational force, the sun itself would explode. The sun is the result of a delicate balancing act between the force of gravity, which tends to crush the star, and the nuclear force, which tends to blast the sun apart. Without gravity, the sun would detonate like trillions upon trillions of hydrogen bombs.**

However, let's come down from the lofty heights, the mind-boggling mathematics of light-year distances and energy systems in the macrocosmic universe, and recognize that in the biological realm of microorganisms a similar confrontation between opposing 'microbe powers' is also going on – one which again ensures that a positive balance between opposing organic forces is maintained. In *My life with the Microbes* (1954), Nobel Laureate Selman A. Waksman writes: *It is usually not recognized that for every injurious or parasitic microbe there are dozens of beneficial ones. Without the latter, there would be no bread to eat nor wine to drink, no fertile soils and no potable waters, no clothing and no sanitation. One can visualize no form of higher life without the existence of the microbes. They are the universal scavengers. They keep in constant circulation the chemical elements which are so essential to the continuation of plant and animal life.*

Think for a moment about the constants in nature: the effective numerical count which governs the strength of gravity, or the charge of an electron, or the weak and strong nuclear forces. A slight difference in the numbers and atoms would not hold things together, stars would not burn... no life forms would ever have appeared. For the universe to persist, the physical laws which regulate the forces involved must be marvelously

---

* Michio Kaku, Hiperspace (New York: Oxford University Press, 1994) p.15.

well adjusted: a situation which, at the very least, should ensure that one keeps an open mind when it comes to the matter of deciding whether the cosmos is the result of blind-chance happenings, or a creation of the kind of preternatural, supreme intelligence which Aristotle called the Prime Mover.

We have used the term 'spirit' in a pretty broad sense throughout these pages, but always with the implication that it represents a psychical power - operating beyond natural or known physical processes - yielding insights of transcendent importance and providing one with the *will* to act. The theologians talk about the will of God - the supranatural creative will of a supreme beyond-the-world Intelligence.

But in human terms, the revealing insights and *willpower* engendered by a surge of what is commonly and casually referred to as 'the human spirit', can hardly stand comparison with the powers attributed to the concept of any deity; yet it certainly manages to stimulate the mental performance of the brain's neurons wonderfully well, enabling one to achieve ends which, initially, seem to represent an impossible undertaking. Such accomplishments are sometimes described as 'superhuman' - suggesting that a 'supercharging' of consciousness takes place once the vitality of spirit takes over.

The notion of spirit as a central force is as fundamental to theology as it is to psychology - a factor which makes for some difficulty when attempting to discuss the concept of God, or account for the autonomous, inner directives of thought and feeling which rise unbidden to consciousness. For in both instances one must accept the fact that spirit, whether it describes the force moving heaven and earth or directing human 'being', is an intangible entity, quite incapable of being observed, measured, or sensed. Consequently, to illustrate the possibility of its existence one can only use the philosophical language of metaphysics. The objective terminology of science has nothing to go on - nothing it can describe.

It is not that difficult to consider the idea that, *individually*, we are more than just the sum of our biological parts -that a non-biological factor in the form of an energizing and informing spirit-intelligence is at work within the psyche - because there will likely have been times when many of us have been inexplicably moved to feel, think, and act in a way that is quite a-typical, more than usually significant, and totally unrelated to the affairs of the moment. But to readily surmise that a whole universe of energy and matter might have come into existence through the will of a supreme spirit force - an unadulterated, sheer, spirit-power - is altogether more exacting. An open mind, a flexible imagination, and the presence of some lurking intuitive feeling that it might be so, are all required if serious consideration is to be given to the Prime Mover proposition. For the difficulty of comprehending the scale and nature of such a cosmic event, so far removed from any personal experience, renders the whole supposition ineffable. But then of course, there is always faith - that indefinable and unquestioning belief in divine providence as the guiding power in the universe - that likes to dog us.

In any event, hypotheses such as these lie beyond the capability of science to either confirm or dismiss. The very immateriality of such a force as spirit ensures that it cannot be revealed and defined by any scientific method: through the observable, experimental and mechanical means we employ to discover and identify the physical forces responsible for life and phenomena in general. Yet I would say that the art and science of mathematics, of the equations created by those working in the realms of physics and quantum physics come the closest, so far as science is concerned, to touching on the mysterious possibility that such an abstract power as spirit is a force to be reckoned with. Nevertheless, it is well to remember that an intuitively held belief in such a godlike omnipotence at work behind the surface of appearances is to be found throughout human history. Over three thousand years ago the architectural marvels of Egypt were inspired by it and testify to its powerful influence.

And the idea has persisted throughout human history - particularly well expressed by the classical Greek philosophers and playwrights - that human beings harbor an indwelling drive to attain a state of moral and spiritual wholeness: *'a spirit superior to every weapon,'* writes Ovid in Book iii of his *Metamorphoses*.

Consequently, to talk about 'leading a double life' is to suggest that not only do we exist as extraordinarily complex physical creatures, subject to the unpredictable play of happenings beyond our control, but that we also live in an abstract mental world of our own making. A duality of existence which, in itself, justifies an Aristotelian, metaphysical question mark. At some point we have to face the fact that the tug-of-war between these two factions - to which consciousness gives equal time - is responsible for what has often been referred to as 'the tragic sense' of human life. For example, we might find 'true love', achieve exemplary happiness.... or discover who and why we are.... but the world has its own ideas as to how life should be lived: random happenings can bring with them misfortune, disappointment, tragedy and suffering. Such is our condition; which led Thoreau to make the much quoted remark in *Walden I*, '*The mass of men lead lives of quiet desperation. What is called resignation is confirmed desperation.*'

Yet these are precisely the psychologic tensions without which we would be simply 'beings', as opposed to 'human' beings. And I would say that the ultimate tension in this tug-of-war between these 'outer' and 'inner' selves, results from the fact that we are driven, both consciously and unconsciously, to bring about an eventual rapprochement between body, and spirit. And to be even more specific, between the biological death of the first, and the possible spiritual continuity of the second. Our 'double life' knows no greater schism. And the building of the bridge can be a lifetime's work.

> *In my picture of the world there is a vast outer realm, and an equally vast inner realm; between these two stands a man, facing now one and now the other, and, according to his mood or disposition, taking one for the absolute truth by denying or sacrificing the other.*
> C.G. Jung, *Modern Man in Search of a Soul* (1933)

To embark on even a cursory study of world mythologies and religions reveals that from the very beginning they shared one belief in common: the notion that beyond the obvious reality of everything manifest in the material world lay an indeterminable realm of otherworldly powers – deities and spirits both hostile and benign, through whose ministrations the drama of human life was played out. The attendant circumstances of birth, destiny, and death were laid at their door. And then, as Nikos Kazantzakis puts it, '*Out of the great God-producing desert*' came the three great monotheistic religions – Judaism, Christianity, and Islam. Born of the vast spreads of Middle Eastern wilderness and their equally vast spreads of starry sky – they each in their own way conceived and held to the idea that the spirit power behind the scenes was invested in but One God, and that the diversity of everything in a state of being must be contained within a *single* creative principle. It is a concept which, with a few exceptions, has persisted for several thousand years in the recorded history of Western and Middle Eastern societies.

Even the early historic records (5000 to 4000 B.C.) of Old Kingdom sages in Ancient Egypt, established the guidelines for a philosophy of life that outlined the principles of harmonious interrelationships that must be maintained between man and man, man and the cosmos', and thus between man and the Creator. "Order', 'truth', and 'justice' must be matched against 'chaos' 'injustice' and 'falsehood'. All opposites must be reconciled or anarchy reigned. And anarchy separated mankind from the creative principle that transcended man. To live in a state of anarchy was to live cut off from the 'theological' truth of Oneness. Such was the philosophical and practical way of life known as *Ma'at* which was seen as created by *God*. It applied no less to the Pharaoh than to the lowest member of Egyptian society.

The earliest historic evidence I believe we have we have of this moral and otherworldly code, can be found in some tomb paintings from both Old and New Kingdom periods, and certain passages translated from the letters of pharaohs, urging the recipients to put the moral principles of goodness, justice, fairness and tolerance above natural desires and self - aggrandizement. As Cyrilin Aldred wrote in his book *The Egyptians*, *'For The Egyptian, the good life consisted in achieving Ma'at'*. The high ranking Ptah-hotep wrote for his son: *'Truth is good and its worth is lasting and it has not been disturbed since the days of its creator....'*

However, if we turn to the archeological record we can reach much further back in time to find indications of ritual practices - of rites that could have been intended to mediate between the forces of the natural world and those active in a supranatural world beyond. There is one particular excavation which takes us far back in time - to a prehistoric Neanderthal group - which forces us to confront the question as to when notions of a material, time-bound world, coexisting with an invisible, timeless spirit-world.... found their way into the so-called primitive consciousness. (There are those anthropologists who consider Neanderthals to have constituted a separate branch of the *homo erectus* line distinct from ourselves.)

Between the years 1953 and 1957, Professor Ralph Solecki of Columbia

University carried out cave excavations at Shanidar in northern Iraq where he found the remains of both adult and young Neanderthals. On a later dig in 1968 he unearthed what appeared to be a burial site – a grave which yielded evidence that flowers had been used in the burial. Pollen grains from flowers indigenous to the region were found throughout the grave, showing a particular concentration in the chest cavities of the dead. Dating techniques – not entirely reliable in dealing with events occurring over 50,000 years ago – put the burial date at around 62,000 BC. The question then naturally arises as to whether the flowers betoken a ritual burial – one in which they serve the same symbolic function they do today - expressing the hope and wonder of renewal: a springtime-like re-blossoming and continuity after the sleep of winter, the passing of death. In his book, subtitled *The First Flower People,* Solecki offers – and I would say supports – the view that the Shanidar burials provide the first signs that a belief in the spirit's or the soul's* survival beyond the grave was present in Neanderthal life at an early date - before the arrival of *Cro-Magnon* man, a Caucasoid type of human who is often regarded as the precursor of our own species. Given the fact that skilled archaeologists in the field, working for extended periods over the same site, can come to gain a special 'feel' for the evidence they are uncovering - experience intuitive deductions that spring from both the nature of the facts they are gathering; and from the a pervasive, underlying resonance of historic events that sometimes seem to linger at isolated and undisturbed sites - I have no difficulty in accepting Solecki's conclusions.

It is not difficult to imagine how the bursting of buds into flowers could come to symbolize the thoughts and feelings aroused by the idea of regeneration. For after winter comes spring. And with the arrival of spring the land reawakens to life in the burgeoning of plants and flowers of all kinds - delighting the eye, warming the heart, trailing hope and optimism in their wake. Small wonder that flowers are still used in our own funeral rites – their symbolic potency working subliminally to soften the pain of bereavement. Even in climates where seasonal differences are not extreme and exotic blossoms are constantly present in all their ephemeral beauty, the very fact that they wither and resuscitate continuously stands

---

* Thoughts on distinctions to be made between spirit and soul constitute the subject of Chapter 20.

to support the very human hope of some form of persistence, some form of metamorphosis when life ebbs. As symbols of love, sympathy, and reunion, flowers still have few equals. (A passing thought: how would the nurserymen and florists of the western world fare if some ancient-of-days flower ritual at the graveside had not persisted into our own time? If we did not, albeit unconsciously, continue to accept the idea of the rites of passage from one form of being to another? And even the giving of flowers to another on special occasions serves as an expression of gratitude, of caring – of a reaching out from the heart.)

The Shanidar cave burials raise the interesting thought that when it comes to dealing with the issues of life and death, we share a common sensibility, an awareness of life's values, with the Neanderthals – those much disparaged beings who have, in the past, been so frequently written off as mere 'grunters'; yet who, ironically enough, seem from the scant evidence available to have been, in many ways, more sensitive beings than those who overran them - namely, us.

I can almost hear the mutterings of disapproval – perhaps even the indignant and heated rebuttals – emanating from those resolute materialists who believe that every aspect of human consciousness - even the most abstract of thoughts and the most spiritual of feelings - can be reduced to survivalist mechanisms of a purely biological nature serving the processes of natural selection. For if one comes to reflect on the sheer psychological complexity and authority of the inner motivations and insights which drive one through life, it is difficult to believe that this subjective side of consciousness simply reflects the principles and needs of biology. Many of us throughout our known history have been moved to follow highly personal directives, and act in ways which did not serve a purely physical wellbeing - even in the face of death. History is full of examples of men and women who found that the power of faith in principles and beliefs, was stronger than the raw compulsion urging their bodies to live. One thinks of the early Christians facing the lions in the days of Rome... and wonders how the instinctual and natural biological imperative to survive was overcome by a state of mind that transcended it.

"It's surely obvious" - a behavioral psychologist specializing in biofeedback said to me the other day - *"that primitive man conjured up a spirit world, and thought of an afterlife, out of fear and ignorance, that's all. Thank God,"* he

added - seemingly unaware of the irony - *"that we've finally become a more rational species."*

Have we? I hadn't really noticed it myself. The Grand List of Twentieth Century Human Follies & Perversions - (not forgetting those of the preceding three or four thousand years) - hardly supports such a view. Putting moral sensibilities, values, aside for the moment, there is nothing rational - nothing sane and reasonable showing intellectual and emotional stability - behind the genocide practiced by evil and irrational men against millions of innocent people; nothing rational about the extremes to which fanatics of any religious or political persuasion will go anywhere in the world; nothing rational about religious intolerance when it would seem obvious that if any absolute Divine Intelligence, exists, it must be one and the same for all of us - not 'custom made' to suit any specific cultural heritage - the *'My God's better than your God'*, kind of idiocy; nothing rational about the cult of violence which pervades the world - the phenomenon of 'road rage', the mindless anger of fans at sports events, the hate displayed in the streets of racially divided cities, and even the sadistic nature of many computer games. And surely reason must be brought to bear in controlling world populations; in preventing the rape of the planet's resources, and the damage done to the geophysical environment in order to satisfy the needs of the super-consumer nations. And reason demands that the 'god of development' be brought to its knees - that the proliferation of suburbs, together with their green and thirsty golf courses, should not extend into desert lands where it is known that in due course the water will run out. The extent of human perversity is not encouraging, does not altogether support my psychologist acquaintance's belief that we are *'a rational species'*. For reason is traditionally defined as the exercise of logical thought and judgment, leading to logical conclusions that serve the general good.

Also, if the rites of those primitive societies which display overtones of a 'naturally' religious nature, sprang entirely from fear and ignorance - a condition occasioned in large part by the prevalence of sickness and the prospect of an early demise then what about contemporary religious rites? Nowadays, much of that sort of fear and ignorance has been alleviated in many parts of the world by tremendous medical advances resulting in a greatly increased longevity. Insofar as ignorance is concerned, science and philosophy have extended the compendium of factual knowledge, and the boundaries of thought, throughout civilizations both ancient and modern. And particularly so today as science races ahead. Is then, the frequently heard parting expression nowadays, 'God Bless...' uttered out of fear and ignorance? Do clergy and shaman alike go out to 'bless' fishing fleets around the world out of fear and ignorance? Do the members of a rescue team, searching for victims of a terrible terrorist act in one of the world's largest cities, stop as one man when a body is found, remove their hard hats and bow their heads, unmoving for a couple of minutes... out of fear and ignorance? And do we continue to put flowers on the graves of our dead in this age of global high technology, out of fear and ignorance?

Prayers, blessings, invocations, and acts of faith, have been part of the fabric of life in human history, and continue to be the way people - of many religious persuasions - attempt to engage a spiritual resource. (The ridiculous thing - and here the behavioral psychologist has a point - is that the world's monotheistic religions frequently fail to see that the power of spirit, if defined as the radiations emanating from one true Godhead must manifest in the same way, and represent one essential spirit-source for *all* who pray. To indulge in a competition between faiths - claim that a particular monotheistic religion is more valid than another - suggests a measure of unsurpassed stupidity and irrationality.)

I find it thought provoking in a country like the United States, where pursuit of the 'good life' can be an end in itself, that God's help should be invoked so readily in times of personal and national emergency. And while one might be justifiably cynical about this, it seems obvious that a belief in the efficacy of prayer does, in general, exist - even in a 'high-tech' society, where 'success' is largely measured in monetary terms; where the computer is king providing all the answers; and where hours may be spent passively as slave to the 'entertainment' provided by television and film. Such a way of life can ensure that the 'inner voices' become redundant, more or less silenced by the volume and frequency of the outside competition. Consequently, is it not surprising that a mystical belief in the efficacy of an inner spiritual power called 'prayer' should persist?

Obviously, the human spirit does not completely go to ground; become eclipsed by the crowding-in of events that come with the extroverted life. In times of personal crisis when, as is said, 'one's guard is down', its intuitively sustained voice can break through the dominant hold on a consciousness dependent on a constant need to be occupied by outside happenings. For without experiencing, at some level, what Cicero described as, *'the inner knowledge of things human and divine and of the causes by which those things are controlled'*, how could individuals resort wholeheartedly, without apparent pretense, to such a mystical activity as prayer?

It is not only Christians who pray for the souls of friends and loved ones departed; for spiritual support to carry on alone when bereaved; for those suffering intolerable physical pain in the belief that they may find relief; and for the distressed and needy throughout the world. (And how often will those who pray offer a silent prayer for the wellbeing of their own soul?) The fact that this goes on in prosperous parts of the world where educational opportunities flourishes, and science advances, makes nonsense of the belief that prayer results from primitive fears and states of ignorance. I think it more reasonable to suggest that it is practiced, given whatever form it takes, in response to an intuition persisting throughout many millennia: a subliminal apprehension that a transcendent and potent energy-principle, responsible for everything in a state of *being* is 'out there', beyond whatever laws of physics are known to pertain at the time.

It is generally accepted that there are many more forms of radiant energy affecting earth than those presently identified, and it has been suggested that with our highly developed brain we are capable of responding physiologically - of creating images in consciousness, that is - due to the *resonance* of the most potent among them. This is the energy we conceive to be that of spirit in order to distinguish it from the more obvious forms of physical energy. But to speculate further... how might the brain be capable of picking up the transmission of such spirit-energy resonances? And here I find myself envisaging a 'special forces' unit of hard-wired neurons serving as a highly sensitive 'antenna' - a mental unit which, as a result of prayerful meditation (conscious or unconscious), becomes especially empowered to both receive and utilize the superphysical powers of spirit-energy.

> *The centre that I cannot find*
> *Is known to my Unconscious Mind;*
> *I have no reason to despair*
> *Because I am already there.*
> W.H. Auden: *The Labyrinth*

I have seen the transformation wrought by the intervention of prayers – automatically uttered by men in war - and the strengthening of will, resolve and courage which enabled them to overcome the sheer biological fear of death during eight long hours in the air. And in more normal times I have known instances when the prayers involved in spiritual healing have seemingly saved a life when every medical prognosis declared the situation hopeless. The *Old Testament* statement, *'Man doth not live by bread alone'*, is surely referring to this kind of spiritual (as opposed material) sustenance.

But if, in looking at the world, it seems obvious that all men do not view spiritual sensibility in the same way. For prayers have been said, and continue to be said, by those requesting divine help in pursuing political and secular ends. Such invocations of spiritual power often ask for assistance in the ruthless killing of other mortals, and come from those who define 'God' in anthropomorphic terms.... rather than recognizing that they are attempting to communicate with an indefinable spirit-force not of this world. Throughout history, the worst offenders in this regard have been extremist and fanatical religious movements whose adherents pray for the destruction of those whose articles of faith differ from their own. Yet it would seem to be a matter of common sense to realize that the human spirit is the *human* spirit - an individual yet universal life principle to be distinguished from the physical body living naturally in time, and seen to be linking each one of us to a supreme and transcendent power, whether called God, Jaweh, Allah, Great Spirit... whatever. And the message delivered by the most mystical and holy followers of the world's most mystical faiths, is that we should strive to embrace virtue and wisdom, goodness and morality: the enhancement of life; not its destruction. If prayer is then to be true to its spiritual roots, and to work its extraordinary effects, it must be an intercession to the powers of benevolence and light; rather than invoking the forces of hate and darkness:

> *If God listened to the prayers of men, all men would quickly have perished: for they are forever praying for evil against one another.*
> Epicurus (341-270 B.C.): *Fragments, Physics 58*

Two scientists - Andrew Newberg, a radiologist and professor at the University of Pennsylvania, and the late Eugene d'Aquili, psychiatrist and anthropologist - came together in the early 1990s to work on d'Aquili's theory that brain function was responsible for, '...*a range of religious experiences, from the profound epiphanies of saints to the quiet sense of holiness felt by a believer during prayer*'. These are the words used by Vince Rause, writing in the Los Angeles Times Magazine, to describe the purpose of the experiments carried out by d'Aquili and Newberg.* He goes on to describe, clearly and precisely, the methods they employed and the results they obtained; and rather than paraphrase his essay - which I feel would diminish the import of his words - the central and distinctive passages of his report are reproduced here untouched:

> *They used an imagining technology called SPECT scanning to map the brains of Tibetan Buddhists meditating, and Franciscan nuns engaged in deep, contemplative prayer. The scans photographed blood flow - indicating levels of neural activity - in each subject's brain at the moment that person had reached an intense spiritual peak.*
>
> *When the scientists studied the scans, their attention was drawn to a chunk of the brain's left parietal lobe they called the orientation association area. This region is responsible for drawing the line between the physical self and the rest of existence, a task that requires a constant stream of neural information flowing in from the senses. What the scans revealed, however, was that at peak moments of prayer and meditation, the flow was dramatically reduced. As the orientation area was deprived of information needed to draw the line between the self and the world - the scientists believed - the subject would experience a sense of a limitless awareness melting into infinite space.*
>
> *It seemed they had captured snapshots of the brain nearing a state of mystical transcendence - described by all major religions as one of the most profound spiritual experiences. Catholic saints referred to it as 'a mystical union with God. A Buddhist would call it 'interconnectedness'.*
>
> *These are rare experiences, requiring an almost total blackout of the orientation area. But Newberg and d'Aquili believed lower degrees of blockage could produce a range of milder, more ordinary spiritual experiences, as when believers 'lose themselves' in prayer, or feel a sense of unity during a religious service. Their research suggests that all these feelings are rooted not in emotion or wishful thinking, but in the genetically arranged wiring of the brain*
>
> *"That's why religion thrives in an age of reason," Newberg says. You can't simply <u>think</u> God out of existence, he says, because religious*

---

* Vince Rause: *Searching for the Divine* (Los Angeles Times Magazine, July 15, 2001).

> *feelings rise more from <u>experience</u> than from thought. They are born in a moment of spiritual connection, as real to the brain as any perception of 'ordinary' physical reality.*
>
> *"Does this mean that God is just a perception generated by the brain, or has the brain been wired to experience the reality of God?" I ask.*
>
> *"The best and most rational answer I can give to both questions," Newberg answers, "is yes. "And "no'.*

Earlier in his report, Rause had said to Newberg, "You're saying your research proves this higher reality exists?"

"I'm saying the possibility of such a reality is not inconsistent with science," Newberg replies.

PRAYING and LOVING

To pray in true spiritual vein is essentially a mystical experience. Yet in an increasingly secular and science-oriented world the mystery inherent in the experience of prayer has not suffered the disparagement that might have been expected. Neither has the mystique inherent in loving. To pray, and to love profoundly, are essentially internal meditative states - and both can bring about a trance-like suspension of the routine flow of sense impressions together with our behavioral responses to them. Prayer leads one into an inner mental realm where thought and feeling are directed beyond the temporal and physical aspects of life. *Selfless* love brings one to feel extraordinarily deeply for a fellow human being, dog, horse…. allowing one to step away from the world and live, as they say, in the heart: and *'The heart has its reasons which reason cannot know.* (Pascal: *Pensees* IV). Such are the transitions from one level of consciousness to another that can be induced by prayer and love. To lose oneself in prayer, and become spiritually empowered through the constancy of love, are the most overwhelming and effective ways to become whole as a human being. For they bring together the physical and spiritual aspects of our condition. The great French poet, novelist, and playwright Victor Hugo (1802-85) described this human dichotomy in the Preface to his play *Cromwell*, as follows: *'Thou art twofold, thou art made up of two beings, one perishable, the other ethereal, one enslaved by appetites, cravings and passions, the other borne aloft on the wings of* enthusiasm and reverie - in a word, the one always stooping towards the earth, its mother, the other always darting up toward heaven, its fatherland.'

If I were to become a practicing psychoanalyst, two signs would be prominently displayed in the office. One would read, **What does prayer do? It takes us beyond the nature of our animal selves**; the other, **What does love do? It takes us beyond the ego of our human selves.** Or perhaps it might be more effective if one's patients were introduced to the following line from the English novelist George Meredith (1828-1909), found in *The Ordeal of Richard Feverell*: 'Who rises from Prayer a better man, his prayer is answered'. Or were brought face to face with the statement quoted in the pages of the *Goncourt Journal* - (compiled by the brothers Edmond and Jules Goncourt, writing in Paris between the years 1842 and 1870) - which reads, 'I believe that love produces a certain flowering of the whole personality which nothing else can achieve'.

At this point, I find myself turning to regard the brain and wondering just how the neurons manage it. How, on the one hand, they bring the existential reality of things in nature's world to register as perceptions; while on the other they induce meditative states of mind that remove us from temporal and factual realities. It seems to me to be little short of miraculous that electrical and chemical activity in billions of the brain's nerve cells can engage the psychical states which have us adrift - loving and praying.

But if you, the reader, are uncertain about the close psychological relationship between praying and loving, I would ask you to consider the following verse from A. E. Housman's, *A Shropshire Lad*:

> *If truth in hearts that perish*
> *Could move the powers on high,*
> *I think the love I bear you*
> *Should make you not to die.*

The words of Alfred Edward Housman, English poet and classical scholar who died in 1936, speak volumes in revealing the profound nature of the attachments we make in the name of love. For he writes about it as an experience that takes us to a metaphysical level of truth - living as much by the heart as by the head; more by feeling than by thinking. But the extent of our capacity to *'be in the loving'* - (as a native Australian, living on Melville Island in the Darwin Gulf, expressed it to me one day) - causes us to form the most powerful emotional bonds of unselfish devotion. The other day a man died in a storm-lashed river trying to save his dog. He could not wait to pray, but acted and gave his life. The dog was saved. Would his act be any more commendable, or seem any more reasonable, had he been attempting to save his child? The emotional pain suffered if the object of love is about to be lost does not always discriminate between a child and a dog when absolute love is involved. And for the one who is left, the grief can persist throughout life, whatever material comforts and distractions are to hand.

Over a hundred years have elapsed since Housman wrote *A Shropshire Lad* in 1896, and lifestyles, together with philosophical attitudes to life, have changed dramatically - particularly during the last years of the twentieth-century and those heralding the start of the twenty-first. During this period the technological revolution in the communications industry has invaded our sense of personal space and diminished our ability to enjoy private moments of time. Nowadays it's very easy to 'pass the time': instant telephonic contact with somebody, somewhere, is literally just at one's fingertips; as is passive entertainment at the touch of a button - distractions which enable those with little inclination to reflect on either the depth or shallowness of life to just drift along. Which was certainly not the case in Housman's day when the average person had to live much more intimately with his or her worries, hopes, thoughts, dreams... Today, having ready recourse to electronic and mechanical devices, it is all too easy to fritter time away and avoid putting the brain to work facing issues. Idle talk on cellular phones is a way round the boredom of being 'stuck' with one's own company (even when simply walking or driving from one location to another); or one can attach oneself to a computer, keep occupied by 'surfing the web'.

In such a cultural climate Housman's romantic and mystical imagery may well be seen as representing latter-day flights of poetic fancy. That is, until a child is lost, a lover gone, a beloved parent, friend, or pet no longer around. And then the spiritual power of his verse - telling of how a profound love embraces a spiritual truth - may provide some hope and comfort for the contemporary reader.

I have concentrated in the final pages of this chapter on the themes of Love and Prayer, seeing them as representing aspects of consciousness that stand in the most marked contrast to the material nature of our day-to-day life. For in surveying the fields of literature and philosophy.... two psychological drives appear to have constantly engaged the imagination of the world's thinkers: the human need to love and be loved - and the invocation of divine authority through prayer. They both have seemed to invest life with a convincing sense of purpose and worth - alleviating doubt and concern as to why on earth one should be alive at all. Take a stroll through an anthology of poetry, for example, and notice the predominance of poems inspired by both the idea and the experience of love. *It is love, not reason, that is stronger than death,* wrote Thomas Mann. And the following verse from *Light* by Francis William Bourdillon expresses the transcendent quality that love can bring to human life, and the darkness that descends when a loved one is lost. The wonders of high technology are of little avail in helping those who face this kind of suffering.

> *The mind has a thousand eyes,*
> *And the heart but one;*
> *Yet the light of a whole life dies,*
> *When love is done.*

Pray, for all men need the aid of the Gods, writes Homer in the *Odyssey*. And, 'Our prayers go out for all those who have lost loved ones. . .' intones the British Prime Minister publicly as news is received of the first casualties in war. To be followed almost word for word by the American President a few hours later. Almost three thousand years of Western history since Greek civilization and still - as Victor Hugo put it in *Les Miserables* - '...the soul is on its knees'.

Loving and praying work very closely together as forces of the psyche: *He prayeth well, who loveth well/Both man and bird and beast...*wrote Samuel Taylor Coleridge. They are both ways of reaching out beyond our solitary state to approach levels of existence that have a whiff of the transcendent about them; (perhaps more than a whiff for those who attain, through prayer, a state of meditative calm that, as is said, *'passeth all understanding.'* There is an inherent loneliness attending the human condition: a sense of our isolation as individual entities that is from time to time unavoidable - for everyone of us is governed by an independent mind operating in a way peculiar to itself, rendering us completely accountable to ourselves. And this is an essential part of our humanness.... however much we embrace an extroverted view of life: indulge in our much touted right to pursue a life of pleasure, or go through the motions of substituting one new 'adventure' or possession for another. Even under these circumstances the disquieting realization creeps in that truly one is going it alone. Particularly as one ages. It is one of the hallmarks of 'growing up' that the awareness of something 'missing' in life - some experience that would complete the circle - becomes more pressing. And the sense of 'waiting...', waiting for the key piece of life's jigsaw puzzle to show up, becomes more insistent. More than ever at this juncture one must try to find the missing piece within oneself by a calm and constant reflection on the course of one's life. Yet the lesson we never seem to learn - despite the expostulations of ballad singers and poets, of holy men and wise essayists and philosophers - is that the surest way to see oneself confidently through to the end is to trust in the refinement of thought and feeling induced by love and prayer: the principal sanctuaries we have along the way, providing us with the means of experiencing a truth greater than our timebound and body-wracked selves. Sanctuaries which allow the human spirit its moments of intercession with a supranatural Providence to which, for thousands of year, mankind has been intuitively drawn. Sanctuaries that ease us through life. Whether or not they augur for a personal continuance beyond time and space is, indeed, another 'matter' (if I may be allowed the pun).

When the celebrated doctor and philosopher Carl Jung - considered by many to be the outstanding healer of the mind and profoundest thinker of the twentieth-century - was featured on B.B.C television in a famous program called *Face to Face*, he encountered a tough, no-nonsense, interviewer in the person of John Freeman, politician and writer. For an hour Freeman questioned Jung on the complexity of the human psyche: on his thoughts concerning the mental roles of mind and brain, his concept of the unconscious, the difficulties of reconciling the biological realities of a body with the immaterial forces of soul and spirit, the processes of individuation... and so on. And gradually, as the hour wore on, Freeman became more and more pensive, more the student listening to the master. The last question he put to Jung was, 'Doctor Jung, do you believe in God?'

'No,' came Jung's reply. 'I don't believe. I know.'

John Freeman went on to become one of Carl Jung's strongest

supporters, persuading him to write a popular book entitled *Man and his Symbols* to bring his sagacity and wise counsel to the attention of a larger lay audience; also to write an autobiography which Jung was loath to do, for he disliked the cult of personality and was averse to promoting personal publicity. But, after much prodding and urging, he wrote *Memories, Dreams, and Reflections* - an account, unselfconscious yet intimate, of the richness of what he called the '... *inner happenings*' in his life. It became a much sought after book that influenced the lives of many and is regarded as one of the most significant personal testimonies of our time.

> *Love is the only satisfactory answer to the problem of human existence.*
> Erich Fromm: The Art of Loving

10

THE BRAIN'S PRODIGIOUS CEREBRUM & CORTEX: THE LATERAL SPECIALIZATION OF THE TWO HEMISPHERES

The brain is unique in the universe, and unlike anything man has ever made.
Robert Ornstein and R.F. Thompson:
from *The Amazing Brain*

IN THE BEGINNING... - three words denoting both conundrum and mystery: the *how* and the *why* behind the existence of the universe and the organic forms of life on this planet. Scientists of many persuasions have been concerned for a few thousand years with 'how' it all came about. While philosophers, theologians, prophets, priests, and shamans, have all attempted to answer, in one way or another, the imponderable question as to 'why'.

In Judeo-Christian societies the Bible's book of *Genesis* has, historically, been generally accepted as the account attributing the whole act of creation to God as the omniscient, one true *supra*-natural Intelligence responsible for everything that is. The religious authority vested in this conviction ensured that strenuous attempts had to be made to dovetail scientific discoveries into the general schema advanced by *Genesis*. For in this theological and transcendent scheme of things 'how' and 'why' could not be separated in confronting the mystery of creation: they worked synergistically as manifestations of God's purpose. But as the Middle Ages gave way to the increasingly secular societies of Renaissance Europe, and science developed apace, this rapprochement between theological doctrine and scientific fact became more and more difficult to maintain. And when the so-called Age of Enlightenment dawned in the 18th century - taking the position that reason and experience are more reliable and credible than dogma or tradition - scientific investigation not only assumed credibility in its own right in asking *how* things came into being, but called into question the supposition of a God-like purpose when it came to asking 'why'. Even so, it was also apparent that the question of 'why' could not be determined empirically and objectively, and so did not come within the purview of science. In the broad advance of natural science and technology it became somewhat of a fringe interest, an exercise in metaphysical wordplay between theologians and philosophers. However, the major scientific discoveries over the last fifty years have revealed such incredibly intricate combinations of energy and matter (at both micro and macro levels) that they merit being regarded as 'stranger than fiction' - discoveries that trigger natural feelings of awe in the rest of us; so much so that the question of 'why' comes naturally to mind.

For science has outdone itself by its very brilliance. It has demonstrated the viability of evolutionary theory; opened up the whole field of nuclear physics; greatly expanded medical and biological knowledge; developed the space-navigation and rocket-propulsion know-how that has taken man to the moon; created the sophisticated mathematics of astronomy and the incredible abilities of equipment like the Hubbell telescope to search the mind-boggling vistas of space; produced the technology by which geologists discover the interior dynamics of planet earth; formulated the mysteries of quantum physics; and, in genetics, computed the staggeringly complicated make-up of chromosomes in the human genome. And all aided by the wizardry of computers.

When I say that science has outdone itself, I mean that it has come remarkably close to achieving results that appear as surreal and imaginative as the creative ideas of science-fiction writers, and that challenge the imagination of the philosophers. For example, here is a statement concerning space, time, and distance in the universe at large expressed in cosmological terms:

> *By human standards, one cubic light-year is an enormous chunk of space, enough to hold the Sun, its orbiting family of nine planets, a million roving asteroids and a trillion comets on giant looping paths which carry them so far out that Mother Sun is reduced to a bright star in the firmament. But on the scale of the known cosmos, this is merely our backyard - just one infinitesimally tiny pocket in a universe of a million trillion cubic light-years.**

---

* Terence Dickinson, *The Universe and Beyond* (Richmond Hill, Ontario, Canada: Firefly Books Ltd. 2010), p. 119.

Now can you get your mind round that? Or are you thinking that they must have got the mathematics wrong? For myself, the awesome dimensions of space-time depicted here are so overwhelming that the question of *why* such cosmic enormity, comes just as readily to mind as *how*. When science has progressed to a point where it touches on the truly profound - as in the case of astronomical physics I have just quoted - and the questions of 'how' and 'why' demand equal attention then, as I mentioned at the outset, the *science of physics* gives way to what Aristotle regarded as the *philosophy of metaphysics:* the speculation that drives mind and brain to wonder about the original and prime cause of the universe and all that lies within it; to ask the questions 'wherefore?', to what end? and, ultimately, to ponder whether is it all the result of chance or design.

I imagine the ghost of Aristotle surveying the marvelously sophisticated revelations of contemporary science and declaring, "*I said as much two thousand years ago: that the more inspired the research, the closer it approaches the heart of the mystery behind the façade of existence. As I wrote somewhere, '... we do not know a truth without knowing its cause'. That's why in my treatise on Physika I presented the chapters on material facts and findings first; and then those concerned with abstract speculations concerning why anything should exist at all, at the end - grouping them under the general heading 'Meta-physika'. I coined the word... for 'meta' signifies 'after' - hence, 'after physika'. Such is the order of mental events in any far reaching creative science: the physical factors are tested and recorded; the unanswerable questions concerning the nature and role of a Prime Mover go round and round on the treadmill of the mind. Yet both mental activities result from the one organ of consciousness - the brain - and are both sides of the same coin of knowledge.*"

The unifying theme running throughout this book is expressed in the words I have attributed to my imaginary ghost of Aristotle - words which simply reflect the tenor of his own writings when alive. His discussions on the subject matter of physics are still viable and used today, and his philosophic insight and keen powers of observation as a natural scientist have ensured his continued fame. But his recognition of the ability of consciousness to deal in both the abstractions of metaphysics and the concrete facts of science, puts him in the camp of my friend, the Flying Bishop who expounded at length on the two sides of life's coin while we were *en route* to Dublin in that rather shaky DeHavilland biplane. During the course of his aerial discourse you will remember that he put forward the basic proposition that we have two sides to our nature: that which lives in the world and responds to the material challenges and opportunities it presents; and that which resides in the mind responsible for deep levels of feeling, the processes of abstract thought, and the faculty of imagination, all conjuring, as Shakespeare put it, '...such stuff as dreams are made on'.

A dual-purpose brain: an instrument 'unique in the universe...'

The story of the brain's evolution is fascinating. The original brainstem sitting atop the last vertebrae of the spine and similar in size to the *complete* brain of a reptile, is the ancient seat of instinct, operating automatically to produce an appropriate response to external stimuli without requiring any decision to be made by the subject; in addition it maintained those bodily functions most necessary for survival such as breathing and heart rate. And there it sat alone, so many aeons ago, the first building block of what was to become the human brain we know today. Around and above the brainstem's 'bargain basement' ganglion of nerve cells, an increasingly sophisticated and complex emporium of new brain 'parts' came into being - an evolutionary process estimated to have taken place over a period of 500 million years. The cerebellum develops, attached to the rear of the brainstem, and regulates posture and muscle coordination as well as serving as the storehouse for certain kinds of memory; (its importance becomes obvious when we learn that it has tripled in size over the last million years). Somewhere between 300 and 200 million years ago, maturing above the brainstem - and now in the center of the present brain - came the limbic system. Most highly developed in mammals, its functions are vital: regulating blood pressure, heart rate, and temperature - governing the body's own environment so that we are not cold-blooded like reptiles. It is also thought to be the seat of a growing range of emotions driving survival behavior such as aggression, evasion, and sexual needs. At its center lies the thalamus which can be thought of as a triggering agent for consciousness - a kind of ignition system, if you like - and a distribution center of certain kinds of external information to relevant areas of the cortex.

Now we come to the crowning - literally and figuratively - evolutionary achievement: the cerebrum and its cortex. It is the largest part of the brain, surmounting and enveloping the limbic system and is thought to have first appeared in our forebears about 200 million years ago. Composed of two hemispheres, left and right, the surfaces of which are covered with a complexly-folded, one-eighth of an inch thick, tier of neurons known as the cortex - which is home to the highest number of brain cells found in any area of the brain - the whole forming an amazing 'command center' which makes human consciousness unlike any other of which we are aware. Connecting the hemispheres is a 'bridge' of threadlike nerves made up of some 300 million nerve fibers - the largest such conduit in the brain - called the corpus callosum; by means of which the directives and information provided by one hemisphere complement the mental resources of the other in the overall scheme of consciousness. It is an extraordinarily intricate neural operation. For our purposes here I will describe it as simply as possible. *Physiologically,* the left hemisphere and its cortex is in charge of information coming in from the right side of the body and determines the glandular and muscular responses to be made; while that of the right hemisphere performs in the same way vis-à-vis the body's left side. *Psychologically,* the left hemisphere is the seat of such faculties as reason, logic, and language, serving to objectively explore the nature of both our own existence and that of the environment in which we find ourselves. While the right hemisphere brings imaginative, creative insights to bear on the rational deliberations of the left; as well as evaluating the degree of importance and meaning they hold for one.

The neuroscientist will tell you that, as yet, we know comparatively little about how the cortex works - that attempting to disentangle how the trillions of neural connections function to bring together the objective and rational powers of intellect with the inner-directed and reflective life of the imagination - is an incredibly challenging, if not impossible, task. Especially as to do so requires that we utilize the very powers of the cortex itself - which raises the question as to how can a subject ever objectively know the nature of itself? And it gives one pause for serious reflection to think of the cortex carrying the load of billions of working neurons, and trillions of axon connections: all that chemical and electrical activity which determines the course of one's individual consciousness during every single minute of the waking day - not to mention serving the subconscious to introduce *'little'* dreams into sleep, and the unconscious to communicate in the form of *'big'* dreams. Only with the cerebrum and cortex finally in place, can it be said that humankind had finally arrived on the scene. Here is Robert Ornstein, talking about the cortex in *The Amazing Brain*:

> *The cortex is the 'executive branch' of the brain, responsible for making decisions and judgments on all the information coming into it from the body and the outside world. First, it receives information; it analyzes and compares this new information with stored information of prior experiences and knowledge, and makes a decision; it then sends its own messages and instructions out to the appropriate muscles and glands....*

However, it should be pointed out that those writers who talk about 'two brains', in terms of right and left, are ignoring the fact that although the two hemispheres are performing different functions, their separation is not radical, for they communicate constantly with each other via the corpus callosum. For example, to experience both abstract and objective *thought* (the human brain's most exalted achievement), or to utilize *language* (some would say just as equally celebrated an accomplishment), both hemispheres are reciprocally active and work together as a single brain. Some of the earliest examples of art found in Western Europe are the Paleolithic cave paintings of animals which date from c. 18,000 B.C. to c. 12,000 B.C. They reveal the left hemisphere at work in the keen objectivity of the hunter-artist's eye - grasping the essential physical characteristics of the animals and the fluid nature of their movements; together with the influence of the right hemisphere seen in the painter's creative ability to project into his work strong personal feelings and symbolic significance. One of the finest examples of Palaeolithic art revealing how the two hemispheres work hand-in-glove is the wonderful rendering of the deer (reproduced here) from the great 'gallery' of animal paintings discovered in the caves of Lascaux in South West France. It is a masterly image, revealing the heightened quality of the artist's objective perceptions that enabled him to portray the elegance of its bodily form and the grace of its motions; the delicate drawing of the head catching all the apprehension of a deer in flight. Yet, at the same time, there can be no doubting how strongly the artist's feelings, his identification and empathy with the animal, and his intuitive urge to imbue his drawing with a symbolic aura, were aroused by the confrontation with his subject. Just look at how he got carried away in depicting the flamboyant exuberance of those antlers! The result is an image that captures what might be described as the essence of deerness (See illustration on Page 160). The English novelist D.H. Lawrence made a famous statement concerning Vincent van Gogh's *Sunflower* paintings which, when amended slightly to refer to the Lascaux deer rather than to one of Vincent's sunflowers, serves to illustrate the high degree of creative sophistication the Old Stone Age artist brought to his work. Thus modified, Lawrence's original comment reads: *...the vision on the wall of the cave is a revelation of the perfected relation, at a certain moment, between a man and a deer...*' 'Courtesy of the cerebra and cortex', one is tempted to add. For it is very unlikely that the Lascaux deer could have come into being had not such interaction between the hemispheres been a fact of Palaeolithic life. It is such a partnership that leads Ornstein to go on and say:

> *Between four million and one million years ago, the fourth (and to date final) level of human brain organization emerged: the lateral specialization of the two hemispheres. These differences in function appeared at the time humans first began to make and use symbols (both language and art).*

'*The lateral specialization of the two hemispheres...*' Each side of the cerebrum finally coming to concentrate on its own specific ends, yet work hand in glove with the other - the particular kind of awareness instituted by the one complementing that established by the other: thus providing us with what we assume to be the most comprehensive range of consciousness on the planet. It was perhaps about 15,000 years ago that the Master of Lascaux made full use of his 'modern' brain. And Robert Ornstein, after vividly describing the workings of cerebrum and cortex concludes by saying, '*It does make you wonder how all this happened.*'

There are two ways in which we have benefited greatly from this evolutionary (and revolutionary) mental development. In the first place we have become intelligent beings - *knowing* beings, empowered to think and learn quickly and readily, and to be skillful, adroit and ingenious in doing things: individuals who can adapt well to novel and difficult situations in life, utilizing the now allied faculties of imagination and reason to achieve essentially practical ends. Clever creatures, having the gift of 'level-headedness', so-called. But then we have to consider that this

Drawings by David Macauly

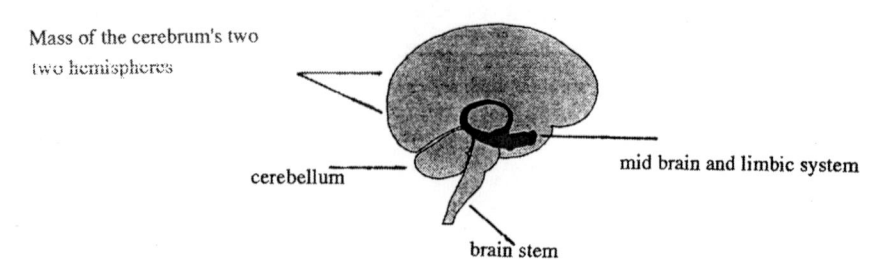

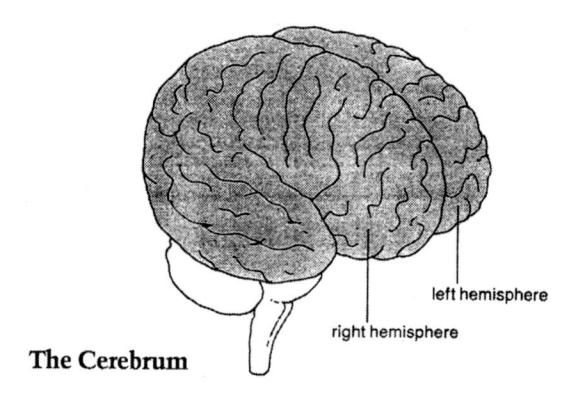

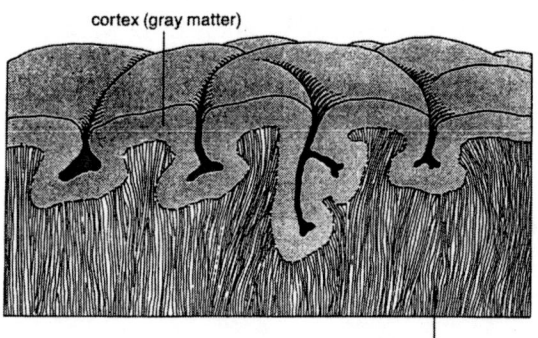

Mass of the cerebrum's two hemispheres

cerebellum

mid brain and limbic system

brain stem

The Cerebrum

left hemisphere
right hemisphere

cortex (gray matter)

A Section of the Cerebrum   nerve fibers (white matter)

Drawing by David from The Amazing Brain

level-headedness (or, if you like, partnership of reason and imagination between the lateral hemispheres), can lead to a still higher level of mental achievement - the attainment of a specifically human quality called wisdom. Now it is possible to be clever and intelligent without being wise, although I would say that levels of wisdom are, more often than not, built on the knowledge gained through exercising a practical intelligence. (Yet it has to be said that there are those individuals who are innately wise, without benefit of conventional learning or life-experience.)

However, wisdom transcends factual knowledge and the intelligent adaption to circumstance that goes with it. It represents an altogether deeper form of knowing - delivering insights of a highly imaginative nature as to the intrinsic worth of things and events, and informing of the right and wrong ways to deal with them for the general good. Consequently, there is both a practical and a moral aspect to wisdom - the kind of measured judgment that deduces not only the most efficient and effective way of tackling situations, but also serves to reveal inner and strongly held convictions as to the most *just* way to deal with situations: undertake the 'rightness' of action that the French 18th-19th century philosopher Joseph Joubert described as being determined by 'the health of the soul'. Such wisdom has, over the years, been described as 'the wisdom of the heart' - a symbolic allusion to the more spiritual insights offered by 'the heart', in comparison to those judgments presented by 'the head', or intellect.

Obviously, as we look back over our history, the working relationship between left and right brain has served us well, particularly in terms of our increasing cleverness. Just look at the heights to which an imaginative intelligence has brought us during the technological revolution of the fifty years. However, it is disappointing to realize that after a cerebral partnership commencing between two and four million years or so - (if the neuroscientists are right in their estimate of the time when this phenomenon took place) - we have not yet managed to collectively attain a commensurate, worldwide, and permanent measure of wisdom. The old proverb, *'Some folks are wise and some are otherwise'*, is a witty way of pointing out that wisdom is not a faculty we all possess. Similarly, it can be said that throughout mankind the ability to think intelligently is by no means uniform - that there are high-achievers and low-achievers across the board. It may be that the differences existing between individuals in terms of their ability to live wisely, and/or intelligently, is due to the level of performance put out by their left and right brains - levels that may range from lower to upper limits of efficiency when it comes to attaining degrees of objective and abstract specialization. Add to this an equally important factor: namely, that the degree of effective communication across the nerve-bridge between the hemispheres - be it good, bad, or indifferent - must play a vital role in determining the quality of the mental balance attained, and therefore the degree of comprehension and judgment reached.

To realize the highest level of consciousness - which I take to be the gift of wisdom - I would think it necessary for both hemispheres to have become specialized to the *'nth'* degree, and be able to communicate with each other through a highly evolved, *super*-conductive, corpus callosum. A union of this sort represents such a fine-tuning of cerebral function that I would say it signifies the highest level of brain evolution thus far. Yet from what we know of our history - and that can never represent an exhaustive assessment - it would seem to be a state achieved by a relatively small number of truly wise men and women.

Writing in 1890, the famous English biologist T. H. Huxley, said: *'The doctrine that all men are, in any sense, or have been, at any time, free and equal, is an utterly baseless fiction.'* The conclusion which may be drawn from Huxley's statement is that - (bearing in mind the degrees of inequality that could pertain in the cerebral fine-tuning I have postulated above) - all evolutionary developments of the brain do not impart a similar degree of mental ability on each individual brain. For there are brains that emerge from the process and bestow high levels of intelligence and wisdom on their owners; others that come to only function minimally in this regard; and those that perform at varying levels between the two extremes. This may have something to do with differences in the neuron count maintained by individuals; or with inequalities in the electrical power generated by brain chemistry. In addition, the laws of heredity are likely to affect brain performance. Also, the effect of differing environmental factors such as climate, family life, education... must also be considered as a factor responsible for determining the organization of each brain and the individual mental characteristics that result. For example, it has been known since the time of Hippocrates that moderate cold has a stimulating effect on brain function, whereas excessive heat (sometimes called 'tropical torpor') debilitates it. And there is another - and I believe crucial - element which helps to determine the ability of a particular brain to reach high levels of imagination and wisdom. A non-biological factor which we have touched on previously - the psychical force of Mind*: a mental power which I see as generated by the activity of a highly active soul and its agent, spirit** - the 'superchargers' of cerebral function.

The Periclean Age of Athens provides an example of how an urban civilization prospered under the leadership of a man possessed of high levels of both reason and imagination - a well-balanced consciousness, but one also inspired by a strong and questing spirit. Pericles (495-429 B.C.), the illustrious Athenian statesman and general has given his name to a period that is generally considered to represent the heights of ancient Greek civilization. A time when, due to his influence, sculptors, philosophers (Socrates, for example), and dramatists, created their 'immortal' works; when Athenian democracy realized lasting political and humanitarian values which continue to be a model for the Western world. And then we have the example of the Roman emperor Marcus Antoninus Aurelius (121-180 A.D.) It is said that when he was a boy attending gladiatorial events, rather than cheering with the thousands when a victorious gladiator plunged his sword into his vanquished opponent, he buried himself even more deeply into a book on moral philosophy. When he came to the throne he was said to be the first philosopher in Western history to assume such a position of power. Throughout his reign he believed that no price was too great if it could buy peace and goodwill, and consistently extolled the powerful combination of reason, imagination, and contemplation in furthering virtue, and in fighting the ignorance which is the root of fear, desire, sorrow... all the negative aspects of this world. Even the Christians, whom he never really understood, acknowledged the spiritual nature of his character - that special quality which distinguishes those men and women whose visionary wisdom has caused the rest of us to ponder the source of their convictions and strength of character. As Emperor, he brought his stoic and moral sensibilities, his imagination and powers of

---

* The relationship between soul and spirit is discussed in Chapter 20: THE DIVIDE BETWEEN SOUL and SPIRIT.

** Similarly, the link between mind and brain is the subject of Chapter 17: MIND AND BRAIN.

reason, to bear on one of Rome's most turbulent periods in its long history: a wise and exemplary exercise of power without which chaos and anarchy might well have reigned. In the words of my friend the Bishop, writing to me after our flight to Dublin, Marcus Aurelius was *'a great soul, working through a great brain'*.

Pericles, Marcus Aurelius, the legendary King Arthur, Joan of Arc, St. Francis of Assisi, Leonardo de Vinci, Voltaire, Florence Nightingale, Jung, Ghandi, Einstein, Jonas Salk, Nelson Mandala, Mother Teresa, the Dalai Lama... just a few of those who come to mind: people whose lives profoundly influenced the practices and ethos of their times: men and women of imagination and sound reason whose presence improved the quality of life and imparted a sense of purpose to human existence. Yet, relative to the millions forming the great body of mankind who plod pedestrianly along, the wise ones always seem to be relatively few in number - those of them we know about, that is.

A HOUSE DIVIDED?

It would seem - certainly at first glance - that the implications of the familiar line in St. Mark's Gospel, *'If a house be divided against itself, that house cannot stand'*, could be applicable to the 'house' of the cerebrum with its divided mental hemispheres. For with left and right brains each 'doing their own thing', it would seem that a natural degree of schizophrenia is built into the cerebrum's functioning, thus militating against any unitary effectiveness, helping only to ensure the development of a split personality. However, the opposite is really the case. The cerebrum is a wonderful example of what has been described as 'unity in diversity', because although the two hemispheres are physically distinct, the mental operations performed by one side are complementary to those accomplished by the other, so creating a relationship of mutual interdependence - a holistic system of enlightenment. And one that would not exist without the presence of the 300 million nerve cells forming the bridge of the corpus callosum. Without its appearance on the scene the 'house' of the cerebra would truly be divided. But this remarkable neural conduit ensures the dynamic interplay between the left brain's ability to introduce objective, sequential (linear) thinking in matters of language, logic, and reason... and the right's more subjective, non-sequential (spatial) forays of the imagination when it comes to insight and creative accomplishment. In addition, without the great number of mental permutations that originate from such an active dialogue, where would be the psychological diversity that results in the forming of character and personality - the marked differences in attitude, manner, and behavior by which we recognize individuality as such?

The 'Deer of Lascaux' for example, was painted to bring about a level of sympathetic magic between man and animal - giving the hunter-artist power over the deer inasmuch as he had now come to possess its image. The Master of Lascaux is thus revealed as a man of great mental vigor, responding to the complementary dictates of both objective and subjective halves of the cerebrum. (Generally speaking, a so-called 'primitive' culture - as the Palaeolithic is said to be - is characterized as one in which raw feelings and irrational fancies and concepts strongly outweigh objective thought and analysis. But I think you will agree that such an evocative work of art as the Lascaux deer can never be considered primitive in this sense.)

'*Without Contraries is no progression...*' wrote William Blake, the English painter and poet in 1790, and although it is unlikely he had specific neurological knowledge of right and left brain specialization, he obviously realized that progress, or lack of it, depended upon the interplay between differing states of mind. And certainly, if those who lead are to advance the life of a society in terms of quality and inventive achievements, they must be driven by dynamic mental interactions between the two cerebral hemispheres. So those who are said to have a 'one track mind' should not be in positions of power; yet all too often we see that such is the case. Right-brain-dominant intellectuals and left-brain-dominant idealists should be given a wide berth when it comes to electing governments. Serious problems beset a society governed by ideologues or fanatics who are fixated on the unchallenged influence of just one side of the brain. Closed minds of this sort pose a real danger for the rest of the world, for they breed a zealotry that leads to hatred and violence - responses born out of an inability to consider any other point of view. So one should always pray for the presence of men and women who are guided by a true bipolar consciousness - one in which knowledge (theoretical and practical) works hand in glove with imagination (vision and wisdom).

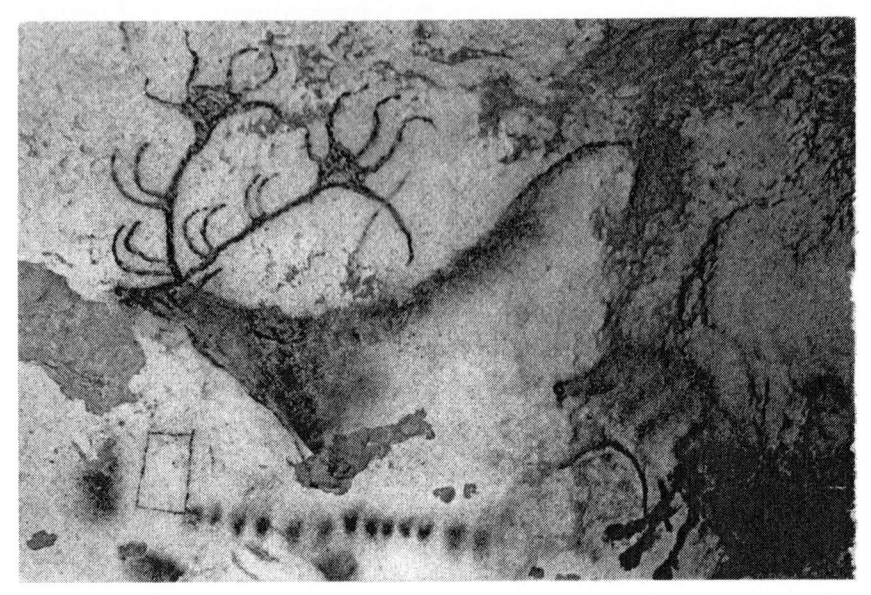

Deer of Lascaux

THE DARK AGES IN WESTERN EUROPE: THE INFLUENCE OF ONE GREAT MAN

The English historian Arnold Toynbee used the phrase, 'The Age of Faith', to describe the years which saw the growing influence of Christianity towards the end of the so-called Dark Ages - the chaotic years following the collapse of the Roman empire west and south of the Rhine and Danube. In the year A.D.331 the emperor Constantine abandoned Rome, leaving the western empire to fend for itself, and established Byzantium (later to become Constantinople) as the capital of the eastern empire. Anarchy and lawlessness continued to sweep western Europe as barbarians from the north - Saxons, Northmen, Franks... - moved into the old Roman colonial territories, constantly fighting among themselves to try and establish supremacy. Europe was the scene of extreme confusion and disorder that persisted for several hundred years, and which only abated to some degree when the Frankish king Charlemagne brought about, by military conquest, a short-lived period of relative order in the name of Christianity and was crowned Holy Roman Emperor in Rome on Christmas Day A.D. 800.

Neither a rule of reason or a spirit of philosophy were much in evidence in the West during the fourth, fifth, and sixth centuries. It was as if the mental life of cerebrum and cortex had been swamped by a resurgence of the spontaneous, destructive, and unreasoning passions lurking in the lower old brain - impulses closer to the days of unregulated instinct than to those that had seen the growth of right and left brain co-operation. (We should not forget the presence of this old and primitive brain that ruled the roost before the crowning edifice of cerebrum and cortex came along. It seems to be always waiting to take over when lust for power and wealth - the adrenal thrill of competition and conflict - cause us to 'lose our heads'; or when we are overtaken by disaster, find ourselves facing the void.) And Europe was indeed a void after Rome's downfall: a vacuum into which swept the savage and relatively unenlightened Germanic tribes who - until exposure to the prevailing remnants of Roman law, philosophy, and civic organization - displayed little of that collaboration between practical intelligence and inventiveness that marked the best days of Rome; produced philosophers and statesmen like Cicero and Marcus Aurelius, poets like Ovid, and engineers such as the men who built the Colosseum and Pantheon in Rome, and the great aqueduct of the Pont du Gard at Nîmes.

Toynbee's designation, the 'Age of Faith', refers to the growing influence of Christianity in the general confusion and lawlessness of the Dark Ages. For with the establishment of the first monasteries, the influence of early Christian thought began to slowly permeate all classes of society, revivifying the faculty of reason and engaging the imagination through the introduction of a religious faith. The historical scholarship of the monasteries helped to bring about an appreciation of a civilized past which encouraged objective reasoning when it came to dealing with local problems of civic and economic order; while the revelatory vision presented by Christian belief and propagated by the monks, kindled the religious imagination of the people and brought hope and faith into their lives.

Common sense on the one hand and a mystical element on the other - a powerful mental combination, bringing a practical resourcefulness and an inner sense of a transcendent reality to combat the political instability of the times, famine, disease and the prospect of an early death. Yet I imagine that if in A.D.520 a poll had been taken in the Subiaco region of central Italy (where a group of small monasteries was already established), to ascertain the greatest benefit the monastic presence had bequeathed to the people, it would reveal that the easing of misery and grief and the consolation and hope advanced through faith in the Christian message, was voted the greatest blessing. For at the time this imagined poll would have been taken - although Toynbee's Age of Faith was in its infancy and just commencing to lay its cloak of spiritual protection over Europe - it was already evident how effective the monasteries were becoming in providing the ordinary man and woman with a belief in their own individual value in the eyes of God; despite their lowly position and unfair treatment suffered at the hands of others, together with the merciless predations of nature. Furthermore, it was a belief that laid before them the prospect of their place in the Kingdom of Heaven. These years leading up to the end of the first millennium after Christ - to about A.D. 1000 - represented, in Toynbee's opinion, the finest, most genuinely spiritual period in Christian history: the Benedictine centuries, as they have been called.

One individual can be credited for introducing the kind of monastic life and doctrine that helped people endure the hardship and uncertainty of the Dark Ages. His Rule - as it came to be called - stipulated that his followers should bring practical, rational, means to deal with the material issues of life, and spiritual counsel to nurture the soul.

It was in or about the year A.D. 520 that the abbot Benedict - later to become Saint Benedict - left the group of small monasteries at Subiaco some thirty miles east of Rome and set out for Cassino with a few faithful followers. Monte Cassino, a prodigiously lofty hill, craggy with rocks, lay about eighty-five miles southeast of Rome in a remote region peopled by herdsmen who were still pagan at the time. There, atop the mountain, he built the first monastery of what was to become the illustrious, worldwide order of Benedictine monks. (The large, citadel-like building it ultimately became was reduced to rubble by the Allies towards the end of the Second World, but was completely restored at the end of hostilities.) Benedict's small cluster of low, stone and wood buildings comprising the original monastery prospered and become a place of spiritual retreat, even drawing the Gothic king and warrior Totila to visit Benedict in 542. The abbot's policy of applying reason and honesty to deal with existential problems, and of bringing spiritual insight to bear in furthering hope and faith for the individual, was ideally suited to a world at a loss for principles and beliefs. Also, it was at Cassino - and throughout the Benedictine monasteries that spread across Europe in the following years - that classical learning was saved from extinction: preserved to influence the resurgence of learning and law in the forthcoming Middle Ages: the years following A.D. 1000. And I would say that the Benedictine ideal - although weakened by an increasing secularism and desire for economic and political power in many of the monasteries during the early Middle Ages – inspired the incredible engineering and powerful spiritual symbolism that characterized the great Gothic churches and cathedrals of the thirteenth, fourteenth, and fifteenth centuries: extraordinary examples of imaginative minds applying the science of mechanics to create the most evocative and mysterious of spiritual environments. Work on the great cathedral of Saint-Etienne at Bourges in central France commenced in 1195; the dedication of the completed Gothic masterpiece was on May 5th 1324. The exterior and interior photographs reproduced here reveal the incredible skill and inspiration of the master stonemasons whose genius worked such engineering wonders in constructing these vertical lanterns of stone and glass: shaping an interior space through which light, color, and sound combined to evoke a mystical atmosphere almost certainly guaranteed to arouse and lift up the spirit. Seen from the outside, the interior magic is, literally, brought down to earth. The rational mechanical principles that enable the load of high stone-vaulted roofs to be carried away from nave walls weakened by increasingly large and highly placed windows - walls which could not possibly support the weight of the roof - become immediately obvious. 'Arms' of stone - flying buttresses so-called - carry the downward thrust of the roof out beyond the slender walls, and transfer it to the ground along rows of massive, free-standing columns. Faced with this intricate visual spectacle of stone ribs in series, cutting diagonally through space to support the high nave walls, allows one to comprehend how the whole thing manages to stand up. Yet on moving inside one still marvels at the monumental stability of the structure. Reason is defied as one wanders, lightheaded, and with little sense of time, senses dragged from horizontal *terra firma* to a high-pointed termination of space that viscerally excites - renders one - especially when the sonorous tones of the great organ reverberate in this mini-cosmos - a stranger to earth.

Saint Benedict's Rule and method only became widespread about a century after his death. And then, only gradually, did the Benedictine

Cathedral of Bourges - external view from *Archeologia*, Dijon. Zoom Studios

Cathedral of Bourges - internal view from *Archeologia*, Dijon. Francois Thomas

Order become recognized and extolled for ministering to both the material and spiritual needs of the people. Perhaps one could say that it had finally 'arrived' when Charlemagne - who was crowned Holy Roman Emperor on Christmas Day, A.D.800 and not known for his patronage of monasteries and monks - could ask if there was indeed any other Rule worth following. Yet Benedict himself would surely never have expected that the years between A.D. 600 and 1150 would come to be called the Benedictine centuries; or that a thousand years after its foundation, Monte Cassino would come to be regarded as the cradle of mediaeval civilization, and that he himself would be acclaimed by a pope as, 'the 'Father of Europe'. Even today, moving into the twenty-first century, the Benedictines still follow their Founder's creed in maintaining an enlightened secularism by keeping abreast of scientific and political change. They reaffirm the needs of the human spirit in times of a scientific materialism, by opening their doors to those in need of retreat and spiritual help - and by running missions to bring both a material and spiritual quality of life to those inhabiting the most backward and hazardous of places. They practice what they preach: contemplation to experience the inner life of soul and spirit; practical aid in helping the body along its way.

I imagine the shade of Saint Benedict drifting through the cathedral of Chartres - or that of Beauvais - around A.D.1400, and saying to itself (in whatever way a disembodied shade says anything to itself), 'So, at last... the Rule is cast in stone'.

POSTSCRIPT

> *I lay flat on my back and looked up into the darkening sky. How sad it would be, I thought, if we humans ultimately were to lose all sense of mystery, all sense of awe. If our left brains were utterly to dominate the right so that logic and reason triumphed over intuition and alienated us absolutely from our innermost being, from our hearts, our souls. I watched as, one by one, the stars appeared...**

So wrote Jane Goodall after spending one of her last nights at the Gombe Reserve in Tanzania - the place where she dedicated so many

---

* Jane Goodall, *Reason for Hope* (Warner Books), p. 177.

patient years to following the lives of individual chimpanzees. She became totally accepted by them in their natural habitat, and is recognized as the authority on the social structure of their communities; particularly with regard to the caring role of the females who, as mothers, hold chimpanzee families together.

I end this chapter with her words. No need to say more. They constitute the perfect epilogue.

11

THE SNAKE and the GREAT ROUND

*All are but parts of one stupendous whole,
Whose body nature is, and God the soul.*
Alexander Pope, *An Essay on Man,* 1.

This is not exactly the sort of statement one expects to find in the writings of a poet-philosopher of the English 18th Century Age of Enlightenment. The lines, coming as they do from the pen of a committed rationalist, bring with them an element of surprise and therefore compel one to take note. For to write of God as the soul of the universe, is to reflect more the religious attitude of the early mediaeval Age of Faith than the relatively secular rationalism of England in the writer's own time. Now if such an aphorism were to be written by Blaise Pascal, the French 17th Century mathematician and religious philosopher – who was convinced of the powerlessness of reason to solve man's metaphysical problems and, instead, exalted faith and mystical experience in its own right – one would not be taken aback. But, coming from Pope, such words testify to the enduring and compelling attraction of the concept of duality – of material nature and enlivening spirit.

In the Beginning, it might be said, was the Snake: a creature which is distinctly linear in form, having a head and a tail, a beginning and an end. Moreover, it is a reptile which glides silently and swiftly on its underside from one cover to the next, secretes itself in the undergrowth or beneath and between rocks, sometimes a deadly killer, sometimes not – a mysterious animal which has long kindled a variety of opposing ideas in the human imagination: of energy and fertility, the powers of destruction and darkness, the guardian of the springs of life, the possessor of arcane and fateful knowledge, the harbinger of death… The serpent is one of the leading players in the drama of the Garden of Eden - the treacherous troublemaker, bringer of misfortune. But as the snake grows it sheds its old skin, and emerges regenerated, reborn you might say, sporting a new, fresh-looking, epidermic sheath. Seen in this context the snake now becomes the symbol of renewal - a renewal in some cultures associated with the idea of spiritual rebirth. So we have a creature characterized by mystery, playing symbolic roles ranging from those that signify life-threatening eventualities to others that indicate positive and life-affirming possibilities.

Yet this is not the complete story in the history of symbolic 'snake-power'. Possibly as early as the Old Stone Age it would seem that the elongated snake evoked associations with the male penis, and was seen as an early kind of phallic symbol of masculinity and virility. And then the realization followed that if the snake contorts its body to bite its own tail, it becomes a circular snake and so is no longer representative of the extended male phallus, delineating instead the roundness and openness of the vulva, the external female genital organs. But the symbolic associations of the snake-as-circle go beyond triggering ideas concerning gender differences. For with the realization that the circular transformation of the male-signifying snake is the *means* by which the female vulva is delineated and brought to mind, it is understood that what is now symbolized is the sexual union between male and female - that physical state of 'oneness' in which both are submerged to briefly lose their separate identities. A union resulting in what is still regarded - even in the light of all contemporary advances in biological and genetic science - as the miracle of a new human life: a child on whom ceremonial or formal rites have been traditionally practiced in many societies to set him or her 'right with the Gods', so to speak - thus investing sexual 'creativity' with religious overtones of one kind or another. The wedding ring - the successor to the circle-snake (with head and tail eliminated to remove any appearance of a beginning or end) - is not simply a sign to tell the world that one is married, but to stand, like the ancient tail-biting snake, as the symbol of a union that pulls men and women together in both a physical and psychical bond. It is an alliance sought almost instinctively because it can have the effect, for each partner, of lessening the sense of separation from some central truth or purpose underlying human existence that many human beings feel.

Some of the earliest finds of this snake tail-biting symbol are small talismans from late prehistoric African and Mediterranean communities - small works of art given the common name of *Uroboros*, a term which denotes 'a combining'. Variations on the primitive configurations of the uroboros are found in many parts of the world, yet the symbolic implications remain constant - namely that they embody the union of opposites; not only at the human level between male and female, but also at the cosmic one in a universe where many different forces are held in balance through the operation of some Eternal principle. When men and women join together to become 'one', the balance sought is between their differing biological and psychological natures, their opposing masculine and feminine psyches. And at some point in the history of Western cosmology, the belief that *everything* is held in a dynamic state of being through the balanced tensions of contraposed forces, came to be known as, The Great Round - of which the uroboros became an early symbol.

To arrive at the concept of *wholeness,* and envision it as a circular configuration, represents an imaginative leap which may have been triggered by the presence in the sky of the golden orb of the sun - the source of the light and energy sustaining life - thus imparting occult significance to the perfect circle or ring: a configuration which, as we have noted, represents a circular continuity where neither beginning or an end can be discerned, and where all dualities and opposites inevitably become part of the One. For at all times, the line of the circle traces an unbroken course equidistant from the center, the point of truth where resides the omnipotent creative force.

If the snake-circle is the oldest form of The Great Round, its successors are to be found everywhere. The best known are the circular mandalas (designs to aid contemplation) in Hinduism and Buddhism; and the elaborate 'wheel' or 'rose' windows of the great Christian Romanesque and Gothic cathedrals of the West, in which an increasingly delicate array of stone-tracery 'spokes' emanating from a central hub hold between them luminous areas of mediaeval stained glass. The Hopi Indians of the American West use the hoop in ritualistic dances celebrating the oneness of Heaven and Earth. And then, of course, there is the Christian halo, and the Chinese *Yin* and *Yang*.... The halo, for example – that historic and ubiquitous symbol of holiness in the Christian faith is part of the same tradition - a Christian version of that ancient snake coiling to connect with its own tail.

It is not, semantically, a far cry from 'wholeness' to 'holiness': the connotations of entirety and perfection are common to both words. Many years ago I came across a *square* halo hovering over the head of a saint in a ninth century Byzantine church somewhere in north-central Greece. I have no idea whether or not any other such examples exist. And I discovered that it was apparently used to denote that its recipient, being a holy man - a 'whole' man, if you like - was canonized before he actually died – before he became pure spirit and so could not be eligible for the award of a circular halo.

One of the most universally known images symbolizing the union of the two most constant opposites envisaged by the human imagination – those of spirit and matter, or heaven and earth – is China's *Yin* and *Yang*. This historic duo is described as follows in the Chou record of rites and customs – the *Chou Li* – as early as the eleventh century B.C.

> *The Great One separated and became Heaven and Earth. It revolved and became the dual forces....*

It is a design worth studying, growing in symbolic significance the longer it is observed: a mandala-like drawing, graphically and starkly conveying the continuous flux of two distinct entities which, together, constitute a whole – *Yang* representing the spiritual, uplifting, heavenly powers of light; *Yin* the hidden, dark forces of earth, of the material world in which all things ultimately die. William Barrett, in his wonderfully informative book, Irrational Man, interprets the symbol perfectly. Writing about the 'Platonic celebration of reason' set against the less rational processes of the intuition, he writes:

> *We are a long distance here from another symbol of light and dark which early mankind, this time the Chinese, handed down to us: the famous diagram of the forces of Yin and Yang, in which the light and the dark lie down beside each other within the same circle, the dark area penetrated by a spot of light and the light by a spot of dark, to symbolize that each must borrow from the other, that the light has need of the dark, and conversely, in order for either to be complete. In Plato's myth first appears that cleavage between reason and the irrational that has been the long burden of the West to carry, until the dualism makes itself felt in most violent form within modern culture.**

And so, with Barrett's statement we are returned to face the music of our own mental discord: the dissonance created by *'that cleavage between reason and the irrational'* which he lays at Plato's door. Nevertheless, The Great Round of the cosmos goes warily on its way, maintaining a dynamic

---

* William Barrett, *Irrational Man* (New York: Doubleday & Company Books, 1962)

Uroboros

Yin and Yang

balance between energy systems and physical forces, while modern entrepreneurial man exploits the planet's resources without showing too much concern for nature's balancing act in managing her ecosystems. Neither is he often any wiser when it comes to realizing the breadth of thought his two-sided brain provides, being so easily led by just one or the other. The kind of irreconcilable duality to which Plato refers would become tempered into an all-comprehensive state of mind.... if the ideas and feelings propagated by both left and right brains were given equal time and credibility; seen as both sides of the same coin.

THE ROUND of HUMAN BELIEFS

What would we do without Shakespeare? For once again I find that Hamlet's words to Horatio relate succinctly to the theme of this chapter:

> 'There are more things in heaven and earth, Horatio,
> Than are dreamt of in your philosophy.'

Webster's Dictionary devotes half a page to the numerous definitions of the word 'round': one of which - 'the whole range (the *round*) of human beliefs' - is particularly relevant here. For it leads me to think that inasmuch as the early shamans, poets and philosophers referred to the dynamic interplay between all of life's forces - cosmic and human - as The *Great* Round.... so can one talk about the mental interchanges between the cerebrum's hemispheres as The *Lesser* Round - seeing the brain as a microcosmic mental 'universe' engaged on its own circuit of cerebral events, forming its own concepts of the world and intuitively searching for meaning in order to form a *whole* understanding of truth.

Two questions invariably arise at this point. The first is whether the brain is now, finally, structurally complete. And the second is whether an organ of such extraordinary cellular, chemical and electrical complexity has resulted from random biological happenings, or has come about by design. If the latter, one may speculate that its evolution was planned by a supreme Intelligence at work in the universe - behind the scenes, as it were - such as Aristotle's Prime Mover, God, Allah, or the Great Spirit of the American Indians....; or take a less definitive position and, as a quantum physicist said to me, 'conclude that it appears the universe has conspired to create us.'

The first question is obviously one we cannot answer. As to the second, I would agree with the gist of his remark, and argue against the likelihood that chance events can account for the evolutionary development of such an extraordinary phenomenon as the brain. I take the view that the organic edifice rising above the nub of the ancient brain of instinct is so complex - both in terms of its overall physiology, and in the coordinated interrelationship between its parts (still imperfectly understood) - that it is difficult to see how the resulting mental powerhouse could be brought about by random biological, chemical and electrical happenings. Granted, the whole process took a few hundred million years, but it is thought that only during the last two to four million years was the brain functioning to provide the kind of human consciousness considered to be unique on the planet. If we study *probability theory* - an increasingly popular mathematical exercise today - the possibility that the grand 'architecture' of the brain might have developed purely fortuitously, appears statistically, mathematically, to be extremely remote.

The most obdurate evolutionists assert that all evolutionary developments serve the basic purpose of increasing a species' chances of survival by providing it with the ability to adapt to changing environmental circumstances. And inasmuch as the standard definition of intelligent adaption is to be objectively and imaginatively resourceful when facing new environmental circumstances, this is fair enough. But if we regard the wide range or *round* of human beliefs - many of them embracing abstract ideas and considerations - it would seem obvious that the evolved brain goes far beyond serving only the needs of physical survival. The incredibly broad spectrum of awareness with which we have become endowed provides reasons and purposes for living that surpass even the most desirable prospects of longevity.

It is legitimate to ask how effectively we use the faculties of reason and imagination offered by this brain of the hemispheres. In terms of scientific achievement I would say we have advanced a hell of a long way. The last fifty years have seen terrific advances in all areas of medicine and science - discoveries obviously resulting from the full cooperation of left and right cerebra. But looking back over the violence and suffering that characterizes the twentieth century - and which is spilling over into the twenty-first.... mass graves of slaughtered victims in the Balkans; hapless civilians bombed in Iraq; merciless forms of terrorism and religious extremism worldwide.... we seem to have become little the *wiser* despite our great scientific achievements. For we either fail to comprehend, or choose to ignore, the right brain's message concerning our common humanity: the lurking, shadowy awareness of life's brevity; the fact that we share a similar kind of grief and despair at the loss of love, or of a loved one; the wish to live in the world, peacefully and personally fulfilling our individual destinies and bringing up our children to do the same.... and that when it comes to dying, we are all in the same boat.

There are indeed men and women sufficiently moved by such humane impulses to work worldwide in the service of the less fortunate, and who stand as examples of the wisdom, kindness, and goodness required to bring about at least some global fellowship. Their altruism, compassion and motivation are largely right brain inspired. The Rule of Saint Benedict would sit lightly on their shoulders.

Regrettably, it seems that the imaginative and philosophical powers of the right brain get short shrift in the day-to-day workings of consciousness. The stressful pace of contemporary life requires the constant utilization of the left brain's objective and logistical capabilities to get one through the day. There is little time (and ultimately little inclination) to indulge in any reflective looking inward. Yet the normal brain can handle such objective left brain activity without necessarily restraining or inhibiting the right brain's contemplative pursuits which can remain 'on hold'; ready to surface as soon as one steps off the treadmill of the daily round. One can indulge in a reverie when having a shower, lying in the bathtub, walking the dog... And the right brain comes into its own when engaged in the kind of conversation with others that calls for the articulation of personal points of view - discourses embracing a variety of topics that demand one takes a position, practical or philosophical. Such verbal encounters demand a knowledge of the facts and the expression of personal opinions, and are unlikely to take place when one is falling asleep before the television. Or when 'surfing the net' on a laptop.... forgetting that one's own cerebral computor can take one further afield, imaginatively, if one sits quietly and gives it a chance.

Many of the up-and-coming generations are already slaves to the kind of left brain dominance imposed by information-gathering and problem-solving technology. On my increasingly fewer visits (by choice) to talk at universities, I find both undergraduate and graduate students disinterested in subjects which have no immediate bearing on their job-oriented studies. The old educational ideal of an inter-disciplinary education seems to possess as little interest for students as it does for faculty and administration. Many officials in the big universities see the teaching of computor literacy in the gathering of facts and the solving of scientific, social, or economic problems as their goal - rather than the kind of teaching performed by a learned and impassioned individual who will inject personal points of view into the subject matter. The concept of the university as a place where inspired teaching brings students to realize something of life's 'Great Round', seems - for the time being at least - to have been abandoned by many large institutions. Hopefully, we can still look to the smaller colleges and universities to serve their charges well by providing them not only with a wealth of accurate information, but also ensures that their most personal curiosity and imagination is challenged in the process.

Unfortunately, far too many elementary and secondary school systems provide little or no incentive to 'jump-start' this process. Add to this educational deficiency the brain-dulling effect of television and video games when children are *out* of school - together with the apparent dearth of conversational interactions between youngsters and parents who seem to be unaware of their role as mentors - and it should come as no surprise that far too many university students appear to be subjectively brain-dead when they arrive on campus. Far too many want to simply learn the facts without 'chewing them over' in the privacy of the mind's reflective sanctum. It is not enough to qualify them as being 'educated'.

Loren Eiseley, in his book *The Firmament of Time* - talks of the perceptive philosopher who wrote, *'The special value of science lies not in what it makes of the world, but in what it makes of the viewer.'* And he goes on to say: *'The Renaissance thinkers were right when they said that man, the Microcosm contains the Macrocosm. I had touched the lives of creatures other than myself and had seen their shapes waver and blow like smoke through the corridors of time. I had watched, with sudden concentrated attention, myself, this brain, unrolling from the seed like a genie from a bottle, and casting my eyes forward, I had seen it* *vanish again into the formless alchemies of the earth.... I had struggled, I am now convinced, for a greater, more comprehensive version of myself.'**

As a lad I remember being challenged at the age of thirteen by the much revered - as well as gently feared - classics master, G.M. Lyne. We were reading Caesar's *Gallic Wars* and I was desperately hoping that I would not be called on to translate, when...'*Collier, second paragraph, page 22... come on boy, on your feet...*' I stammered my way through the translation, suffering the wit of old Lyne in the process and the muffled giggles of my compatriots. What is referred to as a pregnant silence followed my rendering. And then a long drawn out *'Mmm...* from G.M. *'Well, let's say, Collier, that you've mangled syntax, vocabulary, and idiom in a remarkable kind of way. But in some strange way it seems you have the gist of it ... so equally to the point, do you agree with Caesar or not?'*

'How do you mean, Sir?' from myself.

'*I mean, Collier, what do you think of Caesar's harangue - which you have so ably translated for us - so far as the Gauls were concerned - was it right or wrong for the occasion... likely to be productive or counter productive in his dealings with a subject people?*'

I thought hard: 'Both, Sir.'

'<u>Both</u>, Collier? How 'both'?'

'Well...'

Good start, Collier, but go on...'

'*Well, he told the Gauls what they could and couldn't do so they weren't free anymore, but they got the benefits of Roman law and order in exchange... Sir.*'

'*So you're a law and order man, Collier: prefer to live securely under the Roman yoke, than be free as a bird without it? Jolly good: at least you got the gist of Caesar's politics, if not the finer points of the Latin tongue.*'

And so the lesson progressed. Others were called upon to translate and to struggle with the content of Caesar's remarks. It was never enough for G.M. that one should comprehend the words and the grammar simply as an exercise in translation. And that was how my schooldays went in every subject. I was expected to learn the facts, and then be able to express an opinion concerning the thoughts they generated in my little head. When I got home, the process of being pushed to express myself in terms of

---

* Loren Eiseley, *The Firmament Of Time* (New York: Atheneum Publishers, 1960)

what I had learned in school that day was continued by my stepfather. But nowadays the educational system tends to be reductionist and analytical, particularly in the teaching of literature, if not in the arts in general. The method often employed is to take the work apart, and play intellectual games in assessing the semantic implications of each word or phrase: taking the book or the poem to pieces without even probing its symbolic and expressive significance - the feelings and personal thoughts it engenders.

At the moment I think a case could be made to show that, viewed *en masse*, we are regressing rather than progressing in right brain sensibility. If such an imbalance persists then evolution - or whatever power may be responsible for our arrival on the scene - will, for the time being at least, have wasted its time on us. And any sense of purpose that might be provided by the full exercise of our magnificent brain, be denied us.

The irony is that it seems *Homo neanderthalensis*, living between say, 200,000 and 40,000 years ago, was an altogether kinder creature - one, as the distinguished American anthropologist Loren Eiseley put it, of *'a curious gentleness that we know now had long ago touched the vanished Neanderthals...'* It would still seem to be in dispute among paleoanthropologists as to whether the Neanderthals were a sub-species of our own (*Homo sapiens*), or a close but distinctly separate species. Yet it would appear that they were vanquished by our own forebears, the taller and more erect hominids of the Upper Paleolithic period – territorially minded men; the early men of conquest: ourselves, some 30,000 years removed. The evidence of a humane Neanderthal society extant in northern Iraq some 60,000 years ago, leads one to think that these supposed savages were – when it comes to the practice of violence – more humane than many of the following, relatively ancient, supposedly more 'human' civilizations.

Perhaps *Homo neanderthalensis*, was, compared to ourselves, on the right track. Nevertheless, it should be remembered that there have always been those in the history of *Homo sapiens* who have advocated tolerance and gentleness. In the Roman Senate, about 50 AD, Seneca reproved his colleagues, saying:

> *Let us ask what is best, not what is customary. Let us love temperance. Let us be just. Let us refrain from bloodshed. For none is so near the gods as he who shows kindness.*

12

CONSCIENCE & MORALITY: CHARACTER - FOR GOOD OR ILL

Good is all that serves life, evil is all that serves death.
Good is reverence for life... and all that enhances life.
Evil is all that stifles life, narrows it down, cuts it to pieces.
Erich Fromm, *Saturday Review*, January 4, 1964

I find myself wondering how Erich Fromm and Friedrich Nietzsche would have got along together had Fromm been able to reach back through time, and consult with the great man on a broad spectrum of philosophical and psychological issues. Whatever disagreements might have resulted from such a meeting, I feel pretty sure that the German philosopher – who died in 1900 – would have voiced his approval of Eric Fromm's statement you see above. For Fromm's dictum concerning good and evil neither makes mention of God or Satan, salvation or damnation, sin or virtuousness... or mentions the moral imperatives deriving from the ideologies of the world's major religions.

The title of Nietzsche's book, *Beyond Good and Evil* – published in 1886 – directs attention to the author's belief that, as the century approached its end, it was necessary to go beyond the Church's traditional moral attitudes, beyond its specifically 'religious' way of understanding the problem of good and evil. For by the mid-nineteenth century the Anglo Saxon and largely Protestant countries of Europe had developed a rigid and puritanical attitude to life – one to which society, at least on the surface, appeared to subscribe. But this was the time when significant breakthroughs in physics and mathematics, as well as in the new infant 'science' of psychology, began to reveal the complexity and authority of the forces at work behind the scenes of the natural world: the physical systems which order the cosmos, and the psychological ones which determine the course of an individual human life. It was an inspiring period in European history for an intellectual of Nietzsche's calibre, and a time demanding that the conventional concepts of 'good' and 'evil' be secularized - be seen as two highly significant notions brought to mind by a sophisticated consciousness for a specific reason. Namely, to cause us to recognize the differences between desires and practices that work positively to both enhance the quality of human life, *and* ensure the wellbeing of the natural world... and those that do exactly the opposite. Whereas the Church would have it that awareness of right and wrong exists predominantly to serve one's spiritual needs in determining the nature of the soul's progress in this life and beyond. (Yet I suspect that Fromm, in any such discussion with Nietzsche, would make little distinction between existential and spiritual values in this regard.)

Nietzsche regarded such theological strictures as working against life rather than for it; particularly the assumption that bad things result from succumbing to the seduction of an imaginative intellect, the appeal of the senses, or the compelling drive of strong feelings – while good things result from avoiding these various 'temptations' by maintaining an ascetic and disciplined way of life. The doctrine that one was on the way to hell in a hand basket unless mental, sensory and emotional stimulation were kept in check, was anathema to Nietzsche: ridiculous and untenable. It negated the rich variety of experiences offered by the menu of life, and made accessible through the broad spectrum of consciousness. The Church, after all, should argue that God – 'working in His mysterious way' – had provided us with the mental appetite to partake of a smorgasbord of life-sampling options, and to recognize which were beneficial and which harmful: thereby providing a challenging obstacle course for us to negotiate in learning to practice the good that, in Fromm's words 'serves life', rather than employ the evil that 'stifles' it. For is anyone going to seriously disagree that it is largely through the experience we gain, and the choices we make in running life's course, that moral and altruistic values make themselves known; character and personality take their shape?

If I understand Nietzsche correctly, he is saying that if religion does not understand that in order for someone to grow in spirit and character, he or she must plunge into life - then religion is an ass.

However, we face an interesting paradox concerning this selective interplay between oneself and what life has to offer. For the brain which kindles both curiosity and desire - inviting us to 'live it up,' so to speak – is the same brain which is responsible for counseling us as to the right and wrong ways to go ahead and do it. This ability of the neurons to present life as a multi-layered cake, but then tell us how to eat it, is really quite extraordinary. Once again, the question has to be asked: how is their chemical and electrical activity able to bring to consciousness such abstract and subjective opinions as those 'voiced' by this autonomous mental authority we call consciousness? It is one thing for the brain's neural electrochemistry to be responsible for the senses, and inform of the presence and form of tangible things and events in the outside world, dealing in fact, with *facts*. But conscience is not a material thing able to be scrutinized by the senses in order to become known. It is a mental impulse: a spontaneity of thought and feeling influencing our response to life's experiences - and one has to wonder how these same neurons bring to our notice, without benefit of the senses, and, presumably via the same system of electrical and chemical mechanics, *qualities* of experience as distinct from facts. The traffic signal of conscience shows green if involvement in a particular offering of life is likely to be generally beneficial; red if harmful effects appear unavoidable. What part, if any, does the unconscious play in this mental phenomenon? Or is there a metaphysical element involved - the influence of soul, spirit, or God?

THE NATURAL ORDER

> *Yet before language and self-knowledge brought meaning and order to our sense perceptions of the world, awakening imagination and curiousity to ask 'how this' and 'why that'... the Ancient of Days Universe was out there. And the laws of physics, chemistry, biology... as we know them today, together with the early mechanistic forms of life, were all in place. It was one creation, and it was not until three or four million years ago - or so the scientists say - that early man saw himself as an individual - a distinct entity within it.*
>
> In a letter from the Bishop, 2 February 1953

My erstwhile flying companion is referring to the time before the right and left hemispheres of cerebrum and cortex, bridged by the nerve fibers of the corpus callosum, came to operate in their own particular way - *lateral specialization* so-called: creating a partnership between the faculties of language, logic and reason (left hemisphere), and the mental imagery we consider more imaginative, intuitive, and abstract (right hemisphere).

Before such specialization was achieved our earlier ancestors would have been functioning with a more basic consciousness - with a brain that essentially regulated the body's physiological systems and provided instinctual responses to environmental challenges; yet which also, at some point in the earlier stages of cerebral evolution, could have facilitated learning from experience, and even the ability to anticipate the outcome of events - not to mention the ability to experience certain fundamental, emotional responses to life.

For we have examples of primate behavior indicating that a consciousness is at work which gives the lie to the frequently heard derogatory expression, 'dumb animals'. Jane Goodall, with her many year's experience observing chimpanzees in the African wild, leaves us in no doubt that they are highly cognizant of what is going on around them. Living in natural groups or 'tribes' they cohere socially, and in terms of the family unit it is the mother who holds things together: the maternal link between a mother and her young reveals a high level of caring and nurturing, the result of a strong emotional and tender bond. Both mother and child experience real and sometimes debilitating grief when death intervenes. They also use sticks as elementary tools; and seem able to anticipate what is likely to happen in terms of cause and effect. Generally speaking, they 'look out' for each other and know their territory and its resources. But she also records instances of violent, even berserk behavior on the part of some males. ('Ah! there's the origin of the macho male syndrome,' I hear you muttering to yourself.) It is not difficult to introduce the word 'love' to describe the maternal side of chimpanzees' world; even to describe the affection which some of them obviously gave to Goodall herself - anthropomorphic projections notwithstanding.

But let us return to the prime question posed by the Bishop. What is required of a consciousness if a creature is to see itself as an independent entity: a singular being, possessed of its own particular essence, distinct from its environmental context, and thus able to perceive the 'outside' world in all its physical manifestations as a distinct and separate phenomenon? And the answer seems obvious enough: namely, a high degree of self-awareness. Now experiments involving gorillas and chimpanzees have revealed that some will ultimately come to recognize themselves in the mirror. And that others are capable of pushing certain buttons that will select and deliver - and so signify their personal choice - of either, say, a banana or piece of chocolate. Yet to simply be aware of their own chimp-like presence, and develop an appreciation for certain kinds of treats or food, stands fairly low on the scale of self-knowledge. The important issue is whether the animal is thinking, 'This is *me* in the mirror - *I* exist as *me*, quite differently from the others'; or, '*I* am a connoisseur of bananas (or chocolate), savoring their taste and texture in a way peculiar to myself '. The '*I*' and the '*me*' in these contexts signifying a mental recognition of '*my*' distinctly personal physiognomic contours and '*my*' psychological characteristics. And there are those primatologists who tend to think that some chimpanzees and gorillas are aware of themselves in this way, and others who would not go so far.

Jane Goodall's observations over the years would seem to leave little doubt that the caring behavior of female chimpanzees goes beyond the automatic, severely practical responses of instinct, and suggests that a mother is aware of her special identity as the bearer and supporter of her infant - of this creature that is 'flesh of my flesh', so to speak - and acts maternally in an individual and intimate way. Yet whether one can say that such behavior is proof-positive of a humanlike self-awareness is another matter.

In his poem *In Memoriam*, Tennyson wrote of '*Nature, red in tooth and claw*', thereby suggesting that animals, driven entirely by instinct, can be only killers or victims. The famous 'Mrs. Ples' skull found at Sterkfontein in South Africa in 1947 is virtually complete and is of an adult female. She was about four feet tall, of the genus known as *Australopithecus*, lived about two million years ago, and possessed a brain cavity very substantially larger than that of a chimpanzee; which likely puts her within the period when cerebrum and cortex were finally 'up and running'. I remember holding her skull in my hand in Pretoria's Transvaal Museum, and inquiring of the museum's Director how she might have come by the neat puncture at the forefront of the skull. "She was most likely killed by a prehistoric leopard," he said. "The hole is consistent with that made by a leopard's tooth." I suddenly felt very close to 'Mrs. Ples' - (the name taken from the famous palaeontologist Dr. Robert Broom's genus *Plesianthropus*) - and tried to imagine the circumstances of her demise as Tennyson's words, '*Nature red in tooth and claw'* came to mind. Standing there, her skull in my hand, provided a stark reminder of the elemental nature of her world where the predatory creatures reigned more or less supreme.

I wonder how far she had progressed as an individual, and could be thought of as living beyond the time of the Natural Order? How much the search for food and drink, defense of territory, procreation, aggression - survival behavior in general - was still governed, even in her day, only by the efficient and automatic response of instinct. Or was she capable of functioning independently, determining for herself the course of action to be taken - one of the first members of the Human Order?

Yet the destructive and predatory violence associated with early primate life in the Natural Order cannot be attributed to any malicious intent on nature's part. Nature has no moral awareness, makes no judgment as to whether a particular earthquake, drought, or plague... or a kill by leopard, tiger, or bear... is a just or unjust event, resulting in good or bad consequences. I think we can be pretty certain that the leopard (or perhaps saber-toothed tiger) surely did not think when it attacked 'Mrs. Ples' that it was committing a dastardly and immoral act: it was simply driven by the terrible and mechanical proficiency of the instinct to kill, eat, and survive - immediate and direct action unencumbered by conscience or feelings for its victim. And the forces behind a natural happening such as a tornado, resulting from the violence of air movement induced by violent thunderstorms, can have no awareness of themselves as malevolent and destructive powers. We are the ones who coined the saying, *An ill wind bloweth no man good.* Nature herself knows nothing of the good or ill effects of her energies - of what we, anthropomorphically, call her 'moods'.

THE MOVE TO THE HUMAN ORDER

There are those who argue that our newly found ability to think and feel in terms of the moral standards set by a working conscience - courtesy of the brain's evolved cerebrum and cortex - can only be seen as purposive in one respect: namely, that such mental activity helps to ensure the survival of the species. For they point out that once early man came to consciously evaluate the harmful or beneficial affects of his actions, it is likely that he would quickly learn to apply thought and reason in order to deal more effectively with life's practical problems. In either event, the automatic and unthinking reactions of instinct which were so often violent and destructive, would be countered, resulting in the preservation of life, an increase in longevity, and the general advance of society.

On the other hand, one can take a less material and more metaphysical position by suggesting that the moral and caring imperatives presented by an evolved consciousness introduced mankind to a new level of personal existence - to a subjective and inner self quite distinct from the ego-driven person oriented to life in the outside world, and given to the satisfying of physical appetites and needs. Over time it would seem as if a new channel of consciousness had opened providing access to a central psychical authority - an inner being generating ideas and feelings that were not necessarily linked to perceptions of the world, and who spoke for an inspiring 'spirit-force' emanating from an otherworldly source that ultimately came to be named, more or less universally, the soul. And over the years philosophers have pondered the question as to whether we only became fully human once the mysterious spiritual energy of this soul was known to consciousness.

Welcome, *Homo sapiens* - or perhaps I should say, *Homo humanus*.

Yet in following this train of thought we have to face once again the inevitable and ultimate question. How do the extremely abstract values and concepts such as conscience, love, compassion, charity, beauty... - so often associated with the spiritual influence of a non-biological entity such as the soul - make their forays into consciousness? For these ideals represent positive and highly satisfying *qualities* of life we create from within ourselves, and are not to be confused with the images of external *things* or *events* perceived via the senses; yet we are able to comprehend them no less clearly, name them, and even think of them as 'facts' of life. But if such visionary precepts as to the truly significant ways to live are not to remain faintly stirring in the unconscious (which might be the case with other higher animals) they must, if we are to become aware of them, be able to make use of the biological channels of communication that only the brain can offer.

We have approached the questions posed by this dual nature of consciousness - serving, as used to be said, 'body and soul' - in various ways throughout these pages; and have consistently taken the position that the most highly inspired human achievements in all fields of endeavour spring from an unconscious source of 'knowing' - one that lies beyond the objective capabilities of the senses alone. In contemporary parlance it could be said that we have access to an intelligence system of undefined origin, 'broadcasting' on a wavelength of consciousness known as the 'intuition channel'. Such a theory implies that the brain, while organizing the mental structure of such outstanding insights - and serving as the conduit without which they would never come to mind - may not be the *a priori* source of their origination. We may one day discover that consciousness does not reside solely in the brain.

Over the last hundred years the knowledge we have gained of the physical and dynamic nature of this planet and of the cosmos to which it belongs, has grown dramatically - while within the last fifty years medical science has made prodigious strides. Yet the Greek scientist and philosopher Aristotle who coined the word *metaphysics* - signifying *beyond* or *after* physics - and who would surely delight in contemporary scientific achievement were he to be around for a day or so - would be particularly fascinated by the theories of what is sometimes called the 'new' physics, quantum mechanics, that is - for here he would find exemplified the philosophical and scientific conclusions he had in mind when he introduced the term 'metaphysics' over two thousand years ago. In 1927 the German physicist Werner Heisenberg developed a new type of *quantum mechanics* and formulated the principle known as that of *indeterminacy:* showing that in the microcosmic realm of atoms, electrons, neutrons, protons and photons... things are going on - forces are at work which do not conform to the established laws of physics; which defy reason and logic, challenge common sense; indeed, may well be described as non-sense. The new physics indicate, for example, that particles (as electrons) seem to act as organic entities, possess their own brand of intelligence, and react unpredictably to an observer's attempts to discern their position and momentum. In fact, it proves impossible to determine such a particle's position and momentum simultaneously. On being exposed to these developments our ghostly Aristotle might well have smiled knowingly, murmuring, 'Well, I told you so. Remember? After physics can go no further... what then?' And he might have gone on to wonder why the Judaic, Christian, and Islamic religions of this twenty first century in which he found himself a visitor, did not take their cue from the enigmas and dilemmas attending this new physics. Why they did not use the findings of quantum physics indicating that science is unable to establish absolute truths concerning the workings of matter and energy - thereby lending credence to their own beliefs that a transcendent, metaphysical power lay behind the grand design of the cosmos. After all, it is essential to the cause of religion that some ineluctable mystery pervades the existence of the cosmos; cloaks our own brief presence on the scene. However, our distinguished visitor could return to the Greek Netherworld content with the knowledge that the monotheistic religions, having taken man himself to be the focus of metaphysical attention, invested him with an indwelling soul - a power certainly beyond the scope of physics to define: more elusive even than the particles of quantum physics, for neither its position *or* its momentum (energy), can be ascertained.

The Elizabethan poets of the late 16th and early 17th centuries such as William Shakespeare, Christopher Marlowe, Walter Raleigh, John Donne, Andrew Marvell... came to be known as the 'metaphysical poets'. And understandably so. I can think of no other period in English history when the poetic imagination was so caught up in the certainty that supreme values reside in the soul. Consequently, the surpassing experience of love, the inspiring urge to scientific discovery, the intuitive mental imagery that finds its expression in the practice of the creative arts, and the enlightening felt-thoughts that spring from philosophical speculation... all speak to the activity of 'great souls'. And I would ask those of my readers who have never heard the music of the 17th century composer John Dowland, or the sounds created by the 18th century geniuses Thomas Arne and William Boyce, to rush out at the first opportunity and buy recordings - that is, if they wish to reach within themselves, and journey beyond present time for a while. Which might be the kind of experience Shakespeare had in mind when he wrote in *Sonnets CVII*:

The prophetic soul
Of the wide world, dreaming on things to come.

The soul has been defined in a thousand different ways by poets, dramatists, philosophers, shamans, priests... but Webster's Dictionary provides two particular definitions that help one capture the essence of what they all have to say. The first describes the soul as, 'the actuating cause of an individual life'; the second as, 'the spiritual principle embodied in human beings'. Both of these characterizations convey the numinous quality that inevitably attends the idea of the soul as what some have called the 'immortal guest' - a resident without form or substance able to offer us, from time to time, a presentiment of truths pertaining to a reality lying beyond the space-time continuum in which we presently exist.

Such communications constitute a remarkable phenomenon - similar to the inspired reaches of consciousness we have discussed previously when a supremely physiological brain manages to intuitively formulate highly abstract and creative images. Inevitably one wonders how the brain is able to use its chemical and electrical energy to bring imaginative insights to mind; and particularly how this energy is able to provide a channel of communication between an intangible something called the soul and our very real physical entity of a brain.

Nowadays, in this material and overly rational age, I am not sure how many of those caught up in the technological and materialistic Western way of life will pay more than lip service to the age-old belief in the soul. But at least this discussion could engage their interest if they consider how the brain's biophysical energy-systems - regarded from a quantum physics point of view, that is - may be seen as the means by which the soul's influence registers in consciousness. First, it seems reasonable to suggest that the soul, as an immaterial force within the human psyche, could only come to affect our thinking and feeling once the brain had developed to its present highly evolved and sophisticated level - that is, when the potential energy level of a hundred billion neurons was in place. In other words, with such sheer chemical and electrical neuron-power available, could the evolved cerebrum and cortex have become capable of 'receiving' energy waves (or particles) of unknown origin carrying what might be described as 'spiritual intelligence' to the amorphous repository of the soul? And of then converting this psychic energy to the electrical energy necessary for the brain to create its own conscious imagery of spiritual values

At this (final?) stage in its development the human brain was about three and a half times larger than the brain of the chimpanzee, our closest relative in evolutionary terms. Even so, the Natural Order's rule of instinct was down but not out - is still lurking beneath the surface of the Human

Order that supplanted it - always waiting to take over when passions run high, as we know only too well. Consequently, we have a foot in both camps - causing Benjamin Disraeli, the British politician and author, to remark at a meeting at Oxford in 1864, 'Is man an ape or an angel? Now I am on the side of the angels.' I suppose the question is, will it ever be so? And a cursory answer would be that as long as runaway *egos*, lost to reason, goodwill, and conscience revert to the 'fight or flight' directives of instinct, it will be so. But more about this in the following chapter, *Mirror, Mirror on the Wall*. (Yet one can't help wondering which side Disraeli would choose were he alive today, knowing of the extermination programs fostered by the likes of Lenin and Stalin, Hitler and Himmler, Mao Tse-tung, Pol Pot... of the fratricidal massacres in Africa, the Balkans... He might have come down on the side of the greatest of the apes, the gorilla - called by some observers, the gentle giant.)

'SO HUMAN AN ANIMAL...'
(With acknowledgement to René Dubos)

I often wonder just where 'Mrs. Ples' stood in the long process of transition from chimpanzee to human being - whether she had acquired a *fully* operational cerebral cortex and the interactive organization of neurons to make her thoroughly human: how much was she aware of herself as an individual person, trying to determine her own destiny, make judgements, and develop strategies to stay alive. For she would likely be speaking some kind of rudimentary language to express her thoughts and feelings, and perhaps be imaginatively stimulated through the symbolic power of early forms of ritual. If such were the case she would have gained some measure of freedom from the Natural Order's rule of instinct. So I think of her as a person, an early prehistoric member of the Human Order, no doubt aware and fearful of the dangers surrounding her, yet likely having no technical defence against nature in the form of the leopard. And while holding her skull in my hand, I found myself hoping - rather ridiculously I suppose - that her death had been instantaneous.

In distinguishing the fundamental differences between the mental capability of our nearest primate relative, the chimpanzee, and that of ourselves, the difference in brain size is obviously significant. The larger mass of the human brain, for example, accommodates a cerebral cortex which is physiologically dominant, exceeding in amplitude the total rest of our entire central nervous system: thus generating the psychological complexity of human consciousness with which we are concerned throughout these pages. But when it comes to discussing the mental differences *inter species* - to account for the fact that not all men and women are possessed of the same levels of intelligence or imaginative insight, or can draw on equal depths of feeling or moral strength - it has been shown that factors other than brain mass play just as vital a role.

David Wechsler, the designer of the Wechsler-Bellevue Intelligence Scale, came to the conclusion in his *Measurement of Adult Intelligence (Baltimore, 1941)* that ... *'heavy' brains have generally been those of men of genius and there would seem to be some correlation, though not a great one, between size of brain and mental capacity.'* He goes on to mention that, *the largest undisputed brain weight on record is that of the Russian novelist, Ivan Turgenev, at 2,015 grams.*

However, *structural* peculiarities in the individual brain may also account for an individual's ability to perceive life and the world more acutely, revealingly, and compassionately. The brain of the French satirist and novelist Anatole France, while found to weigh only 1,017 grams - lower than the normal weight of 1,360 grams for a man of his height - nevertheless possessed an unusual configuration. L. Guillon, writing in the *Bulletin Academie Medical, Vol., XCI (1927)* described it as follows: ...*the cerebrum had marked asymmetry; the convolutions were long and tortuous, and the foldings were unusually complex. These and other peculiarities provided considerably more gray matter than the brain weight would indicate. Moreover, he was eighty when he died, by which time normally a good deal of brain shrinkage has occurred.* So here we have the case of a highly creative writer who was awarded the Nobel Prize in Literature in 1921, and whose mental powers were obviously not affected by the overall smallness of his brain - whose perceptive and inventive capabilities resulted from the amplitude of gray matter required to form such a convoluted and meandrous cortex. (It would be interesting to know how the obviously significant spread of cortex overlaying Turgenev's large brain, compared with the area covered by the folded configurations of cortex overlaying Anatole France's much smaller cerebrum.)

There is still a third factor to take into account when considering the dissimilarities between animal and human consciousness, and that is the pattern of neural *'wiring'*. For in examining chimpanzee and human consciousness, researchers find that regions of the human brain are 'wired' differently from those of the chimpanzee brain - particularly in the area processing language where complex neural connections allow us the great advantage of linguistically defining, and thus comprehending, all aspects of experience. However, while the term 'wiring' is a convenient one to use when discussing the electro-chemical activity of the brain, its connotations to conventional electrical wiring and connections give a pretty simplistic picture when it comes to depicting the fantastic complexity of the brain's neural pathways.

In his *Meditations* (Bk. ii), that most reflective and humane of Roman emperors, Marcus Aurelius wrote: *This Being of mine, whatever it be, consists of a little flesh, a little breath, and the part which governs.* In one sentence he sums up all I have discussed in the preceding pages. For when one reads the *Meditations* it becomes clear that the author considers *'the part which governs'* to consist of more than simply the mental faculties of perception, reason, and analysis leading to objective judgements - but also must include the moral and spiritual insights that spring from an intuitive consciousness and urge that we live in accordance with some standards of virtuousness. And he believes - as he makes clear - that our ability to find a purpose to existence, to know a personal destiny, depends on how well we balance and coordinate these two dimensions of consciousness in the process of becoming 'whole'. The *Meditations* is still in print today: as a declaration of human values and ideals it has appealed to readers across the centuries, and new editions appear at regular intervals.

Marcus Aurelius, in company with such other great Romans as Cicero, Senecca, Ovid... was well aware of how men's attitudes and actions may work for good or ill in the course of events. He knew only too well how to size up the character of every man coming before him in the course of the day: his assessment being based on an intuitive perception of a councillor's or petitioner's moral reliability. He writes: *Begin the morning by saying to thyself, I shall meet with the busybody, the ungrateful, arrogant, deceitful, envious, unsocial. All these things happen to them by reason of their ignorance of what is good and what is evil...*

I wonder what he would have thought had it been suggested to him that the unique blend of character traits possessed by each one of us - the character differences of which he writes and of which he was so well aware - could be accounted for by variations in brain size, structural configuration, and the pattern of 'wiring' in neural interconnections. And particularly how he would reconcile this biological information to his belief in the indwelling presence of soul and spirit - two influential forces in consciousness about which he writes frequently and for which it is difficult, if not impossible, to advance a biological explanation.

Knowledge of the brain's physiological complexity would certainly have intrigued this sensitive and wise man. After all, a hundred billion neurons and trillions of neuron-connections...! He would surely have been amazed - possibly more so than we who have grown up with modern science's staggering mathematical hypotheses - at the sheer volume of neurons involved, and by the astronomical numbers of permutations available in terms of their possible interconnections: their arrangement and rearrangements capable of producing a seemingly infinite number of individual 'consciousnesses' - every person responding to life in his or her special way. And given the keen workings of this Emperor-philosopher's mind, I feel sure he would have advanced a sound philosophical argument to account for the manifestation of such immaterial factors as soul and spirit within this maelstrom of neurons we call consciousness.

Looking back in time we have no sure way of knowing when thoughts and feelings pertaining to the rightness or wrongness of things, or concerned with attributing degrees of perfection to this or that experience of the world, first stirred in human consciousness - when *ideals* of this nature first began to influence human attitudes and behavior. (Yet if flowers were used for symbolic reasons in the cave burials at Shanidar some 50,000 years ago - as we have mentioned previously - it certainly suggests that Neanderthal man lived with imaginative ideas and ideals in addition to coping realistically with life's practical problems.) The earliest written statements (of which I have knowledge) that testify to the influence in consciousness of thoughts concerning goodness and truth, are found in the Maxims of Egyptian pharaohs from as long ago as the Third Millennium before Christ. Here, translated from hieroglyphs by Boris de Rachewiltz, and from his Italian by Guy Davenport, is an extract from the *Maxims of Ptahotep*:

> *If you desire your conduct to be well free from evil, keep away envy, that grave and incurable sickness. It makes intimacy impossible, makes the sweet friend bitter, alienates the faithful from their master, corrupts father, mother, and separates the wife from her husband. It is a bundle of all kinds of faults, and a basket of all that is shameful. He lives long, the man whose rules of behavior are just and in accordance with the true way. Thus he comes to his wealth, while the envious come only to the grave.**

So here we are, reading Ptahotep some five thousand years later, and the question that comes to mind is how seriously the ideals he expresses would be taken nowadays? How many parents in Western society would write such an epistle to a son setting out in the world?

Ptahotep is a good example of a man displaying the wisdom of a fully-fledged mind. One has the sense in reading his words that, like Marcus Aurelius more than two thousand years later, he was an idealist who would inspire as a visionary and philosopher; yet also a realist who would impress as a good military leader and civil administrator. It is difficult to think of either of them 'loosing their cool', as is said nowadays. For a calm response, waiting for the cogs of intuition and reason to mesh, is one of the marks of a finely balanced, right and left brain consciousness. Yet for many of us there will be stressful moments when we are not capable of such composure and deliberation - times when instinctive and automatic impulses-to-action seize the opportunity to break through and take over: as we have noted, these elemental responses rooted in the Natural Order are never entirely lost. Their dim memory-shadows are stored as subliminal imprints in the cerebellum, able to rise to the surface when rising passions become ungovernable and overwhelm the control of reason - squelch what we think of as 'civilized' behavior. Such latent volatility of instinctive behavior testifies to the potency of a residual well of raw emotion thought to remain hoarded in the brain's limbic system - surging to provide a similar kind of 'adrenaline rush' to that which would have given rise to

---

* Boris de Rachewiltz: Maxims of the Ancient Egyptians (translated from the Hieroglyphs & Guy Davenport translated from his Italian).

the unconstrained and unpremeditated response of the days when instinct was in command.

Both the cerebellum and the limbic system were relatively early additions to the original brainstem and so antedate the arrival of the cerebrum and cortex. Yet neuroscientists say that the cerebellum has tripled in size over the last million years, and this certainly provides food for thought: causing one to wonder if such a development might already be allowing age-old memories of the days of 'nature red in tooth and claw' to become more subliminally evocative today than they have been in the past. It is to be hoped not. For as one surveys the clash of political and religious ideologies around the world at the present time, it becomes evident that reason, compassion, conscience...fight a losing battle against the rudimentary emotional urge to destroy that which is antithetical to one side or the other and taken to be a potential threat. We see tides of hate, anger, suspicion... and waves of killing. The will to reason is overcome; the consciousness of those intractable individuals who initiate and maintain such conflicts seems to be in a state of devolution.

Some of the anthropologists and historians who have studied the cave art of Palaeolithic man - particularly the paintings found at Lascaux and Altamira in south western Europe and thought to have been made about fifteen thousand years ago - have been led to feel that the Old Stone Age hunters, who were probably also the artists, did not kill the animals so necessary for their survival *'without mercy.'* It is not a sentiment we can attribute to those who today so readily and easily destroy human beings *en masse* without mercy or pity. As civilizations have grown so has the savagery to which we have subjected our own kind. It is a sobering thought, and one which militates against taking an overly optimistic view of consciousness' evolution - even though it is equally obvious that men and women of conscience and remarkable moral character have influenced world events; and arguments could be advanced that the number of institutions and individuals working for the general good are steadily increasing. Nevertheless, there are moments when the hypocrisy and mendacity of politics and the naked aggression fired by religious and national causes, compels one to wonder if the idea of a purposive evolution leading to a wiser and kinder Human Order throughout the world is not mere wishful thinking.

The admonition commonly found in the many codes of moral conduct set up by societies worldwide - and which can be summed up in the three words: *'Do No Harm'* - has no place in the rabid zealousness of fanatics and extremists of whatever ilk. For these three words illuminate a guiding humane principle expressed wherever individuals or nations have felt for the wellbeing of animals, humans, and the planet itself: experienced a sense of responsibility to maintain, if not improve, the life of nature and the world - a sentiment unknown to a primordial consciousness where all the neurons could do was to trigger the quick urge to act unimpeded by moral or pragmatic considerations.

The novelist Joseph Conrad gave the title *Heart of Darkness* to a story of human cupidity in 'darkest' Africa. His central character, a man whose life is devoted to the pursuit of unbridled power and wealth who will go to any lengths to maintain his sinister authority in a little 'kingdom' far from civilization in the heart of an equatorial jungle where primitive tribes make up his 'subjects'. Devoid of conscience he is the personification of evil; and Conrad draws him so convincingly that some readers must surely be left wondering as to how a character of such inner psychical bleakness could come about. For what we have is a man who has regressed to, or never known anything but, a conscienceless state: an individual governed by strong subliminal, a-human impulses that create a pathology of evil. The telling aspect of the story is that he represents a level of inhumanity against which even the primitive attitudes and behavior of the native peoples can be seen as more representative of human sensibilities. The parable of the novel is that an overwhelming greed and lust for power can arouse the old imprints in the pre-cortical cerebellum to produce a completely psychopathic personality. Yet, paradoxically, resulting in behavior made even more vicious because an unrestrained and gross - yet evolved and controlling ego-self has assumed automatic command.

Conrad has created a person for whom the words of a Marcus Aurelius or a Ptahotep would have little meaning; and it is difficult to read *Heart of Darkness* without pondering just how one particular fully-developed brain of a hundred billion neurons (give or take a few) can lead to the formation of character which takes one person to follow the low road into darkness - while another contemporary brain shapes the kind of character which has someone else following the high road into the light. Inevitably,

Conrad causes one to reflect on the central role the human soul may play in this matter: on the influence it's residency may have as a spiritual force inspiring the build up of character that seeks to do no harm. And on what may happen should the soul retreat, leaving one bereft of a spiritual center, knowing indeed only the heart of darkness.

> There is a kind of character in thy life,
> That to the observer doth thy history
> Fully unfold.
> Shakespeare, *Measure for Measure*

How does one discern the kind of character a person possesses? And what is meant by 'character' anyway? During the Second World War the most severe condemnation made of a Royal Air Force aircrew member's character was noted on his personal documents as L.M.F. - lack of moral fibre: the official jargon for being afraid. But we were almost all afraid, most of the time; yet I have to admit that there were those amongst us in whom one could detect no such anxiety - and I do not believe that they were just very good at hiding it. We talked about this phenomenon among ourselves and I remember Squadron Leader Beauchamp saying to me that it was really a matter of nerve or will - of possessing the willpower to transcend 'gut-fear' - and that some of us were better at it than others; that those who couldn't manage it were designated L.M.F. And that this was not a *defect* in their character, rather the *nature* of it. That the same man who could not control fear at 15,000 feet, when the aircraft was caught in the searchlight beams and subjected to a heavy anti-aircraft barrage of exploding shells might, in civilian life, risk his neck to save a drowning child. Even now - so many years later, and recalling the drift of our conversations - I remember that in our often heated discussions we all pretty much denounced the injustice of judging and damning a man's character so arbitrarily.

Shakespeare was well aware that the uniqueness of each individual character - its distinctive psychological texture - is determined by a particularly personal, closely interwoven and tremendously variable mix of thoughts and feelings: an intricate web of mental responses set in motion and shaped by the impact of life's existential happenings; and by the subliminal, unconsciously held inclinations (with which we are likely born) to adopt particularly patterns of feeling and thinking by which character

– and some would say destiny – is shaped. And there are those geneticists who suggest that genes play a role in determining such psychological traits by carrying these innate predilections. Yet is the great playwright correct in suggesting that the nature and quality of this inner self – the core of a person we associate with character – is perceptible to others? There are indeed times when one looks someone in the eye and intuitively senses how trustworthy that person would be, or whether they are of a cruel or kind disposition... And when faced by a preponderantly good person, or by one given chiefly to malice, such opposing and dominating characteristics can usually be discerned – evident in an individual's general manner, in some twist of the mouth or lack of steadfastness of eye. Shakespeare raises an intriguing question here, and one we pursue in the following chapter, *Mirror, Mirror, on the Wall.* In the company of Herbert Read or Bishop Trevor Huddleston, for example, one immediately knew their strength of spirit and conscience. It was written all over them. Never could they have become killers in the mould of, say, a Hutu fanatic or a Saddam Hussein... Lurking vestigial, violent influences would have little chance of prevailing against their moral toughness, compassion, and gentleness of manner. They were men uncaring of the trappings of worldly success and power.

I said earlier that we have no sure way of knowing when, in our history as a species, the development of this psychological individuality we call character would have become a fact of life for every human being. However, if we understand that the cortex evolved to become the brain's chief intuitive and imaginative source, then the presence in consciousness of moral standards, of the urge to love and protect... and of notions pertaining to such abstract forces as soul and spirit, can possibly be attributed to a time somewhere between two and four million years ago when some neuroscientists conjecture that the cerebral cortex became fully operational and authoritative. Consequently, it is reasonable to suggest that a benevolent character results from the ability of the evolved cortex to respond positively to the soul-spirit force – whereas a malevolent one is less able to do so and remains – more or less intractably – under the age-old influence of instinct.

In *The Principles of Psychology* (1890) the American psychologist and philosopher, William James, comments on the persistent influence of millions of years of instinctual behavior before the brain achieved its cerebral maturity. In Chapter 24 he writes:

> *The hunting and the fighting instinct continue in many manifestations. They both support the emotion of anger; they combine in the fascination which stories of atrocities have for most minds... the pleasure of disinterested cruelty has been thought a paradox and writers have sought to show that it is no primitive attribute of our nature, but rather a resultant of the subtle or other less malignant elements of mind. This is a hopeless task. If evolution and the survival of the fittest is true at all, the destruction of prey and of human rivals must have been among the most important... It is just because human bloodthirstiness is such a primitive part of us that it is so hard to eradicate, especially when a fight or a hunt is promised as part of the fun.*

This is well put and illustrates the validity of the 'nature red in tooth and claw' image we used earlier as one way to account for the constant prevalence of human violence. For, as you can see, James considers the *'primitive attribute of our nature'* to be primarily responsible for the cruel and aggressive side of our behavior - attitudes that can be overcome by recognition of the power for good, the growth of compassion. He does not suggest that the urge to violence and the brutal commission of atrocities arises from a *permanent*, built-in predisposition of the human brain and psyche. Whereas Joseph Conrad would seem to subscribe to this latter view, suggesting that mankind is inherently wicked, despite attaining the higher levels of consciousness that can bring him 'to see the light', as it were. Yet in writing *The Heart of Darkness*, he raises - but does not pursue - a third possibility to account for the existence of malevolence and evil - namely, that it may exist in its own right.... an autonomous, *supra*natural force present in the world independent of human existence.

> *The belief in a supernatural source of evil is not necessary; men alone are quite capable of every wickedness.*
> Joseph Conrad, *Under Western Eyes* II, iv.

Many societies have believed that we are prey to preternatural evil forces, working to bring about our individual spiritual and moral downfalls by enmeshing us in a tempting web of self-indulgence - titillating appetites and stimulating pleasures in which we will thrash around to excess; or by encouraging the human desire for personal aggrandizement in the form of material success and lust for power. In either event, the aim, it might seem, is 'designed' to involve us in a struggle to develop an individualistic spiritual strength in the face of such challenges. The Satanic influence - as presented in the teachings of the Christian church, for example - is seen in this light as the negative force enticing one to act so as to commit spiritual suicide.

The word 'evil' is found in the earliest of writings - Egyptian, Greek, Roman... Look back a few pages for instance, to the letter written so long ago by the Egyptian pharaoh Ptatahotep to his son. In the fifth century B.C. the Greek dramatist Euripides wrote the line, *'I am overcome by evil'*, in the play *Medea* - the Greek word for evil being *kakois*. While the Roman statesman Cicero (106-43 B.C.) wrote in *Philippicae* that, *'Every evil in the bud is easily crushed...'* - the Latin word for evil being *malum*. And there is no reason to think that the concept of things evil was not present long before it appears in written records. Many Australian aborigines - the indigenous peoples who some anthropologists think have continuously occupied parts of the vast island continent for over 40,000 years - still hold to their ancient views of what constitutes right and proper behavior - the way to live in order to counter negative or evil forces.

Yet even though there is a general belief in the opposing forces of right and wrong, and in the importance of recognizing the one from the other.... many differing thoughts emerge as to the moral quality that best serves the essentially 'good' way to live. For as Will Durant wrote in *The Story of Philosophy*...

> *Morality, said Jesus, is kindness to the weak; morality said Nietzsche, is the bravery of the strong; morality said Plato is the effective harmony of the whole. Probably all three doctrines must be combined to find a perfect ethic; but can we doubt which of the elements is fundamental?*

Yet more than three 'doctrines' go to form the characters of those human beings who exemplify the best amongst us. I am thinking of Sophie Scholl, the young German martyr of whom I wrote earlier and who, in her brief life, displayed the kind of spiritual, compassionate, and aesthetic sensitivity that encompasses just about every moral and creative attribute one can bring to mind: perception and vision, a natural loving of the world, courage and strength of will, and a conscience that informed her every move. Her kind of human *being* brings the bald statements of the philosophers to life. Imbues their 'doctrines' with a living truth. She corresponded frequently with Fritz Hartnagel, a close and long-standing friend, disclosing her innermost thoughts and feelings, and I refer you to a letter she wrote to him some nine months before her execution expressing her moral despair and solitariness living in Hitler's Germany:

> *My dear Fritz,*
> *We're having some really glorious early summer weather. If I had the time, I'd stretch out beside the Iller, swim, laze, and try to think of nothing but the beauty around me. It isn't easy to banish all thoughts of the war. Although I don't know much about politics and have no ambition to do so, I do have some idea of right and wrong, because that has nothing to do with politics and nationality. And I could weep at how mean people are, in high level politics as well, and how they betray their fellow creatures, perhaps for the sake of personal advantage. Isn't it enough to make a person lose sleep sometimes? Often my one desire is to live on a Robinson Crusoe island. I'm sometimes tempted to regard mankind as a terrestrial skin disease. But only sometimes, when I'm very tired, and people who are worse than beasts loom large in my mind's eye. But all that matters fundamentally is whether we come through, whether we manage to hold our own among the majority, whose sole concern is self-interest – those who approve of any means to their own ends. It's so overwhelming, that majority, you have to be bad to survive at all. Only one person has ever managed to go straight to God, probably, but who still looks for him nowadays...?*
> To Fritz Hartnagel: Ulm – May 29, 1940*

The Sophie Scholls of this world are not in the majority. Yet in her short life she presents us with an example of the clear-sightedness, moral strength and nobility of character which represents the apogee of humanness. One would hope to possess a little of such fortitude and goodness.

Most of us in western societies struggle along the moral way with varying degrees of success – some attaining goodness, some having goodness thrust upon them, others seeing the light only from time to time, yet others never having a glimmer. Sir Thomas Browne writing in his *Religio Medici* in 1660 AD declared, 'There is another man within me that's angry

---

* Inge Jens, *At the Heart of the White Rose: Letters and Diaries of Hans and Sophie Scholl* (New York: Harper & Row, 1987).

with me.' I wonder how many of us are familiar with this *'man within'* who speaks with varying degrees of firmness and exerts varying degrees of influence; how many of us come sometimes to accept his (or her) counsel, and at other times dismiss it. But when it is a matter of great moment – when a course of action involves the suffering and violation of others, or compromises principles of honesty and truth – then to turn away from the voice of conscience is to discourage the spirit of good and diminish the true self. And I remember the Headman of the Sakkudei telling a film crew as he led his tribe out of the jungle in Java, *"We must always behave well so that our souls will like to stay with us."* (The tribe, about which I write later, had always lived in the forest, and was being relocated by order of the Indonesian government in order to facilitate logging operations; the Headman's remark was in response to a question about the possibility of militant resistance by the Sakkudei).

What a contrast between the words and actions of the leader of the Sakkudei, and the tirades and predations of the men of violence who move through the pages of history. On the world scene today there are many for whom the random taking of life is seen as the only way to redress wrong, propagate their beliefs, or attain positions of power. Media coverage of bombings and other acts of terrorism ensures their visibility on the world stage, and perhaps for this reason those who perpetrate such outrages appear to be more diverse and prolific than I remember in the past - proliferating in factions and groups whose adherents have come to know no other way of life, and whose minds are closed to the way of reason through dialogue and mutual compromise. They are able to bend conscience - (if indeed it should murmur from time to time) - to justify both the ways and means by which they murder, and the terrible suffering they inflict on others. Some two thousand years ago the Roman statesman Marcus Tullius Cicero wrote in *De Officiis*:

> *Reason and speech, which bring men together and unite them in a sort of natural society. Nor in anything are we further removed from the nature of wild beasts.*

A most reasonable statement in itself; and one against which the havoc wrought by extremists and fanatics in recent times is seen to be spiritually and humanistically bankrupt. The faculty of reason has not advanced to 'bring men together' in a 'natural society' as Cicero might have hoped would happen as civilization advanced. Were he around to be consulted he would probably agree that while we have made giant strides in knowledge we have advanced but little - if at all - in wisdom. And I think it likely he would observe that religious absolutism breeds the most unreasonable, ultimately evil kind of terrorist - the one who believes that only his or her religion reveals the true way to God, thus justifying the killing of disbelievers without a scrap of concern for the sanctity or value of human life as such.

The Sakkudei Headman possessed a serene and commanding presence, and I remember thinking at the time that his 'man within' would be more likely to offer calm and wise counsel when crucial decisions had to be made, rather than make the angry protestations of Sir Thomas Browne's conscience. Also, in watching him and listening to him as he emerged from the trees, my impressions were of a man who stood head and shoulders above many of the leading world figures of the time in wisdom and nobility of spirit - the one, I suspect, being inseparable from the other.

As I write I am reminded of the great 19th Century French painter Eugène Delacroix's comment in *The Journal* to the effect that *'The Muse is a jealous mistress. She leaves you on the slightest sign of infidelity...* ' by which he implies that if the serious artist gets totally caught up in the social life of the outside world - or becomes completely enamoured of the kind of new sensations and experiences which gratify the externally oriented side of the ego - then the inner world of creative inspiration is lost. Similarly, when there is neither time (or inclination) to mull over the happenings in life, indulge in the sort of healthy introspection that brings the essence of an event to mind, then one is less likely to differentiate between this or that experience in terms of relative worth and significance. Such inner blankness can give no voice to conscience; nor to the imaginative self-revelation of which Wordsworth speaks in that line from his poem *The Daffodils:* "I gazed - and gazed - but little thought/What wealth the show to me had brought...' Delacroix's Muse is indeed a jealous mistress. Don't hope to discover her if living on the 'fast track' as it is called - constantly on the move seeking change and the temporary stimulation offered by new things, people, places...

'*Conscience was born when man had shed his fur, his tail, his pointed ears.*' So wrote Sir Richard Burton, the intrepid 19th century British explorer and orientalist, in his book *The Kasidah*. A graphic and poetic statement which implies that it was present as an active force within the human psyche considerably earlier than we have suggested. Essentially, conscience breaks through to consciousness as a synthesis of involuntary and internally generated feelings and thoughts - what Herbert Read described as 'felt-thoughts' - 'whispering' their conclusions concerning the beneficial or harmful consequences (to oneself or others), of the attitudes one adopts and the actions one takes: judging them against standards of truthfulness, loving, compassion, charity, selflessness, virtuousness.... all those sensibilities which are often taken to be *passé* in today's world. The image of conscience as a voice - *'that still, small voice'* sounding quietly from some innermost recess of the mind - is the one most universally used. And while conscience is capable of being defined in many different ways, there is no denying that it acts as a constraint, as an urge to virtuousness testing the moral and spiritual will. At its most potent it exerts a powerful, transforming inner force inspiring one to act, heroically at times, in putting the safety and wellbeing of others before thoughts of one's own survival... yet always for the preservation of one's own integrity as well as for the general good. This linking of conscience to a spiritual center is found worldwide. For instance, it is seen as the spirit-directed *'fount whence honour springs'* as the Elizabethan dramatist Christopher Marlowe put it. *'Virtue is to the soul what health is to the body,'* said the French writer and moralist La Rochefoucauld in *Maximes Posthumes*, published a few years after Marlowe's death.

The dictates of conscience usually present themselves as vague feelings of unease or, conversely, as a sense of moral certainty - both of which somehow manage to convey *a priori* knowledge that such and such a mental attitude is good in that it 'serves life', or evil in that it 'cuts it to pieces'... as Erich Fromm declares. It has to be admitted that this kind of silent communing between conscience and consciousness is a puzzling kind of psychologic transaction. For we have seen that it appears to be entirely unwilled, occurring spontaneously when particular thoughts and feelings suggest a course of action which is especially significant in that it invokes a moral judgement. And inevitably the question follows as to whether this persuasive power of conscience is learned or innate. I take the view that it is an innate psychological faculty, and that its effectiveness depends upon the potency of spirit with which a human being comes into this world. For

I have known men and women whose childhood and adolescent years - the most impressionable stages of life - were lived in squalor and deprivation seemingly bereft of any moral example; devoid of contact with people who had little if any positive influence to offer. Nevertheless, they turned out to be reliable individuals, strong in spirit when it came to dealing with life's vicissitudes, and fully deserving the appellation, 'a *good* person'. While there are others who were privileged in their early years to be brought up in the most benevolent and advantageous of environments, yet who seem to lack all manifestations of conscience and compassion.

I have always thought of conscience in the same way that I think of inspiration: as a guiding force arising intuitively from the unconscious. And although it can be argued that its beneficial effects serve - in an evolutionary sense - to aid in the preservation of the species, more significantly it brings to the individual intimations of truths that surpass the drives to survival to which the body so strongly holds.

> *I fully subscribe to the judgment of those writers who maintain that of all the differences between man and the lower animals, the moral sense of conscience is by far the most important... It is the most noble of all the attributes of man.*
> Charles Darwin, *The Descent of Man* (1871)

Whether Darwin would go so far as to consider the idea that the 'noble' attribute of conscience may be perceived as the voice of spirit, I seriously doubt. Unless he were to read this book and agree with the Bishop (my companion on the flight to Dublin, you will recall) that the overall reality of human 'being' is dualistic, and that conscience comes with the 'spirit side' of the coin.

'*I only know,*' wrote Ernest Hemingway in *Death in the Afternoon*, 'that what is moral is what you feel good after and what is immoral is what you feel bad after.' I suppose most of us recognize what he is talking about. For when the murmurings of conscience regarding crucial moral issues are overridden, we feel that we have not served ourselves well; have taken a step in the wrong direction, not kept faith with the prescience of significant truths - however enigmatic and mysterious their source may be. And if we see nobility as signifying integrity of character and charitability of spirit, then I would think that a strong and active conscience must be its guiding light. Frans de Waal, a scientist who has studied captive chimpanzees for many years at the Yerkes National Primate Center at Emory University in Atlanta, observes that they live, as a society, according to certain behavioral rules: they display a sense of fairplay in their daily round with each other, 'feel' for victims of violent or unjust acts, and are concerned that good relationships within the society are maintained. Such attitudes governing a communal existence may well be thought to represent the beginnings of a rudimentary morality. Yet *'... the most noble of all attributes of man'* of which Darwin speaks are a far cry from the chimpanzees' 'sense of fairplay.' 'Soul', 'spirit', and 'conscience' - the words we have given to the most 'noble' of our attributes - identify mental experiences which owe their origin to inspired insights giving rise to the most creative and altruistic acts. Are there Mozarts or Mother Teresas lingering unrecognized in chimp circles?

Which brings us back to the neuroscientists' view that once the two hemispheres of the cerebrum were in place and the cortex fully operative, mankind came into possession of an advanced and highly imaginative aspect of consciousness - a state of awareness bringing insights into spiritual dimensions of human existence. A phenomenon that can reasonably be thought to have been concurrent with the onset of linguistic and ritualistic modes of expression which - as we have previously mentioned - some neurological researchers think happened between four and two million years ago.

I find my thoughts turning to the remarkable energy-field created by the electrical and chemical activity of the mature brain, and wonder if this powerful cortical network of billions of neurons could have brought into being a highly sophisticated 'receiving system' - the 'new' cortical brain serving as an antenna-like receptor capable of picking up transmissions of radiant energy - electromagnetic waves hitherto inaccessible to the older brain and capable of being converted into the thoughts and feelings that shape the noblest of our attitudes: energy emanating from some cosmic source of ultimate and supreme intelligence. And when casual mention is made of the human spirit as a resource on which we can draw in times of trouble, I find myself wondering if this is not another way of referring to the surpassing spiritual reach of the especially finely-tuned and highly efficient cortex we have come to possess.

This is a prospect that must appear ludicrous to those who accept only what science can submit as 'hard evidence' - or who see all aspects and experiences of life resulting from chance encounters between the known forms of matter and energy. Yet there are also many of us who, in experiencing the dualistic enigma of intuition working hand-in-hand with reason, and learning of the phantasmagoric complexity of the cosmos at macro and micro levels, find that our curiousity leads us to search for answers as to why and wherefore life beyond the scientifically provable operations of nature. Which was Aristotle's position when he coined the word 'metaphysics'.

It is interesting that Darwin considered 'the moral sense of conscience' to be the most significant difference between ourselves and 'the lower animals'. For to accept a mental authority that conjures up abstract value judgements such as right and wrong, indicates that Darwin gave no less credibility to our imaginative and intuitive powers than to the rational and analytical strengths by which we gathered information through the senses.

In pondering the effect on the individual of the first stirrings of the 'voice' of conscience, I imagine that the thoughts and feelings such an inner voice provoked, must have given the impression that some profound influence was at work behind the scenes, evoking a sense of a profundity to life: a level of mystery that induced thought as to life's purpose and origins - what we now would regard as a form of metaphysical contemplation - that, over time, helped to contribute to the development of countless religious theories and philosophies. Consequently, it can be argued that it was with the advent of *conscienta profunda* that the concept of God entered human life. There are two ways of looking at this proposition. Either the brain itself, once it was physiologically complete with cerebrum and cortex, imagined, invented God. Or that with its sophisticated cerebral cortex now capable of absorbing an unknown form of radiant energy, God was then enabled to intervene directly into human life - an intervention by which we became, in René Dubos' words, 'so human an animal'.

> *Then a sense of law and beauty,*
> *And a face turned from the clod –*
> *Some call it Evolution,*
> *And others call it God.*
> W. H. Carruth: *Each in His Own Tongue*

Yet it is obvious that conscience works at different levels for different people; can be moribund on some occasions and active on others. There are those of great resolve who always struggle to keep faith with the values it represents. Yet when its counsel runs contrary to personal desire or advantage I suspect that many of us find ways to justify 'putting it on the back burner', to use a current idiom - 'the spirit is willing but the flesh is weak...' But then it must also be recognized that some individuals seem bereft of any values or impulses of conscience – seemingly experiencing no moral promptings that would cause them to reflect on whether their attitude or their actions would do harm and cause suffering, do good and bring pleasure. It is one thing to be privy to the soundings of conscience, yet deliberately choose to suppress the advice and go ahead anyway and compulsively gratify any desire – but quite another to live in a bleak and silent inner world where no counsel is offered at all.

Could such a silence be the result of a neurological problem? A malfunctioning or incapacitation of cortical neurons thereby rendering consciousness unable to respond to the communications of conscience? Or might it result from a spiritual rather than a neurological weakness - say the possibility that soul and spirit are somehow so innately and inexplicably constricted as a combined force, that conscience is insufficiently empowered to effectively register moral persuasions at the level of the cerebrum and its cortical processes? Or perhaps, given the uncertainty of guesswork in theorizing how any mental understanding of abstract moral principles becomes a conscious event, we would do just as well to adopt poetic images like that offered by John Wolcot, taken from *The Lousiad: Conscience, a terrifying little sprite, that bat-like winks by day and wakes by night*.

I have met three men in my life whom I would deem to be 'saints' - men who possessed such a strong aura of goodness and wisdom that people, on coming into their presence, suddenly became tranquil and quiescent as if released from the social façade they daily presented to the world. You have already met my onetime flying companion, the Bishop and, like him, neither Sir Herbert Read, the English poet and critic, or Father Trevor Huddleston of the Community of the Resurrection in Mirfield in northern England, ever seemed to be aware of the ineffable quality of grace and moral authority they displayed. In their presence one felt 'safe' - in every sense of the word: protected from any internal or external vicissitude. During the First World War Herbert Read saved the lives of many men in his regiment after their positions were overrun in a German advance. Their situation behind the new German positions seemed hopeless, but Read, the only surviving officer, kept morale among the survivors high; had the men moving at night unseen by the enemy, until finally they regained their own lines. His was always a serene presence: the embodiment of a quiet spirit of confidence in running life's journey and of faith in the ultimate outcome - personal characteristics not unlike those displayed by Shackleton in keeping the men of the *Endurance* alive through two Antarctic winters. Yet there was something else about Read for which it is difficult to find the words - a transparent honesty which, as a fellow poet might say, 'encompassed him all around'.

It is a quality which comes across strongly in his war poems - testaments which stand with the poetry of Wilfred Owen and Rupert Brooke as the most powerful images to emerge from the bloodbath of the World War I trenches. But his friend and literary colleague Graham Greene, in writing Herbert's obituary in 1968, did find the words to convey the absolute integrity of the man. He said that when Read entered a room in the course of a literary gathering - during which the usual critical gossip about personalities and their accomplishments, or lack of them, was well under way – the malicious backbiting would wane, die out, because, *'All goodness, all truth, had come among them.'*

And Trevor Huddleston? Well, he is the kind of priest you would want to have with you at the end, his presence and his Blessing guaranteeing you passage. In the 1950's his was the most powerful voice in South Africa - an implacable conscience crying out against apartheid. His book, *Naught For Your Comfort*, is a most compassionate plea for human equality under God. His faith and spirituality were palpable. There could be no dissembling in his presence.

My flying companion The Bishop, Herbert Read, and Trevor Huddleston constantly linger in my mind as examples of human greatness; yet of an eminence that transcends the customary view of genius as the display of great creative gifts in music, science, art, or literature. Their distinction lay in conveying a dimension of human existence that I can only describe as a combination of great wisdom and sublimity. Thomas Campion the

Elizabethan poet expressed it in the words, *'Soul is such a Man'*. The latest edition of Webster's Dictionary characterizes such a quality as 'nobility of soul'. Not surprisingly, perhaps, greatness of this order has long been associated with magnanimity of soul and the eloquence of conscience seen as its emissary. A long time before the Elizabethans, Lucius Annaeus Seneca (4 B.C.? - A.D.65), Roman statesman, philosopher and dramatist, wrote: *'Why, then, is a wise man great? Because he has a great soul'*.

There are those who would agree with the Marquis de Sade that conscience and morality are delusions serving no 'higher' purpose. For inasmuch as nature possesses no conscience, knows nothing of moral standards - and as we are existentially creatures of nature - notions of good and evil are simply irrelevant creations of our own freewheeling brain. And others who will argue that one of the main planks in this discussion - the concept of soul as constituting the essence of *human* being is, in any event, purely fanciful, in that it is not scientifically demonstrable. And we have to face the fact - made more obvious nowadays by the instant relay of news (with pictures) - that worldwide there are many who seem able to compromise the moral standards they claim to uphold in order to suit their particular financial, political, ideological or religious agenda.

History shows that human beliefs concerning the nature of what constitutes evil and what good have much in common, despite differences in culture, time or place. Murder, adultery, and sacrilege appear to be the 'big three', constantly on the lists of 'deadly' sins - transgressions recognized as damning and supremely immoral across the religious and philosophical spectrum. Yet increasingly in the contemporary world we see the proliferation of *"interpretations'* of conscience that compromise traditional and widespread views of right and wrong to serve the political or sectarian needs of the moment - making it extremely difficult for groups or nations to work together for the general good according to the basic moral principles of compassion and fairness: ideals which, as Erich Fromm states, 'serve life'. And in this he takes a classical Greek position, believing that we, as *human* beings endowed with conscience and intelligence have a duty - indeed are required specifically - to act as the stewards of nature and, in so doing, look to our own good psychic health and safeguard her very existence. It represents an ideal view of the role we should play in life, yet few of us are completely up to it. The Italian author and statesman Niccolo Machiavelli

(1469-1527), summed us up correctly when he wrote in his *Discourses on Livy* that *'Men are not altogether bad or altogether good'*. Such is indeed the condition in which most of us find ourselves. And in one important sense - given the way consciousness works - this is how it has to be. We can only be aware of one ethical standard if we are presented with its complete opposite. If we had no idea of what is meant by 'doing harm', we would have no idea of what is signified by 'doing good'; and if such were the case we would have no judgements to make in terms of how we act - have no choices to take as to whether we travel the high road or the low road. Life would be 'mindless', you might say: there could be none of this 'by their actions you will know them' kind of sermonizing. And without having to ponder the good or ill that can result from our actions and make decisions accordingly, 'character' would be a meaningless term. As meaningless as the idea that we have a 'self' (other than the face in the mirror) to discover; a personal destiny to shape.

Let us return for a moment to the historical record. In studying the earliest of written statements from the Old and New Kingdoms of Egypt, of Mycenaean and Homeric Greece, and the first historic Kingdoms of Asia Minor and China - we find a certain number of pharaohs, kings, prophets, poets, philosophers, and statesmen expressing similar views concerning the right and proper way to live. Common to them all is a belief in the need to separate the false from the true, both in the manner of one's thinking and in one's conduct - in order to try, overall, to do no harm to others or to the principles of harmony and balance by which everything is maintained in a state of being. And for this we need the direction offered by conscience. *'There is measure in all things; certain limits, beyond and short of which right cannot be found'* and, *'Yes, there's a mean in morals...'*, wrote Quintus Horatius Flaccus (65 - 8 B.C.) - the Roman poet and satirist otherwise known simply as Horace - in Book I of his *Satires*.

In Plato's *Phaedrus* - a dialogue between Socrates and his friend Phaedrus - the discussion introduces the now well known metaphor of the charioteer and two horses, one black and one white, to symbolize every human being's 'chariot journey' through life. The charioteer represents the soul as the 'supreme commander', bearing the light of reason by which we become responsible for our respective destinies; and which ultimately brings us to bridge the irrational and rational aspects of consciousness and experience them as two sides of the same coin - thus providing us with an overview and understanding of the form and direction life is taking. Now it is the black steed that represents the irrational side of consciousness, embodying all aspects of our involuntary, imaginative life from the most positive and creative to the darkest and most destructive; consequently, this is the horse that demands the charioteer's constant attention and supervision. While the white horse, standing for the faith, hope, steadfastness, and courage generated by the power of spirit, can always be allowed its head.

In terms of this illustration I wonder how many of us currently pursuing the chariot-course will manage to achieve the balanced integration between the two horses that brings us successfully over life's finishing line. For those of us who do, the Welsh poet Dylan Thomas' words, *'The wire-dangled human race'* will never have been all that applicable for, as charioteers, we will have driven the black and white horses as a team.

> *A peace above all earthly dignities,*
> *A still and quiet conscience.*
> Shakespeare, *Henry VIII*.

LACK OF WONDER: 'GOOD AS DEAD'

> *The fairest thing we can experience is the mysterious. It is the fundamental emotion which stands at the cradle of true art and true science, He who knows it not and can no longer wonder, no longer feel amazement, is as good as dead, a snuffed-out candle.*
> (Albert Einstein, *The World as I See It*, 1950.)

It is a sobering thought to realize that since Einstein made this statement five years before the start of the Second World War, the sense of mystery he feels to be lurking behind the scenes - of phenomena in general and his own existence in particular - does not seem to be much in evidence during these opening years of the twenty-first century. And this despite the remarkable cosmological, biological and paleontological discoveries made by science. It is difficult to understand how anyone can remain apathetic to wonder, unmoved by awe, when confronted by the Hubble telescope's photographs from space revealing such an incredible scale of *macro*cosmic grandeur; in contemplating the unbelievable *micro*cosmic complexity of the human genome; or when reading a news report that biologists at Cambridge University have come to the conclusion that birds - with their small chip of a brain - can understand each other's intentions, utilize tools more effectively than chimpanzees, and understand the workings of cause and effect at the level displayed by a three-year old child.

We live in a time when sophisticated instruments and machines play important roles in daily life to an extent that Einstein would likely have not envisaged in 1934. Electrical technology has revolutionized the way we do things and the way we spend time. It has led to the harnessing of the energy produced by nuclear fission; put satellites into orbit to provide practically instant worldwide communication systems; created a new market economy known as 'consumer electronics' through the production of ever more sophisticated home visual and auditory systems, personal computers, mobile cellular phones... and all kinds of electronic accessories designed to make time-demanding tasks easier and more intriguing. For example, our whole way of life has been transformed by the technology that allows the electronic information imprinted in a plastic card to be approved by an orbiting satellite to allow *instant* credit - facilitate the *instant* purchase of just about anything. It is an invention that has turned us into a society living on credit, buying things whether we can afford them or not. But science is the big winner in this age of electronic marvels. Research has advanced by leaps and bounds over the last fifty years in the understanding and applying of electrical and electromagnetic principles to create such instruments as radio telescopes, sophisticated computers capable of controlling space-rocketry and revealing the incredibly complex code of the human genome... not to mention the invention of the Magnetic Resonance Imaging machine - a marvellous instrument that produces computerized images of internal body tissue induced by the application of radio waves; and is to medicine what the radio telescope is to astrophysics.

One of the most significant of all these developments has been to speed up life, shrink time and space. At the macrocosmic level this has been achieved by bringing the furthermost reaches of the cosmos within our purview. Whereas at the human level it has been accomplished by facilitating almost instant communication between individuals - whether the distance involved be one mile or one thousand - and between governments and commercial interests worldwide. Everywhere people are talking on their wireless phones as if they cannot bear to be incommunicado for very long - must constantly be in touch with something or somebody. Yet barely two hundred years ago distance was the factor determining how quickly and how frequently one could communicate with someone who may well be 'dear' but not 'near'. To be within shouting distance was to be close enough to allow some level of 'instant' conversation to take place, but when the 'dear one' was far away a seeming eternity of time could intervene - depending on whether the other was a three-hour horse ride away or a month's sailing time distant.

Such radical developments in electronic technology have done much to change both the way we regard life and the way we live it. A particularly significant psychological casualty is what was once called 'the virtue of patience' - a response to the uncertainties of existence we can ill afford to loose. Time and patience walk hand in hand - for to be patient it is necessary to be unhurried by the flow of time and events and wait, judging and responding calmly as things play themselves out: *'There is nothing so bitter, that a patient mind can not find some solace for it'*, wrote the Roman statesman and philosopher Seneca in *De Animi Tranquillitate*. And, indeed, tranquillity (a mental force in its own right) frees one to play the waiting game and, in so doing, encourages curiosity and wonder. But one has to work hard at developing patience in a society where time is at a premium: we move at speed, distances are slashed, things happen in quick succession, and the logistical details of day-to-day living become ever more pressing and complex. It is all too easy to become impatient and restless. And one prays for quiet moments. The intrusive noise (music?) blaring out from sound-systems in restaurants and supermarkets, or the din of city traffic, and the constant loud-voiced-chattering that passes for conversation... all militate against one finding that mental space where it is possible to become serenely patient, reviewing things calmly, accepting things as they are without complaint, *waiting*... trusting one's judgements.

Waiting nowadays has become an art. Waiting until one can afford to indulge the latest desire, buy that big-screen television, for example; waiting for the moment when it is possible to sit down and write a personal and interesting letter to a friend, instead of reaching for the cell phone; or waiting expectantly for the half-hour when it will be possible to sit and quietly meditate, listen to music... 'All in good time...' is the maxim of the patient man or woman. It is (or was) a practice of the aborigine peoples of Australia to go 'walkabout' from time to time - leave the habitual daily round and walk long distances as a ritualistic way to induce periods of patient contemplation and so renew themselves... *'perchance to dream'*, as Hamlet might have put it.

Shakespeare, in writing *Hamlet*, allowed the Prince of Denmark short respites from the dramatic events shaping his destiny which were closing in around him - moments when he is permitted to patiently gather his thoughts together and consider the options open to him. His most famous soliloquy is often quoted as an example of such an internal dialogue - a patient summarizing of the facts in which he calmly gathers his thoughts together and considers his options:

> *To be, or not to be, that is the question:*
> *Whether 'tis nobler in the mind to suffer*
> *The slings and arrows of outrageous fortune,*
> *Or to take arms against a sea of troubles*
> *And by opposing end them...*

But patience and Hamlet-like introspection are not the marks of an extroverted society - not the means by which decisions are made when the instant gratification of a desire is thwarted, or instant solutions to problems are not to hand. For in becoming avid consumers, measuring personal progress in life according to the number and variety of expensive possessions owned, constantly desiring to be on the move, it is difficult not to be psychologically driven by the imperative, command-like overtones of the word 'Now'. Patient reflection is out; immediate satisfaction is in. The notion that one should live for the moment - 'seize the day' and live 'the *good* life' (visit the Mall) - is propagated subtly by the advertising industry, and more directly when official pronouncements on the state of the nation's economy constantly make the point that a high level of consumerism is required to keep the country economically afloat. Whereas to live 'the *meaningful* life' has nothing to do with the accumulation of possessions or with how busy one is professionally, domestically, or socially. To find meaning it is necessary to gain some quiet time and become pensive, allow the mind to filter out the dross and illuminate the truths pertaining to one's own existence - or at least the relative truths - that revitalize both body and spirit. Yet this requires some self-discipline. It is certainly easier to keep the radio on 'for company' and so avoid silence and solitariness, join a group, live vicariously - (the television world of soap opera, situation comedies, and twenty-four hour news is always there at the push of a button).

Such is the nature of the post-Einsteinian world. A constant round of looking outwards. Science adventures into the far reaches of space; commerce becomes global at the expense of local industries; national politics is increasingly influenced by international considerations at the expense of regional needs; while a high degree of mobility, the urge to shop and the need to be entertained... all help to shut out the inner life of the mind. Far too many of us live in the belief that 'reality' lies beyond ourselves, and accordingly chase the experience of 'the Now' and 'the New'.

Foremost among the early inventions that changed the ways we think about space and time, and permitted us to gain more freedom of movement, cram more activity into a given period then ever before, are the steam engine, the internal combustion engine, and the jet engine. While in the present Electronic Age it has to be the computor: an instrument capable of flying aeroplanes or simply paying one's bills. It is incredibly efficient in organizing myriads of facts to reveal the sequential relationships between disparate events - knowledge of which can lead to the solution of hitherto insurmountable problems in mathematics, physics, astrophysics, medicine... And memory - to a layman like myself - is nothing short of miraculous; a staggering amount of information is stored therein - data banks, so-called, from which any one fact can be readily retrieved: a veritable electronic Encyclopedia Britannica. Yet there is a downside to the tremendous advantages bestowed by this marvellous invention. Through its widespread use, particularly in primary and secondary education, it has effectively turned the whole process of learning into a fact-gathering exercise. A generation at least of younger people in the world's developed nations have access to that vast pool of information known as the World Wide Web. All the facts they could possibly need to know are to hand at their fingertips. This process of collecting data is both facile and fascinating and computors have become indispensable educational tools. However, if used unintelligently, indiscriminatingly, they thwart rather than enhance the faculty of comprehension - of understanding fully the implications and essential meaning of a particular piece of factual information. With such a vast range of definitive statements so readily available it has become all too easy to regard the data, as presented, to be sufficient in and of itself: the be-all and end-all in the process of learning and research. Granted, computor skills are required to search for the information, but having found the right definitions and explications it is surely not enough to simply quote or paraphrase them without digesting the facts as given and putting one's own 'slant' on them; revealing their appropriateness to the topic under discussion. In my experience as a university professor I found even graduate students had difficulty in personally interpreting such information in order to use it in support of their basic argument.

In the first chapter of this book I noted how Paul Valéry, the French poet and philosopher, would frequently give the same advice to the many young, would-be poets who sent him samples of their work hoping for a critique - saying, in effect, that they were quite right in believing that it was absolutely necessary for the serious writer to indulge in a reverie... 'but,' (he went on to exhort them) 'please, please, make sure that it is aided by a little accurate information'. Yet in this day and age he might well find it necessary to say exactly the opposite: point out how important it is to get the facts pertaining to any situation absolutely right, but then emphasize how *vital* it is that they be subjected to the tide of personal thoughts and feelings that rise to appraise their worth and revelatory significance. For, as we have seen, this is the way consciousness follows up *after* the fact - inducing a contemplative state, a form of reverie through which one comes to appreciate the intellectual and emotional wealth certain facts or happenings bring with them; evoking, on occasion, a totally absorbing sense of wonder.

Wonder: a highly-charged state of awareness involving the elements of surprise, curiosity, respect, admiration, astonishment... Yet it would seem that nowadays wonder is out, and *data* alone is in.

But is there not something of a paradox here? For the computor is a formidable instrument in terms of the range and import of the information it presents, and one would think it impossible *not* to be brought to reflect on the wonders and mysteries of the cosmos, and of the incredible complexity of life-forms on this planet - not the least being the marvel of the human mind that invented such a machine in the first place. Yet, unfortunately, having so many facts readily to hand encourages what one might call the 'compulsive collector syndrome' - a form of obsessive computor use to satisfy a magpie-like need to accumulate reams of facts - in the face of which the subjective processes of contemplation - Valéry's 'reverie' - will have little chance to 'kick in'. But it is also the case that even discounting the attitude and activity of the 'magpies', such easy access alone to the world-straddling the 'information highway' known as the Web, will draw in many millions who come to believe that truth and 'reality' are only to be found in the 'hard facts'. There seems to be little doubt that the omnipresent use of the computor in the world's developed nations is fostering the general attitude that by knowing how to access, and gather volumes of intelligence, one can consider oneself to be knowledgeable and wise.

However, it is another matter if one is driven by a strong personal interest in a particular subject, a compelling curiosity pushing one to research the facts concerning *how* it came into being, and *how* it works... and is then led on to experience the kind of wonder that evokes the ultimate question, *'why?'*

'Wonder -- which is the seed of knowledge...' wrote the Elizabethan philosopher Francis Bacon, anticipating Einstein's remarks by some four-hundred years. And if I were to have the last word and expand on the great physicist's statement, I would do so by saying: *'Wonder at cosmic radiation, the structure of a starfish, the delicate beauty of a flower, the devotion of a dog... induces a catholicity of outlook, quietness of mood, an overall gentleness of temperament and, perhaps most importantly, brings one to feel a deferential esteem for things of the world and the forces that maintain them in a state of being; and ultimately to respect, and wonder at, the blend of physiological, psychological, and spiritual energies that determines one's own uniqueness - the mysterious nature of one's own character.*

If the time should ever come - as the Electronic Age progresses and we become more and more its victims (automatons in our own right), bereft of wonder and pursuing a totally extroverted, escapist life .... then we may reach the stage when the word *human* will no longer qualify the word *being*. In which case Darwin's statement that *the moral sense of conscience is the most important and noble of all human attributes, and that to be without it renders us more akin to the 'lower animals,'* will assume the nature of a prophetic truth. There is evidence enough in this, the second year of the new millennium, that points to the total decline of self-respect, and a Caligulalike indulgence in degrading and unwholesome behavior by some of the younger generations in Western society. Writing in a recent issue of the London *Sunday Times*, India Knight - in a commentary entitled *Promiscuity blues* - recounts her reactions to British ITV's new series *Club Reps*, which records the goings-on of staff and 'customers' at the resort of Faliraki on the island of Rhodes. I quote liberally from her report, for I could not equal the graphicness and impact of her words by attempting to paraphrase her comments:

> .... I don't have a problem with the idea of chronic promiscuity... I don't have a problem with nudity or with drunkenness. I'd pass on the offer of getting my tits out for the lads at a disco in Rhodes but hey, it takes all sorts. I don't find drinking so much that you leave pools of vomit in your wake entertaining or sexy, but nor do I think that those that do are necessarily morally deficient.
>
> Why, then, am I so appalled by Club Reps...? Its partly to do with the combination of tits, arses, sick and sex all at the same time, and partly because it confirms my suspicions that there is a generation of young adults, 18 to 30 - who appear to live their lives in a complete moral and intellectual vacuum.
>
> They are quite unbelievably stupid. All of them, men and women, have those thick, inbred faces that are usually accessorised with cheap fags and a Rottweiler. They are also unbelievably hedonistic, in the coarsest sense of the word: not intelligent enough to be decadent or knowing enough to twig that there are more entertaining ways of being hedonistic than flashing your tits or sicking up on your feet...

The author continues... recounting the gross antics of 'beer-bellied males and the serial spreading of legs in mechanical and joyless couplings', noting that all the 'performers' were, 'apparently grimly unaware of such old-fashioned concepts as self-respect.' Towards the end of her account of the depressing spectacle, she writes: 'What is the matter with these people? Who are their parents, teachers, employers? How have they reached adulthood seemingly as mindless as a cluster of amoebae? And when you add conduct such as this - (which, as India Knight points out, might be slightly redeemed if it possessed any vestiges of style or a genuine, caring intimacy) - to the wholesale use of drugs, teenage killing sprees and street muggings... you might well have doubts about the moral progression of, say, the last two generations in some Western countries, and remember Darwin's thoughts about the importance of *'a moral sense of conscience.'* To which I would add, *'... and the vital importance of the capacity to wonder.'*

The English journalist and author, G. K. Chesterton (1874-1936), wrote the following prophetic sentence in his book *Tremendous Trifles*: 'The world will never starve for want of wonders; but only for want of wonder'.

And here is a passage from Graham Swift's recent novel *Waterland,* in which a history teacher who is being pushed into retirement is delivering a wonderfully humorous, intriguing and provocative defence of history as a school subject. He might well have written it - just for me - to end this Chapter:

> *Children, be curious. Nothing is worse (I know it) than when curiousity stops. Nothing is more repressive than the repression of curiousity. Curiousity begets love. It weds us to the world. It's part of our perverse, madcap love for this impossible planet we inhabit. People die when curiousity goes. People have to find out, people have to know. How can there be any true revolution till we know what we're made of?**

---

* Graham Swift, *Waterland* (London: William Heinemann Ltd., 1983) p.178.

13

WHAT PRICE A 'BRAVE NEW WORLD'?

We are mad, not only individually, but nationally. We check manslaughter and isolated murders; but what of war and the much vaunted crime of slaughtering whole peoples?

So wrote Lucius Annaeus Seneca (4 B.C.?-A.D. 65), Roman statesman and philosopher in his narrative *Ad Lucilium*. Two thousand years have passed since Seneca made this statement and here we are, living in a modern world so much geographically larger and diverse than he could have envisaged; benefiting from medical, scientific, economic, and cultural amenities that make life immeasurably easier, extend our mental horizons and enable us to live longer... Yet the indictment he makes above pertaining to the attitude and conduct of his own civilization can just as easily be leveled at our own: we are just as 'mad', individually and nationally, and are still waging wars and 'slaughtering whole peoples'.

Perhaps the best-known book of the English novelist, essayist and critic Aldous Huxley, who died in 1963, is entitled *Brave New World*. In it he expresses his worries about the dangers of scientific progress in general, and particularly - after Hiroshima and Nagasaki - of the perils facing a nuclear, atomic world. The title of the book is both ironic and commiserating, implying that science has done well in meeting its challenges head on, but has, nevertheless, done so in a cavalier fashion – and is now left 'holding the baby', so to speak. That in unlocking the secrets of matter and energy, and of microbiological life, science has opened a veritable Pandora's box, loosing evils upon the world in the form of atomic or germ warfare; not to mention a dehumanising form of social engineering. The Greek Gods would say, if asked, that we have been too presumptuous and self-assertive in going so far, and that such *hubris* will ensure our ultimate downfall. But who takes any notice of the Greeks or their ancient Gods nowadays. The question we now face is whether the world's nations will find the collective will and sanity to face the moral issue - join together despite differences in religious faiths, political systems, economic circumstances - and destroy all stockpiles of nuclear and biological weapons, and pledge that all such advanced science and technology will never be used for military purposes.

Good luck! Too much to expect? Of course. Is there any evidence to suggest that we have attained the wisdom and moral integrity allowing us to claim that we have finally overcome our historic 'madness'? The following paragraph from Huxley's *Views of Holland* suggests that what we should ultimately, and essentially, have come to learn from all our scientific and technological progress is the underlying mystery and impenetrability of our circumambient universe. If such truly were the case, then he might agree that we are some way along the road to sanity:

> *We have learnt that nothing is simple and rational except what we ourselves have invented; that God thinks in terms neither of Euclid nor Riemann; that science has explained "nothing"; that the more we know the more fantastic the world becomes and the profounder the surrounding darkness.*

Although he wrote this fifty years or so ago, it has surely lost none of its relevance. Yet there are some inspiring human achievements in the arts he does not mention such as, for example, the great advances made in the creation and performance of music over the last five-hundred years: the gift of the great composers, conductors and instrumental performers in bringing the light of the human spirit to bear on the *'surrounding darkness'*. Unfortunately, Huxley died before the science of astrophysics and the curiosity and courage of astronauts put a man on the moon; and before medical science had advanced to the incredibly sophisticated levels from which we now benefit – heartening forays into the previously unknown.

But a brave new world, free of the individual and national madness of which Seneca speaks?

Hardly.

We have just lived through a century of unprecedented slaughter. Millions of human lives lost in the jingoistic madness that was played out in the terrible trench warfare of World War I. Twenty million died in Russia as a result of Lenin's purges and Stalin's peasant relocation orders. Six million Jews went to the gas chambers in Hitler's Germany. Military and civilian casualties in the Second World War are again numbered in millions – thousands, for example, dying in Japanese prisoner-of-war camps.

And then came Korea, Vietnam, the 'killing fields' of Cambodia, the tribal massacres in Rwanda, the gassing of the Kurds by Saddam Hussein, the internecine warfare in the Balkans, ongoing violence between Jews and Palestinians... to be followed in the new century by wars in Afghanistan and Iraq, and merciless killings worldwide carried out in the name of Muslim fundamentalism (to which adherents continue to flock), and by violent political revolutionaries across the globe.

And as the century has worn on, so has the destructive power of weapons systems led to more carnage in numerical terms than ever in human history. Ordnance of all kinds, developed primarily by the advanced and industrialized nations (primarily the United States, France, Britain, Germany and Russia) is sold without too much scruple to other countries, yielding high profits to the supplier which add considerably to gross national product totals.

It has been a century of violence which has left its mark on late-coming generations. So much so that children attacking teachers (knives in school), youngsters assaulting the elderly, mugging on the streets, the mindless anger that leads to physical aggression (road rage, for example), the possession of handguns and assault weapons kept in home or car and facilitating random attacks in school or workplace, or drive-by shootings on the highway... have all become subconsciously accepted as a normal way of life. Violence in one form or another would seem to be the bread and butter of the entertainment industry (especially in films made for television), and it certainly grabs the attention of the newscasters; while the video-games industry produces some of the most sordid and sadistic 'entertainments' imaginable – games avidly sought after by youngsters in their teens. It is 'cool' to be tough and unfeeling.

The age-old wisdom that youngsters learn from the examples set by adults has gone by the board. The argument that a child – say between thirteen and sixteen years of age – is not affected by what he or she sees being played out in the context of adult lives is completely specious. *'Example is the school of mankind, and they will learn at no other'*, wrote Edmund Burke (1729-1797), the British statesman and orator in *On a Regicide Peace*. And as a schoolmaster turned university professor I have found it to be generally true that when students find themselves intrigued by a teacher's manner and personality, by the passion of his or her involvement, they learn readily and perform more effectively: because it is in the nature of youth to be influenced by the behavior and special kind of appeal presented by those who are older – even if only by a few years. It is a natural form of emulation that works throughout the animal kingdom. But in the case of humans, the individual's own fledgling character will help determine whether to follow the example presented by a particularly intriguing personality – whether or not it seems 'right', either practically or morally.

Adolf Hitler said that his brave new Fascist World would last a thousand years. It perished after fifteen. He practiced a ruthless philosophy based on his declaration that, *'In starting and waging a war, it is not right that matters but victory.'* Whereas some two-hundred and fifty years earlier that wise humanist and statesman, Benjamin Franklin – whose words carried considerable weight in Europe and America alike – echoed Cicero by writing: *'There never was a good war or a bad peace.'* There can be little doubt as to which of the two statements best serves the spirit of such a concept as that of a brave new world.

Throughout history, and in every part of the world, societies have instituted a variety of laws which possess moral connotations – rules and standards of behavior arising from culturally inspired religious or ethical beliefs; whereas in the case of Moses and the Ten Commandments the moral precepts were seen as being of divine origin, revealed to the prophet on Mount Sinai. But whether institutionally established, or divinely revealed, the penalties for infringing such laws vary enormously. Women adulterers in Islam, for example, could be stoned to death and even today risk being beheaded. Whereas nowadays the Christian church tends to turn a blind eye on adultery and social ostracization barely exists. The damage to life wrought by drug traffickers is considered an evil to be punished by death in Singapore. Not so in the West. Yet in early-nineteenth century Britain to steal a sheep could result in banishment or even death – whereas today in some Muslim countries, thieves may lose a hand. And in the Marquesas Islands a hundred years ago it was morally indefensible for a woman to set foot in a canoe, never mind take to the water in one; the offence was punishable by death.

Conceptions of right and wrong - of what is permissible and what is 'taboo' - have been evident throughout our history. (Taboo – a word taken from the Polynesian *tapu* – connotes the imposition of a sacred prohibition:

a restriction on any form of involvement with people, things, or events considered to be associated with the mysterious powers of spirit.)

Of all the moral admonitions, *'Thou shalt not kill'* is one found in cultures worldwide. Yet it is hedged around by all kinds of exceptions. One may kill in self-defense, in the service of one's country, or for ritualistic purposes. The Aztec priests regularly slaughtered their kinsmen to propitiate their Gods. And we should remember that it is less than a thousand years since the Christians set out to free Jerusalem from the Muslim Ottomans, enthusiastically killing as many as they could in the process – 'saving their souls,' was one of the justifications; neither were they averse to raping and pillaging in Byzantine (Eastern Christian) lands as they went along. The Thirty Years War between Catholic and Protestant Europe in the seventeenth century produced horrible atrocities on both sides – bearing out Cicero's statement written almost seventeen hundred years earlier, *'Laws are silent in times of war.'* He might well have added, *'and this is particularly true of moral law.'* This grim silencing of conscience through the 'creative' redefining of moral principles - whether initiated by those who seek conquest or personal power, or religious zealots out to convert 'the unbeliever' – is responsible for many of the miseries suffered by ordinary men and women who simply want to live as trouble-free, and as personally satisfying a way of life, as possible. Juvenal (c. A.D. 60 – c. 140) the Roman satirical poet, was not very sanguine about the state of his world when he wrote in his *Satires*, *'Earth now maintains none but evil men and cowards.'* I wonder what this very perceptive Roman would think of the state of contemporary Western civilization if he were to travel forward through time. He would undoubtedly by incredibly impressed with our science and the material benefits it has conferred on so many in the modern world – together with the technological advances that have changed the way we think and the manner in which we live. But what would he think about our progress since Roman times in terms of our political altruism, philosophical sophistication, respect for the sanctity of human life, stewardship of the planet and all its creatures.... all those qualities that would justify talk about a brave new world? Is it likely that he would see us as predominantly charitably disposed individuals; as nations living intelligently together on this crowded star... or simply as a more up-to-date collection of *'evil men and cowards'*? He would note the orbiting satellites circling their way around the globe, making naught of time and space, and providing almost instant intelligence about what is going on, and where...and wonder if we used such ability to oversee the world wisely and courageously for the general wellbeing of nations.

For nowadays no developed nation can fail to be aware of the horrors of war or genocide; of the scale of human suffering in times of disease, famine, epidemics, earthquakes, tornadoes, hurricanes, typhoons and floods... Certainly, the fact that so many international relief agencies are on the scene immediately may indicate that a more compassionate spirit is moving us in the direction of a Brave New World. Yet on surveying this often unhappy globe such thoughts as to whether ultimately the power for good invested in the human psyche will eventually overcome the vicious, greedy and destructive side of our nature, sometimes seems altogether too idealistic and vague a hope.

Nevertheless, such utopian ideas have been around for some time. In Shakespeare's shortest, perhaps most optimistic play, *The Tempest*, Prospero – the rightful Duke of Milan, whose powerful wizardry has conjured the storm responsible for wrecking the ship carrying those who usurped him – has created magical situations which have changed for the better the hearts of the men who brought about his downfall as they wandered aimlessly over the apparently uninhabited island: home only to the exiled Prospero, his daughter Miranda, and the deformed subhuman slave Caliban; yet also beset by the spirits – notably the engaging Ariel – working for Prospero in creating the surreal nature of events that result in the transformation of his erstwhile enemies. And it is Prospero's design that Miranda should meet and fall in love with one of the castaways – Ferdinand, son of the King of Naples – thereby contributing to the romantic ambience of the place and the experiences that have chastened and changed them.

In this magical atmosphere, with her father's fortunes restored and seeing those who originally deposed him being truly penitent and taking on a nobility and graciousness of mien and manner, it is easy to understand the spontaneity and genuineness of Miranda's utopian outburst:

> *O wonder!*
> *How many goodly creatures are there here!*
> *How beautiful mankind is!*
> *O brave new world*
> *That has such people in't.*

Yet this is a small island and given the number of assorted 'mariners' who accompany the shipwrecked courtiers, the number of *'goodly creatures'* is no more than twenty. A veritable microcosm of a *'brave new world'*; yet seemingly an ideal and potentially perfect human situation having all the makings of an earthly paradise.

Nevertheless, one wonders how long it will last. Once everyone is off the island and back in the 'real' world will it not become the stuff of dreams – a visionary afterimage of what life could be like in such a beautiful place: no more or less 'real' than the imaginary island of *Utopia* as envisioned by Sir Thomas More in his book of the same name written in 1516? For one is not being unduly cynical in recognizing the vagaries and unpredictabilities of human nature – especially when it comes to assessing the chances of achieving a permanent and perfect rapprochement between twenty individuals driven simply by a newly acquired spirit of good will. As the saying goes, 'Two is company, three's a crowd.' And for a 'crowd' – constituting three or more – to experience the kind of harmonious relationship that plays a key role in any utopian situation, require that a general philosophy of life be held in common. Members should believe in a mutual sharing of experience…. in order to gain a felicitous insight into their own individual natures - and in so doing benefit from the general uplifting of spirit and psychical harmony that, theoretically, a utopian society could engender. But Shakespeare's characters were not truly motivated by such grand ideas – being moved to take on a companiable and beneficent outlook solely through the power of Prospero's magic – and it seems likely that once away from the enchantment of the island they would quickly revert to their own personal ways of life.

The utopian dream is a fragile concept at best. If such societies are to work effectively, individuals must be prepared to modify their personal views in order to accord with the guiding principles that are seen to constitute the greater good: principles sufficiently inspiring to warrant the tempering of individual attitudes and opinions. And although it might appear that this sort of close knit society could easily breed the seeds of fanaticism, such a development would be totally antithetical to the whole utopian ethos. For fanaticism drives the urge to proselytize, first through intellectual persuasion and ultimately by the physical coercion that depends on violence. And any resort to violence deals the death blow to the utopian ideal. Violence, gratuitous or planned, remains the persistent obstacle to civilization's advance to any utopian-like goal. Utopia may appear to represent an utterly dull existence. But none of the authors who have produced their own versions of such a place have suggested that it is a land where all the questions pertaining to existence are solved, and all personal hardship mitigated – leading to a placid, tensionless, and vacuous drifting through time. Boredom exemplified. Instead, life could be wonderfully enriched. For with personal animosities, social, political, and ideological tensions eased…. the way is open for the soul to bring its revelatory insights to consciousness. Aldous Huxley makes it quite clear in *Brave New World* that the reasoned sanity of the common cause – living wisely, peacefully, and in harmony with one's fellow man and with nature…. wary of undue self-interest - would bring a richness and fulfillment to human existence that would go a long way to easing spiritual uncertainty and satisfy the quest for meaning.

Are there any among us who at one time or another have not harbored thoughts about 'getting away from it all' – suddenly feeling the underlying urge to discover the kind of peace that nourishes body and soul: a place where the environment brings a sense of oneness with the whole of the natural world, and where the community allows one to be at peace with oneself; a psychical place where one finds what is biblically presented as, "…*the peace of God that passeth all understanding*'. Yet another way of describing Utopia. The literary record of such longings speaks of small islands, mountains, monasteries… Yet the word 'utopia' is taken from the Greek *ou*, 'not' and *topos*, 'a place': signifying 'no place' – nowhere in the real world, only in the imagination, or perhaps one should say only in spirit. *Brave new world…Utopia…*an escape from discord to harmony; from violence to gentleness; from the prosaic to the poetic; from time to no time at all.

> *Listen: there's a hell of a good universe next door: let's go.*
> e.e. Cummings: *Times One*

Utopia. The Great Escape – almost as if from a prison camp. We are well aware of the lengths to which the imagination can go, but is there not something perverse about a physical brain generating neural activity that would have us dreaming of somewhere other than where we are - rejecting the known world in favor of an imaginative one? Yet this is the sort of paradoxical mental activity undertaken by the brain's left and right hemispheres.

A hundred years ago if some average 'man in the street' – whether living in the West or the East – had been able to foresee the nature of modern life, he might well have thought of it as 'utopian'. It would seem to be a particularly human mental feat to shift attention from what actually *is*, to imagine what might be *better*. "What if…" we say to ourselves….

The idea of Utopia – the perfect place – can be seen as part of a generally held and intuitive apprehension that a place of near-absolute excellence, of harmony, light, and truth – the lion lying down with the lamb 'peaceable kingdom' kind of thing – is 'out there' somewhere. Such thoughts go beyond a conception of Utopia as simply a place offering earthly pleasures and satisfactions. Instead, they verge on the transcendent because we have already experienced the state of 'other worldliness' that accompanies perfect love; a perfect afternoon, sunset, or symphony…. or encountering absolute beauty, be it of spirit or form. All events touching on the perfect state of things when the blinkers are removed. It would seem that we are *unconsciously* aware that a perfect state of being does exist: that in our heart of hearts – (otherwise the soul) – we know this. Consequently, when we are deeply moved, we experience an uprush of the unconscious…. and we dream of things Utopian.

> *I thought I could not breathe in that fine air,*
> *That pure severity of perfect light.*
> Tennyson: *Guinevere*

Ethereal splendor. The structure and dynamics of the cosmos, the very expanse of limitless space, defined and shaped by planets, stars, galaxies… home to light and dark energies…have always engaged the imagination to speculate on forms of life beyond our earth – even inducing thoughts of a Utopia called heaven, - 'a place' representing a level of peace and fulfillment, for which the pleasures and travails of mortal life are seen to be but a preparation. The very concept of heaven represents the perfect solution to the kind of psychical restlessness that has us, individually, never quite content with the present… wondering if some ultimate truth of being is not to be found on the other side of that time-space galactic fence – a place just waiting for our arrival.

When the Welsh poet Dylan Thomas talked about *'The wire-dangled human race'* he was referring to the intellectual and emotional tensions induced by this kind of mental dueling between what *is* and what *might be*. When I last talked to him – over sixty years ago now – he described the double life this caused him to live. As a poet he was positive about life when moved by the fantasies of nature and the love that pulls human beings together. Yet he was also driven to a near-bleak and ironic despair by the play of events that could make the struggle to survive a grim, sometimes devastating struggle. The irreconcilable nature of these opposing experiences led him to celebrate life on the one hand, and negate it on the other. It was either a fascinating and heroic journey - or a nihilistic process whereby Utopian-like ideals were mere disparate follies to divert the mind. Poetry was for him the precarious bridge he was driven to build between the two. And when the bridge was not passable he would drink the night (and ultimately his life) away.

I have talked previously of how one of the greatest scientific minds in Western history was preoccupied with the idea of flight: of how Leonardo da Vinci's dream of flying led him to be the first to lay down the principles of an aeronautical science – of how to build and operate a machine that would remain airborne, gliding with the aid of wind currents and thermals. The apocryphal story is that he did indeed construct a glider that flew for a short distance –leaving Leonardo pretty bruised and banged up after a crash(?) landing. But whether or not was such the case, he certainly devised the means by which man would be able to defy gravity and fly – all he really needed was an engine of some sort to provide lift and momentum. Then he would have been the first man to be *'up and away...'* (Discounting Icarus, of course, whose father, Daedalus' approach to flying was not scientifically sound: the wax fastening the wings he had made for his son melted as Icarus got closer to the sun).

'Oh that I had wings like a dove! for then I would fly away, and be at rest' is the plea in the Bible, *Psalms* 55:6. *'O for a horse with wings!'* writes Shakespeare in *Cymbeline*. This constant cry, this image of flying away to escape time and place, is found over and over again in myth and literature and, as such, suggests a utopian-like quest. Leonardo's urge to fly and conquer the

element of air was no doubt driven by an intense curiosity concerning the physical forces that control an earthbound existence: one more problem for his hungry intuition and intellect to solve. Yet who knows whether or not this nagging need to experience the ultimate freedom by taking to the air, was being subliminally nudged along by the dream to break through the earth-binding barrier of gravity to attain more limitless levels of his being? For in the infinite reaches of sky and space, 'nowhere' could become the 'somewhere' for which his imagination and spirit reached.

Certainly, the sense of psychic release that comes with being airborne, freed from the mundane limitations of an earthbound existence, is an experience described by many pilots in the early days of flying when they felt to be actually 'handling' the element of air, and physically 'fighting' the forces of gravity. To traverse the heavens in a prop-driven aeroplane was to reach a level of being not readily attainable when both feet were firmly on the ground. Ultimate solitude. Ultimate self-realization in a dangerous element (or so the body's apprehensive nerve cells indicated was the case.) Yet strangely enough, the very awareness of such mortal frailty can permit the mind to take a very cavalier view of the body – simply see it as a vehicle for material existence – as if a higher form of one's being had finally come into its own, discovered its only element. Nowhere is this strange and mystical possibility expressed so convincingly and poetically as in Antoine de Saint-Exupery's *Night Flight* (1931) and *Flight to Arras* (1942).

These are flights in which sensory objectivity and imaginative wanderings are not mutually opposed, but enjoy a common sense of reality. The pilot flying alone, physically feeling the 'air-shift' of rudder and elevator movement, scanning the gauges, artificial horizon and compass, and manually trimming the 'set' of the control surfaces...nevertheless finds himself in a mental space where – although these practical functions are being performed efficiently – the neurons find themselves free to indulge in thoughts of a generally philosophic nature as to what is significant about one's own little life: what is *meant* to be significant about it. To what distant horizon – above and beyond worldly concerns – is it traveling? As if consciousness had suddenly gone through the 'time, space, and object' barrier... flowered, come to the full, and reached a mental stratosphere where the abstract themes and images created by the great thinkers, poets, musicians, mathematicians, physicists...could no longer be considered merely imaginative fictions but tangible truths floating in and out of the dark cockpit. Meanwhile, the weather is unpredictable (ice on the wing's leading edge?). No satellite navigational help is available. Radio contact is uncertain. Wireless telephone range extremely limited.

No wonder the motto on the Royal Air Force crest is *Per Ardua ad Astra* – through hardship to the stars. Which is another way of saying that in taking to the element of air in the old days, the mind expands in musing on life in general and one's own destiny in particular. In *Wind, Sand and Stars* (1939), that pioneer of aviation, Antoine de Saint-Exupery wrote: '*I know of but one freedom and that is the freedom of the mind in flight.*' (Yet it must be admitted that when flying a heavy bomber at night over enemy territory, in company with six other crew members, this idealization of flight – the transcending sense of freedom and exultation I have just described – is not inclined to show itself. In such circumstances the soul, and its messenger the spirit, sit it out. Quietly. Almost resignedly. Yet ready. In reserve, you might say).

The Garden of Eden – before the 'Fall' of Adam and Eve – stands as the first testament to the Utopian idea. And the best known of those that follow and still exist, are Plato's *Republic*; Sir Thomas More's *Utopia*; Francis Bacon's *New Atlantis*; Rabelais' *Gargantua and Pantagruel* (in which the kingdom of Utopia plays a large part); Samuel Butler's *Erewhon* (supposed to spell '*Nowhere*' backwards); several by H. G. Wells including *A Modern Utopia*; Aldous Huxley's *Brave New World*; and James Hilton's *Lost Horizon* which tells of the mythical land of Shangri-La where there is no ageing... just a timeless serenity.

> *It is the darling delusion of mankind that the world is progressive in religion, toleration, freedom, as it is progressive in machinery.*
> Moncure D. Conway, *Dogma and Science*

Conway, an American clergyman and author who wrote *Dogma and Science* in the last years of the nineteenth century, was concerned about the cultural gap he saw opening up between the impressive advances made in science and technology, and the lack of a similar movement in the social and philosophical spheres he names as 'religion, toleration, freedom'. He was wanting to see an opening of the mind in directions other than those pursued by the hard sciences in terms of the material world - into the mental processes that determine how we think and feel as *individuals.* He was referring to the expansion of thought and feeling that can release us to live imaginatively - survey the course of our lives to uncover whatever meaning or purpose can be ascertained. Over two-thousand years ago the Roman philosopher Lucretius, writing in *De Rerum Natura,* expressed the Utopian view that the goal for mankind should be to realize *'The highest good at which all aim.'*

Now here *we* are, in the first decade of the 21st Century, living through a time of phenomenal scientific and technological progress. And yet – as we have listed in some detail – the catalog of crimes against humanity carried out within living memory provides a depressing record of our overall progress in attaining *'The highest good...'*

Significant scientific progress in Western countries has been made throughout the 20th century, yet it was immediately after World War II - in the second half of the century - that it went racing ahead to make amazing breakthroughs in nuclear physics, medicine, biology, engineering, and astrophysical adventuring .... spawning technologies that have changed the way we think of our cosmic environment and the way we live. One might say that the 'science-gene' - (if one may allot specific mental capabilities solely to genetic influences) - went on a rampage, accomplishing more in a few years than had been achieved in the previous several thousand. Which leaves one to wonder why the 'wisdom gene' - the Utopian fellow traveler - has not been as effective in bringing about a similar revolution: bringing us to live together with a compassionate and humane sense of purpose.... perhaps even influenced by a spiritual one. For if we were *wise,* we would see that at rock bottom - stripped of the conditioning imposed by geographical and cultural circumstances - we share the basic attributes of a Utopian humanness. We experience the emotions of love, and pleasure in what we find to be beautiful; we wish and strive for happiness; we seek faith in some truth that transcends the biological mechanics of being alive; and we hope to die to some purpose. Tears and laughter we have in common.

Meanwhile, the electronic 'worldwide web' is in place, making possible the almost instant relaying of information concerning significant happenings across the globe: a communications-system knitting the world community together, enabling nations to intervene and quickly dispatch help to areas suffering from natural disasters, or to intervene where populations were being terrorised. And when it comes to commercial issues affecting the growing global economy, computors talk to other computors thousands of miles away.

Are we witnessing here the first tangible signs of a brave new world - courtesy the science of electronics?

There can be little doubt that the computer revolution represents a major step forward in our overall search for answers concerning both the phenomenon of the cosmos, and of ourselves on this little planet. And I would say that the most significant advance in creating a better world has been made in the realm of medical science where, with the aid of computors, seemingly intractable biological problems have yielded to research, and surgical techniques have become incredibly more sophisticated. Even so, there are still parts of the world where men, women - and children in particular - are ravaged by pestilence and malnutrition; where governments neither encourage nor facilitate the arrival of aid. The human factor, alas, is less certain than the electronic one.

The development of electronic gadgetry in general has radically affected the way we live. It imparts an existential immediacy to our existence: things can be made to happen *now* - at the touch of a button - thereby adding a new dimension to the American dream of the good life: the reality of *instant gratification*. The constantly expanding range of sophisticated communication devices allows a person the opportunity to get in touch with just about anyone else - even on the whim of the moment while walking down the street. As a result more and more people - and the younger generations particularly - live a more extroverted life than was previously possible, participating in other people's lives and immersed in things the world has to offer. Solitude is unwelcome and to be avoided. Wherever one happens to be, and at all times of the day and night, the cell-phone is always to hand; the television always available to entertain. Music or messages assault the ear while waiting for a response on the telephone, waiting in the doctor's office, eating in café or restaurant, shopping in stores, captive in airports... But to be always on the go and subject to the dulling effect of constant background noise, tends to take the fine edge off consciousness - and especially when it comes to experiencing both moments of hindsight and foresight: bring the past into the present and see beyond one's immediate self-interests to assess where we are going in terms of any human purpose and destiny.

In his *Preface to the Lyrical Ballads*, the great English poet, William Wordsworth, talks about *'... emotions recollected in tranquility'*: about the need to review, in quiet and solitary moments, the past and recent happenings in his life - a meditative practice whereby the especially momentous encounters with nature, and with the lives of others, come clearly to mind, charged with insights reflecting the deep empathic nature of his own being. It was the way he came to know himself - his sensitivity to the unpredictable and often tragic nature of human life, and to experience the natural sense of piety that possessed his consciousness when faced with the beauty and drama of nature's moods.

In today's computer-driven society such a subjective disposition of mind is generally regarded as 'old-fashioned': there is less inclination to reflect on the depths of one's feelings - on the discovery of new values they reflect. It is an attitude seen by many as an unnecessary hangover from a pre-technological age. To mull over is to look backwards. Whereas to gather data, 'surf' the worldwide web and thus become a 'player' - even if vicariously - on the world stage, is not only to be actively progressing with the times, but is also seen as a relatively exotic and interesting way to live. There is obviously some truth to this, but when the computor becomes the 'be-all' and 'end-all' of existence - a mechanical extension of oneself that becomes a surrogate for consciousness proper - then one's own thought processes, and the acuity of one's own sensory perceptions, are vitiated. Personal idiosyncrasies of mind, and the breadth of its interests and insights are effectively neutralized - and at some point down the road life becomes more robotic, less individually governed.

A brave new world?

In the first chapter of this book I mentioned that Paul Valéry, the French poet and philosopher who died in 1945, would advise the young writers of his day that they must indeed 'indulge in a reverie' if they were ever to experience the literary and poetic vision that marks the good author... but then he would add the rider that such insight must be aided by 'a little accurate information'. Now, in these first years of the 21st century, the situation is reversed: we lay all the emphasis on the gathering of information and Valéry, were he still around, would likely be heard shouting from the roof tops, exhorting would-be writers to remember that in an age when non-stop information leaves little time for reflection, it is absolutely crucial to 'indulge in a reverie' - filter everything through the discriminating sensibilities of the mind in order to separate the significant facts from the trivial; and reveal whatever intrinsic truths concerning human progress - or lack of it - are to there to be gleaned.

Computers do not indulge in reveries. They do not laugh, cry, or render philosophical judgements. They are marvellous calculating machines capable of instant logistical progression in performing their computings, but they are not *conscious* in the human sense. What can they know of love, for example? And yet computer matching - the comparing of psychical characteristics in order to predict a loving compatibility - is fraught with uncertainties and difficult to defend. Personality traits are so differentiated between individuals that they are difficult, if not impossible, to standardize for the purpose of making viable forecasts of human compatibilities. Their mechanical electronic circuitry cannot be compared to the incredibly complex, *physiological* neural pathways by which the human brain-mind responds over time to the constant presence of another person. But when presented with a problem involving all the intricacies of scientific variables a physicist or biologist may have to resolve, or asked to provide vital information pertaining to just about any thing under the sun, computers work to standards of objectivity, and at speeds of resolution, unattainable in the pre-computer age of research.

Valéry's injunction on the need to indulge in a reverie - a plea which can apply to all of us and not to just writers or poets - is *not* therefore completely invalidated by the advent of the computer and should not be seen as an old-fashioned appeal from an earlier and more romantic era. Had the personal computer been invented in Valéry's time, I can imagine him exhorting one to, 'drop these electrical nuggets of intelligence into the deep pool of contemplation, and wait for the ripples of insight to break on the internal shores of consciousness.'

These are not profound observations: but they illustrate the shortcomings of habitual and indiscriminate computer-surfing - an activity which would seem to be driven solely by a voyeuristic, magpie-like curiosity to see what is 'out there' and, having seen, pass on.... while little, if any, personal evaluation as to how significant the information may, or may not be, is taking place. But if it becomes a way of life - of simply passing time in the serialistic pursuit and casual acknowledgement of random bits of information - the avid practitioners of such 'data-skimming' are in danger of impairing their own natural rational, and intuitive analytical faculties. One already notices how difficult it can be to engage some relative 'youngsters' in any conversation that goes beyond discussing the 'facts' pertaining to this or that occurrence. I frequently hear teachers and employers commenting on the short 'attention span' that characterizes students and younger employees.

Education bears a great responsibility in this regard, for computers are used as the primary sources of information in both high school and university. But just how much the educators concerned ensure that students understand that the computer is but a means to an end in researching the necessary facts - and that then they use their *own* brain to marshal the factual evidence, shaping and ordering it to arrive at a personal level of comprehension, interpretation and evaluation - is open to question. Also, one wonders whether the hypnotic appeal of the computer in its 'all-knowingness', can be held responsible for a discernible shift in current educational theory. For there would seem to be a strong tendency nowadays in some educational circles to believe that research competence in accessing and organizing information, and in asking the computer the right questions, constitutes the most important function of learning. And this can well be justified by the fact that computers fly space-shuttles and aircraft; bring diagnostic and surgical capabilities to the practice of medicine hitherto unimagined; and could, ultimately, bring about some semblance of a family of nations... bearing in mind the scientific revelations and human benefits for which they are responsible.

However, I have come across faculty members in many disciplines who do indeed see the computer's answers as the prime authority when it comes to the matter of learning. Because, they say, in this day and age we recognize that education is all about understanding the 'materiality' of existence: have finally come to accept that 'reality' should be, and can only be, measured in terms of the material properties of things, how they work, and their utilitarian value. Therefore, we must learn to master the 'nuts and bolts' that hold nature (and that includes ourselves) together. And for this - in the final analysis - one turns to the 'outside brain' of the computer to glean how to do it.

In the last few pages we have argued that the most important function of education is to bring the pupil to think for him or her self. It is an issue that has occupied much space in the last few pages - and one that applies to us all as students of life whatever our age. A process of listening to oneself mulling over the facts until one can say: 'This is what *I* think; this is where *I* stand...'

As I write... these familiar lines from Wordsworth's *Daffodils* come to mind: *I gazed -- and gazed -- but little thought/ What wealth the show to me had brought.* And I remember the occasion - just before I left the university world - when the Chairman of the English Literature Department was talking to me enthusiastically about the great significance of courses dealing with what he called 'the Structural Analysis of linguistic form in the Great Books': which he explained as the taking apart of the syntax and vocabulary line by line, in order to examine the rule and pattern of a particular writer's style. *'And what about content.... the communication of content,'* I asked him. He looked at me as if I had just stepped straight from the pages of Charles Dickens. *'In contemporary education,'* he said...*'content can really be left to itself. In understanding linguistic structure, comprehension of the content becomes almost automatic...'* I was about to ask him if he thought that Hamlet's psychological turmoil could have been better resolved if he had perused the 'content' of them.... just what did Hamlet *mean* by the '...*slings and arrows of outrageous fortune*'? But thought better of it.

'Perchance to dream...' wistfully declaimed Hamlet.

Fat chance... you might think, if any Utopian brave new world of the future came to be administered by statisticians, analysts, and structuralists.

Both Saint Augustine in his *City of God* - an especially spiritual view of a Utopian-like world - and Sir Thomas More in the writing of his own *Utopia*, discussed how necessary it would be to ensure that the two principal ways in which human beings respond to life were brought into some kind of ideal accord: that those who took a practical and secular approach were appreciated for their pragmatic contribution to life's existential demands by those who possessed a more visionary - even metaphysical frame of mind - and vice versa. Sir Thomas, who suffered a martyr's death at the hands of Henry VIIIth in 1535 - and who was only canonized by Pope Pius XIth in 1935, displayed a saintly character throughout his life. And the following paragraphs in *Utopia* are an example of the reasonableness and gentleness of his nature. They are words expressing the common lot of our humanity: a combination of empathy and rationality married to a deeply moral sensibility. He is talking about the workings of his imagined peaceable world and the decrees of its ruler, King Utopus - words that reveal how far away we are from such a Utopian spirit and intent in this day and age, four hundred and fifty years after he penned them; especially with regard to the kind of religious intolerance that bedevils the world today.

> *This is one of the ancientest laws among them, that no man shall be blamed for reasonings in the maintenance of his own religion. For King Utopus ... made a decree, that it should be lawful for every man to favore and follow what religion he would, and that he might do the best he could to bring other to his opinion, so that he did it peaceably, gently, quietly, and soberly, without hasty and contentious rebuking and inveighing against others.*
>
> *They detest war as a very brutal thing; and which, to the reproach of human nature is more practiced by men than any sort of beasts; and they, against the custom of almost all other nations, think that there is nothing more inglorious than that glory which is gained by war. They should be both troubled and ashamed of a bloody victory over their enemies; and in no victory do they glory so much, as in that which is gained by dexterity and good conduct without bloodshed.*

We are desperately in need of wise men. Men of Saint Thomas More's ilk. We have entered into a nuclear age; girded the world in communication systems; ventured into space and landed on the moon; developed telescopes that can listen in on the dynamic activity taking place in deep space; developed the skill to transplant human organs; learned about the human genome; harnessed electro-magnetic forces to see inside the human body; developed the technique of cloning...

Yet we are still desperately in need of wise men.

Yet even without the wisest of counsel and leadership... I would have thought that in realizing we are all in the same 'life-death' boat together, living the brief span of time allotted, more of us would have the common sense to recognize the nihilistic folly in not coming together in mutual support. For in recognizing that we are all candidates for death, does it not seen reasonable to at least tolerate, if not pool, the differences between us for our general wellbeing - developing the *'There but for the grace of God go*

I... ' kind of attitude? And wish the other well in his or her brief sojourn in time - instead of hating and killing? Such a sympathetic response to our fellow travellers on the road to the embarkation port of no return, might see us all successfully through Customs on the other side when the boat gets in - whatever the name of the shipping line on which we have travelled. If, of course, there is any such boat. And if there is anywhere for it to go.

And those who deny the possibility of creating a tolerant and sympathetic Brave New World by fanatically insisting on the spiritual supremacy of this or that religious ideology or dogma - and show themselves willing to kill for it - might do well to remember that mystical experience is still as mysterious and arcane a psychical force as it has ever been in human history, and that no one religion has an edge on absolute and ultimate metaphysical truth.

> *Men never do evil so completely and cheerfully as when they do it from religious conviction.*
> Pascal, *Pensees*

14

WAR and KILLING: QUESTIONS OF EVIL, RELIGION, SPIRITUALITY, INNOCENCE

An unjust peace is better than a just war.
Cicero: *Ad Atticum* Vii. xiv.

There were twenty-one of us - all serving in Bomber Command, Royal Air Force, during the Second World War - drinking in our favorite pub when Dick Guterman presented us with Cicero. Yet despite heavy losses in the Squadron, eighteen of us did not agree with the Roman statesman.

Revisionist historians writing of the war have described bomber crews as 'barbarians.' They had in mind the burned-out cities of Hamburg, Cologne, Dresden... - devastation that resulted when the Air Council decided that 'saturation bombing' of German cities should be carried out in addition to attacking bona fide military targets. The orders to carry out these night raids came directly from Air Marshall 'Bomber' Harris, the A.O.C. (Air Officer Commanding) Bomber Command and, as far as I can remember, went into effect in mid-1942. The moral issues are - and were - obvious. Yet although post-war historians have labeled Harris as the evil instigator of the 'saturation' policy, it seems that the decision to bomb German cities were initiated by Winston Churchill. Even so, Harris zealously pushed the squadrons to the limit. He appeared to have little concern for the heavy losses in both aircrew and aircraft. There was a raid planned to take place on Kiel on the night of a full moon that would, the meteorologists warned, make every bomber an easy target for German night fighters and anti-aircraft gunners. But despite objections from Station Commanders, Harris insisted the operation should go ahead. Losses were calamitous. If memory serves me correctly, there were 900 aircraft dispatched to Kiel of which some 500 were lost... with seven men to each aeroplane. This change in air tactics was seen as a retaliation in kind after the indiscriminate night bombing by the Luftwaffe of British cities such as Coventry and London. Yet there was a tactical reason – other than the desire for revenge after the Luftwaffe night raids on England – and that was to destroy the morale of the German civilian population in waging the war.

There were bomber crews, including my own, who were not comfortable with this policy: legitimate military installations were one thing, but to engage in the wholesale demolition of cities and the immolation of their citizens - non-serving older men, women and children - was another. One crew on our squadron had the courage to refuse to fly against such targets. I can't remember exactly what happened to them - just that court martials were convened; the findings were marked 'Restricted'; and we saw none of them again. We all knew that their refusal had nothing to do with 'lack of moral fiber', the official line taken by the powers-that-be. Just the opposite, in fact. And it should have been evident to the Air Council that to kill and maim hundreds of thousands of German townsfolk would, like as not, only serve to unite the nation and stiffen their resistance to a now clearly perceived enemy. In our own country, for example, the indiscriminate night raids on major cities - and particularly the Luftwaffe's attacks on London in which thousands died - only shaped and strengthened a resolve never to 'Give In'. (Neither was there any brutal regime in England lying in wait behind the scenes to viciously put down any attempt at civil disobedience.)

During the years 1942 and '43 we lost hundreds of aircraft in night operations against Germany: a time when many aircrew - on our Squadron at least - were seriously questioning the justification, on both military and moral levels, of saturation-bombing tactics. (And yet the philosophers among us were quick to point out the fallacy of thinking one could impose *levels* of morality in war - taking it to be acceptable, for instance, if fifty high-explosive and incendiary bombs are dropped, but not if five-thousand are let loose.) Later, when the war was over and German photographs revealed the results of 'incendiarising' whole cities - engulfing them in firestorms that melted the surface of the streets and anybody on them - the infamous nature of the decision to attack whole civilian populations really hit home. It gave one small comfort to know that we had frequently agreed among ourselves that the role of Bomber Command should be to concentrate attacks on the German armaments factories, railways, harbors, army locations, coastal defences and submarine pens... And although many civilians working there would undoubtedly have been killed, we saw such a consequence as a less morally reprehensible loss of life. Yet I must emphasize that those of us who talked about the issue of saturation bombing did so as critics questioning its worth as a strategy; discussion of the moral aspects, although they were often present in our thoughts, was generally avoided. And understandably so. We lived too closely with death ourselves to constantly worry about moral distinctions - for them or us. We were well aware of the statistics: only one of every four aircrews were likely to survive. It was - as both sides were constantly being told - *total war*. Moral rules were suspended, swamped beneath the harsh realities attending one's own survival, and by the very nature of war itself.

Yet I know also that in terms of the moral issue, many of us experienced the shadowy persistence of the ancient and archetypal belief that the forces of evil abroad in the world must be countered by human effort. And it was the conviction, widely held among many of us, that Germany and Japan represented such forces. We knew we were not fighting for territorial expansion or some nationalistic political ideology, but to save civilization from a descent into a new Dark Age, a motivation that allowed us – in part at least – to accept night-bombing missions, in the belief that we were helping to destroy the forces of evil abroad in the world. On our intermittent 'nights off' when we talked among ourselves - often in our favorite local pub - the conversation would sometimes briefly become 'serious' in discussing the 'why's' and 'wherefore's' of the war. And it was generally agreed that the country had no option but to go to war and, that being the case, the enemy had to be hit as hard as possible. And, if that meant having five hundred or more bombers over a German city to bring the war to an end as soon as possible then the ends justified the means. The greater harm would be to allow moral quibbles to weaken our resolve in overcoming the evildoers.

Now, sixty years on - at the close of a century in which moral scruples have hardly made a dent in the proliferation of 'killing fields' generated by a series of local and international wars - the moral rationalizations we made to justify such annihilating raids on civilian populations in 1942 seems, sadly, naïve and almost whimsical. Looking back, I clearly remember the night when dense ground mist had caused operations to be cancelled, and about twenty of us had piled into three cars and woven a somewhat unsteady course into Lincoln to drink at *The Saracen's Head*.

The night when Sergeant Richard Guterman, D.F.M. (Distinguished Flying Medal) - a brilliant pianist, and a most articulate and natural philosopher who was lost just two months before his twentieth birthday) - held forth on the subject of killing in war.

It always took 'Guters' between three and five half-pints of beer to get going - whether he was commandingly expressing an opinion, punctuating his delivery from time to time by stabbing the air with his glass; or when, with glass safely deposited atop the piano, he would play Chopin études without need of a score - virtuoso performances that always engaged the rapt attention of those present - most of whom had had a pint or two themselves. Dick Guterman was something of a phenomenon. His talents extended to drawing and painting and he was responsible for many of the insignia adorning the fuselages of the Squadron's aeroplanes. He refused a Commission on the grounds that he preferred the relatively relaxed and informal life of the Sergeants' Mess.

On the night in question, save for two civilians, we had the Saloon Bar of *The Saracen's* to ourselves. We talked about the usual things: the last leave, wives, families, girlfriends, sexual adventures and non-adventures... and difficulties encountered on the last operational flights, the personal quirks of certain pilots, flight engineers, air-gunners (among whom Guterman's eccentricities when airborne in the mid-upper turret always occasioned many good-humored quips).

The interruption of these familiar exchanges came in typical Guterman fashion. A Socratic raising of the right arm, glass in hand; a voice pitched to command attention (yet still retaining an intimate and conversational quality), and generally injecting a provocative question totally unrelated to anything being currently discussed. As if he was oblivious to the group and asking *himself* the question. It was a familiar technique; yet usually effective - as it was on this occasion. Attention shifted to Guterman. He was about to speak.

'*Did you know*' - he said, outlining a circle in space with his glass, beer swirling frenetically around the rim - '*that killing in war was not considered to be an immoral act - not seen to be 'murder' as defined by civil law in the Graeco-Roman world - until early in the second century A.D.? Only then did the philosophers and thinkers begin to question this double standard and decide, ultimately, that to kill is in violation of moral law - is murder whether it be in times of war or peace. Yet a good hundred years before the new philosophers came to this conclusion, my favourite Roman, Marcus Tullius Cicero, declared that 'AN UNJUST PEACE IS BETTER THAN A JUST WAR.'*

He paused, emptied his glass, as if to let Marcus Tullius' pronouncement sink in. It was just over a week since Flying Officer Hunt and his crew had refused to take part in a mass bombing raid on Cologne and, after being put on 'open arrest,' had been transported to an undisclosed destination. And I knew that Guters had this very much on his mind. He put his glass down slowly and deliberately, wiped the beer froth from his moustache with the back of his hand, before going on to say, *'And if I was representing Hunt and Co., I'd bring on Cicero as my prime witness for the defence.'* Nobody spoke. We'd exhaustively discussed the pros and cons of the unanimous decision made by Hunt and his crew – and I have outlined in the preceding pages the prevailing opinion that our part in the war represented a crusade against an evil-disposed enemy. I don't think anyone thought this was the time or the place to bring it up again.

'Go and get another beer, Guters…' from Flight Lieutenant Greene. But the Sergeant was not about to be deterred. Was apparently deep in thought… *'But it obviously wouldn't work, would it- the Cicero defence I mean? Would any of us here buy it?*

Not as it stands… of course. But it could be used to make a point: namely that Hunt and the rest of them represent a special kind of conscientious objector and, as such, have legal rights that should be respected. For they are not refusing to fly on military operations – but only against targets which their consciences tell them are not bona fide military objectives. There's bound to be a Court of Inquiry convened at Group* and I'm going to send a petition to the Presiding Officer – in the spirit of Cicero's declaration rather than in its literal sense – saying that we all respect and understand the problems of conscience expressed by Flight Lieutenant Hunt and his crew and, given the fact that they have already completed 16 missions, would it not be possible to keep them together and transfer them to Coastal Command. In other words, would the Court honour their willingness to fly on military missions where civilians do not inevitably become the principal target. I would also say that such a recommendation by the Court would in no way diminish the commitment of the rest of us. For we know only too well that flying against Germany is a matter of England's survival - that in war one must fight fire with fire – in our case quite

---

* Group is Group Headquarters

literally; and that, paradoxically, to succeed, the moral law must be laid aside and conscience silenced in the hope that 'a greater good' will rise like a Phoenix from the ashes we leave behind us.

I will NOT say that the difficult question, we who continue to fly have to ask ourselves, is just HOW MUCH bombing is sufficient to achieve this. Or that for myself, I consider that the saturation bombing we are asked to carry out is 'overkill' and 'undermoral.'

Anyway... the point is... if I write it, how many of you will sign it?' He shrugged his shoulders, turned around and moved to the bar; presented his glass for a refill.

Greene was the first to speak. 'Well, I'll sign, Guters. Not that I think it'll do much good. You know how fixated the top brass are on the lack of moral fibre business. But...you never know...if they get a statement expressed in those what you might call 'loyal' terms from a number of us, it might make them consider re-assigning old Hunt's lot to Coastal.'

'Yeah, we'll all sign,' from Flight Lieutenant Ellis, my flight engineer – the oldest among us (going on 27) and a pre-war career officer whose opinions always carried a lot of weight. Nods and murmurs of assent seemingly from everyone. In the meantime, I hadn't noticed that Squadron Leader Cooper, the Roman Catholic chaplain, had moved from the crowd by the Bar to join our group.

'I would like to sign your letter, Sergeant' – looking directly at Guterman. 'There is an old Hungarian poem: *If you are among brigands and you are silent, you are a brigand yourself.* And then...*But if you would better the brigands, then play a brigand's game.* The implications of these two lines are particularly relevant to the moral dilemma that faced Flight Lieutenant Hunt and his crew – and that indeed faces you all. Take the Nazis to be the brigands of Europe... while we in England did not keep silent in the face of their aggression. We declared war. Yet the irony is that to beat them we have to play them at their own game - which puts us in a real bind. For here's the rub. We have to employ more deadly force than they can muster in order to win - and how do we do this without becoming brigands in our own right?'

'If you ask me can I justify killing either in civil life or war, I would have to say 'yes'... given two circumstances. Either when you are forced to protect your own life or that of those you love - and when you are called to stand against destructive, nihilistic forces that know and care nothing for the good of mankind, or the sanctity of human life. I am not sure how much right I have to be standing here and 'preaching,' as it were. I'm not up in the air as a professional target every other night or so risking my neck. But may I suggest - just among ourselves - and in response to Sergeant Guterman's use of the words *'overkill'* and *'undermoral'*, that Flight Lieutenant Hunt and his colleagues saw themselves as the agents of *too much* deadly force - more than was needed to play the brigand's game and win, while keeping some vestige of moral integrity. You will agree... it's a bloody fine line. But it was *their* bloody fine line when it comes to bombing the Germans *en masse*. And... decide whether the ends justify the means. I want to sign Sergeant Guterman's letter in the hope that the authorities will recognize the moral and military benefits of transferring Flight Lieutenant Hunt to fly in Coastal Command where civilian populations are not the target. The kind of letter that Sergeant Guterman proposes to write expresses the moral case for Hunt very well; yet at the same time affirms the expressed intent of the rest of you to fly the missions as ordered by Bomber Command. I can't see it doing anything but good.'

'Anyway, sign up! And the best of luck to you.'

A few days after Dick Guterman had held forth in the *Saracen's Head*, Flying Officer 'Johnny' Atkinson persuaded him to paint a large, winged figure of the Archangel Michael on the nose of his aircraft, Lancaster L7516. It was a magnificent design: a bronze-gold body, arcing white wings reaching up to the cockpit window on the pilot's side and down to where the bombdoors opened beneath the fuselage - Michael, the Christian soldier-angel, holding aloft a flaming sword, poised to strike, defend the powers of Light against the forces of Darkness. Three days later - after the Archangel had been on his first night operation over Hamburg - Guterman added a long pennant streaming from the tip of the archangelic sword which carried the R.A.F. maxim *Per Ardu ad Astra*, brushed-in with fine calligraphic flourishes.

The loss of Richard Guterman on the night when a German Me110 put a cannon shell through the rear fuselage of our Lancaster, badly damaging the mid-upper turret, was felt throughout the Squadron. We were used to seeing empty spaces at the various dispersal points around the airfield perimeter - spaces normally occupied by a particular aircraft that had not returned from the previous night's operation - and to living with the visual memory of the ten or twenty missing faces a week of the men who flew them. Generally we kept our feelings to ourselves - acknowledging the loss of a friend by an overly nonchalant, 'Pity about old Smithers...' Yet this was not the case when Guters was gone. It seemed alright, even necessary, to talk about him: about his impromptu piano performances that increased in virtuosity as the beer flowed; his humour, wit, and never-still mind that made him wise beyond his years; his anarchic and disheveled appearance and irrepressible debunking of authority; the penetrating cartoon-portraits of friends and acquaintances dashed off at a minute's notice... And, although frequently airsick during a six or seven hour operation, the fact that he managed to shoot down two attacking German night-fighters in the space of ten minutes, added further to his reputation. He was missed by everybody, groundcrews and aircrews alike. His 'Michael' had given rise to good-natured quips on the nature of, 'old Guters' heavenly visions while up there in the mid-upper turret... ' Yet I know that Atkinson's crew were impressed with the talisman occupying the front end of their Lancaster, not only regarding it as artistically a cut above the conventional 'girlie' pinups that adorned other squadron aircraft but as symbolically reassuring - justifying the night's mission in moving against an evil enemy, and providing them with a certain, yet quite irrational, feeling of protection (in fact, the aeroplane was never lost: engines were replaced from time to time and airframe repairs carried out - but L7516 and Michael saw the war out).

There were thirty-two signatures on Guterman's letter to the A.O.C. (Air Officer Commanding) 5 Group. We never knew what response, if any, it had aroused. I asked the Station Commander if he had heard anything about it, but he said he had not.

The S.M.O. (Station Medical Officer), Squadron Leader Hughes, was not an impressionable man. Always cool and collected in emergencies and a painstaking diagnostician, he was a doctor in whom I had complete confidence and trust. On the night a few weeks earlier when German 'intruders' were dropping incendiary and high-explosive bombs on the aerodrome's administrative buildings and living quarters, he displayed great heroism in moving from place to place to treat the wounded and do what he could for the dying, even as bombs were falling all around, setting off fires and destroying buildings. But for his bravery and skill, many more lives would have been lost.

He was with me at six o'clock in the morning in the R.A.F. hospital at Cranwell - three days after returning from the Berlin operation - when Guters died without regaining consciousness after being extricated from the wreckage of the mid-upper turret. The Cranwell doctors had done their best but had been unable to stem the internal bleeding.

Neither of us spoke very much on the thirty-mile drive back to the airfield. I'd had the hope all along that Guters would recover. It did not seem possible that such a dynamic and cogent force could have given up the ghost. Mentally and emotionally it was difficult to accept the thought of a still and silent Guterman. Our loss... Chopin's loss. After a few miles on our return journey the weather clamped down - a typical autumnal Lincolnshire fog blight presented a gray wall to the headlights of Hughes' little M.G., turning the world into a featureless void. It took us almost two hours to travel thirty or so miles.

That evening the aerodrome was on 'stand-down' - no operations - the airfield blanketed in opaque and impenetrable mist. The Officer's Mess was almost empty, most of its members happy to forsake the hazards of flying over the Third Reich for those of driving into Lincoln. The Station Medical Officer and I were alone in the bar, and he had started the conversation by commenting on the general response to old Guters' Archangel Michael on Atkinson's 'kite' (R.A.F. terminology for aeroplane). He remarked on how little scoffing there had been at its emblematic significance in representing the forces of good, even by those disinclined to depart from a totally rational attitude to the job in hand - a fact which I put down to the sheer visual impact of the design: a compelling symbolic tour de force leaving the viewer to wonder if Archangels might not, in fact, exist after all.

After dinner Hughes went on to talk about the early nineteen-thirties - when I was but a lad and he already engaged in medical training - remarking that after Hitler's ascent to power it seemed to him that a cloud of malevolence and hate was gathering over the new Nazi Germany: forces which took over the country, intensifying and exploiting the dark side of the German nature - pushing those serving the regime to commit the kind of vicious and destructive acts against humanity which revealed their total moral collapse.

I remember we went on to talk about the paradox presented by two such contradictory life-driving forces as good or evil, and that Hughes said - speaking as a doctor (he reminded me) - how important it is to realize that human consciousness serves two masters. That of the body-ego-self with all its temporal, *physical* needs and desires, pursuing, piranha-like, all the offerings of the five senses; and the inner self of mind determining the relative value of this or that experience. He emphasized the *psychosomatic* nature of human beings – body and mind complementing each other's needs.

In his view, good and evil forces were the major protagonists in the mental realm of mind. The forces for good inspire feelings and thoughts concerning truth, meaning and purpose in human existence - the moral and spiritual persuasions. While the negative or evil powers drive the grossly egotistical, sensate passions - the sadistic, masochistic, vengeful, murderous urges, the lust for power in all of its manifestations: impulses unrestrained by any awareness of a mutual humanness – of the feelings that bring us together in the spirit of kindness, generosity and benevolence.

'Or you can simplify it all if you like - perhaps oversimplify it...' said Hughes as if thinking aloud, 'by saying that the most significant evolutionary struggle on which homo sapiens has been embarked for 'x' number of years is the conflict between our proto-human, instinctive and amoral nature, and the arrival of the shadowy presence in our consciousness of the psychic energy we call spirit. A conflict that shows little sign of resolving itself, I'm sorry to say.'

He sipped thoughtfully on his whisky and soda, the knuckle of his right forefinger smoothing down his moustache, eyes looking absently into the distance.

'These concepts of good and evil...', I responded - more or less thinking aloud - 'you don't see them as just *ideas* we invent to categorize behaviour which we perceive to be obviously either constructive or destructive... without necessarily drawing moral conclusions from either? After all, the material universe couldn't care less about our values or how we act...'.

'No: I think it is a purpose of our evolution - as a species - that we should develop into creatures sensitive to principles of universal truth, and particularly where right and wrong are involved: moral principles that transcend the mechanics of the cosmos... are central to the human spirit. (The reader may be forgiven a slow smile of incredulity at this point - questioning the likelihood that two R.A.F. officers in the middle of a war would be talking philosophically about moral sensibility. Yet one should not forget that first and foremost, Hughes was a doctor; and I was a would-be artist and writer.)

The M.O. took a few minutes off to light his pipe. 'Did I ever tell you about the exorcism I attended?'

I shook my head. 'Well,' he continued, the smoke from his pipe rivalling the thickness of the fog outside, 'when it comes to dealing with the problem of evil, my experiences on that occasion answered the question - for me anyway - as to whether *we* are – in our biological nature - the sole initiators of it, or whether it is an external cosmic force in its own right.

"If you ask me does evil - as a tangible force - exist... then I will say 'yes.' I've *seen* it...' Another pause. More puffs on the pipe. 'The exorcism I mentioned went on for a whole week. I was called in by the Anglican priest to medically examine the boy, who was just ten years old. The priest then asked me to stay and be on hand should medical help be required throughout the course of the exorcism ritual. The young lad had been in his bedroom for three whole days, snarling, screaming, like a wild beast, refusing all offers of food and ejecting vile streams of what appeared to be bile at whoever approached him, alternately coiling and uncoiling like some sort of venomous snake. His father had locked him in there to contain the disturbance, hoping that the affliction would leave the boy. When I and the priest entered the room, all hell, literally, broke loose. Foul language spat out at us, and physical contortions that you'd think would separate every bone in his body. It was absolutely impossible to medically examine him. And it was the priest who decided that a colleague within the diocese should be brought in - a man especially skilled in the practice of exorcism.'

'We started that evening. The boy was quiet when we entered the room, crouched at the head of the bed like a terrified and hunted animal. Yet as soon as the exorcist - a Canon of Exeter Cathedral possessed of a quiet and benevolent presence - moved towards the bed, the boy's antics and screams increased in violence and intensity.'

'What I witnessed then was enough to convince me that such mental torment and physical distortion, combined with incredible organic eruptions of stuff from within the body and the emanations of indescribable odors, could *not* be attributed to any form of psychosomatic stress with which I was familiar. What I witnessed was a total change of being. There is no physical way the body alone could produce such unnatural bodily phenomena. At times the boy would become bloated as if to burst; and then vomit a nauseous yellow froth or foam, together with seaweed and feathers which covered the whole top of his pajamas and smelled foul. I've never smelled anything in the natural world like it. On another occasion his body became elongated, particularly his neck, and he squirmed around in the bed as if made of rubber. Then he lay so still on his back, eyes staring sightlessly at the ceiling, that I thought he'd died. But then, rigid as a plank, he levitated a good foot above his bed - stayed airborne at least a minute before slowly, very slowly, descending to the mattress. After this he seemed to be in a coma for the rest of the day. His pulse was barely discernible, and what rate I could detect was not, in my opinion, high enough to adequately feed the brain with oxygen. The pupils of his eyes showed no sensitivity to light. I was sleeping next door - the two priests conducting the exorcism stayed in his room - and early the next morning I was awakened by a cacophony of sound coming from the boy's bedroom: screams of obscenities and blasphemies, loud animal grunts, piercing howls, and the crashing of his body as he flung himself against the wall. By the time I had got to him he was being restrained by the priests. And then the room was filled with a vile stench as his body disgorged vast quantities of vomit and excrement - and, this despite the fact that for the last forty-eight hours he would not allow himself to be fed. We were going to attempt to feed him intravenously with a glucose drip that very morning. The whole thing went on for six days. At the end of that time he went into a deep sleep for eight hours. When he woke up he remembered nothing, and though weak from lack of food, his body showed no sign of the ravages that had tortured it. The priests were unsure as to whether a cure had been effected - whether 'the evil and demonic forces' had finally left him. I don't know how he went on after that. Two days later the war broke out and as a Reservist I was out of civilian practice and in the Service.'

'But in that room, and particularly over that bed, hung what I can only describe as an 'aura of evil' which I'm sure was not generated by any person in that room. It was a *visitation* of something disgustingly horrid.'

It is nearly sixty years ago since Hughes and I had this conversation, yet I remember it almost word for word, and can still see the intense look on his face as he talked. (We met briefly, once, at a Squadron reunion in 1956 and said we would correspond, but we never did. He was almost twice my age... and must have departed this life some time ago.)

Now, with so much water under the bridge, I find myself thinking again of that fogbound night in 1942 when we talked about the nature of evil. For I am staring at the latest evil atrocity: a photograph in the daily newspaper showing the victims of a recent massacre in Kosovo - of villagers executed and mutilated by Serbian forces in the ongoing Balkan's war. It brings back memories of the wholesale slaughter of the population of Lidice in 1942, after the Germans burned this village in Bohemia in retaliation for the assassination of Reinhard Heydrich in Prague by Czech patriots. And I remember how convinced Hughes was that evil could manifest as an external, independent force in the world, using people as mediums: 'Satanic' forces, if you like - although he never used the term. Certainly, when it comes to his description of the exorcism he attended, it seems only possible to regard the boy as the medium through which some unholy power was exercising itself. Which, in looking at the hideous mass grave in Kosovo, brings one to ask the question: can those who ordered such terrible acts of mass execution be seen as but the tools of non-human malevolent forces similar to those Dr. Hughes believed ravaged the young boy? It would seem unlikely, given that there were many apparently willing participants – some of whom must have acted out of their own free will.

We have seen that the novelist Joseph Conrad - whose great novel *Heart of Darkness* we discussed in Chapter 12 - would *not* support the notion that human wickedness may result from the direct intervention and control of such supranatural forces: believing, rather, that the people responsible for such atrocities acted in accordance with their own primitive, atavistic nature - the *"nature red in tooth and claw'* syndrome we have previously discussed. Any twinges of conscience they may experience must be either nonexistent or too weak to be an effective deterrent. Are such men - and there are any number of them throughout history - so brain-damaged, spirit-damaged, or soul-damaged that they totally lack the promptings of conscience? Or are the consciences of men who kill under orders overcome by fear of the consequences should they refuse to obey? And if such is the case, does killing 'second-hand', so to speak, allow them to be somewhat morally absolved?

Such questions concerning the enigma of conscience and moral sensibility keep cropping up throughout these pages, for they are issues which impinge on the background of several chapters.

I think again of Hans and Sophie Scholl who were guillotined by the German authorities for campaigning against the immorality of the regime - of the strength of their moral convictions, the profound level of humanness their sacrifice reveals, pointing the way for the rest of us...if we are up to it. And then the pendulum of thought swings to the other side of the moral coin, and closer to the present... Did the Hutu massacre the Tutsi in Rwanda because, collectively, they could not resist the 'aura of evil' abroad in the land? Or was it the hatred for *those who are not 'us'*, and the lust for dominance in their own hearts which inflamed the mob to indulge in such terrible slaughter? They were obviously individuals in whom atavistic and dark passions were easily aroused; thoughts of right and wrong easily dissipated.

These are not questions that yield easily to any definitive 'scientific' answer. And if we turn to psychologists for explanations, they must go beyond just an understanding of the physiological processes of the brain, and struggle to gain insights concerning the complexity of the mental forces driving the mind... the operational center of the psyche overall. Here are the thoughts of Carl Jung on the matter:

> *We must beware of thinking of good and evil as absolute opposites. The criterion of ethical action can no longer consist in the simple view that good has the force of a categorical imperative, while so-called evil can resolutely be shunned. Recognition of the reality of evil necessarily relativizes the good, and the evil likewise, converting both into halves of a paradoxical whole. In practical terms, this means that good and evil are no longer so self-evident. We have to realize that each represents a judgment. In view of the fallibility of all human judgment, we cannot believe that we will always judge rightly. We might so easily be the victims of misjudgment. The ethical problem is affected by this principle only to the extent that we become somewhat uncertain about moral evaluations. Nevertheless we have to make ethical decisions.*

> *The relativity of 'good' and 'evil' by no means signifies that these categories are invalid, or do not exist. Moral judgment is always present and carries with it characteristic psychological consequences. I have pointed out many times that as in the past, so in the future the wrong we have done, thought, or intended will wreak its vengeance on our souls...*

This passage - and those that follow - from Carl Jung's *Memories, Dreams, Reflections** relate directly to the moral dilemma we faced as aircrews engaged in the 'saturation bombing' of German cities. Many of us had silenced conscience by believing that we were serving a 'greater good' - that of bringing the war to an early end by eroding the will of the German people to continue fighting. And in continuing to read the chapter *Late Thoughts*, Jung brings one to realize that the 'greater good' defence could be ethically justified:

> *For moral evaluation is always founded upon the apparent certitudes of a moral code which pretends to know precisely what is good and what evil. But once we know how uncertain the foundation is, ethical decision becomes a subjective, creative act... a spontaneous and decisive impulse on the part of the unconscious. Ethics itself, the decision between good and evil, is not affected by this impulse, only made more difficult for us. Nothing can spare us the torment of an ethical decision. Nevertheless, harsh as it may sound we must have the freedom **in some circumstances to avoid the known moral good and to do what is considered to be evil, if our ethical decision so requires**. In other words... we must not succumb to either of the two opposites...**In given cases, the moral code is undeniably abrogated and ethical choice is left to the individual.** In itself there is nothing new about this idea; in pre-psychological days such difficult choices were also known and came under the heading of 'conflict of duties.*
> (The Bold type, when it appears, is my addition to the original.)

Jung goes on to refer to the question asked by the Gnostics, *'Whence comes evil?'* He discusses the egomaniacal lust for power of political leaders, the manic nationalism and destructive political ideologies that led to two world wars in the twentieth century; together with the terribly destructive powers of the atom bomb unleashed by science... and writes that, *'we*

---

* C.G. Jung, *Memories, Dreams, Reflections* (New York: Pantheon Books, 1961), p. 329, 330.

stand empty-handed, bewildered, and perplexed; when compelled to face - in our own time - the question of how to account for evil. (The earliest written records attribute the query quoted above to Basilides, the great Christian philosopher of the Gnostic sect in the early second century A.D.).

Writing in the decade following the end of the Second World War, with a bellicose Soviet Union and its satellite countries standing against the West, Jung sensed the dark forebodings hanging over Europe and the United States: presentiments of disaster made all the more fearful because we no longer believed in the saving grace of some surpassing parable or myth. '...*we know no way out,*' he wrote, '*and very few persons indeed draw the conclusion that this time the issue is the long-since-forgotten* soul of man.'

My own thoughts on the matter wander about all over the place. Yet they never seem able to abandon the spiritual and mystical core which, I feel, lies at the very heart of the good versus evil problem. For despite the violence which has continually beset the world there have always been people of great character willing to risk their lives in the service of conscience, compassion, and the righting of wrongs. And I find it difficult to understand why the peoples of the world are not brought together in the spirit of sympathy and concern when they observe the emotional intensity of suffering experienced by the survivors of tragedy, particularly by those who have lost loved ones. For the degree of pain which is evident is the same no matter to which side of a conflict people belong, to what ethnic group, or to what religious faith. In terms of geological time our brief sojourn on this planet is scarcely the blink of an eye. The brevity of human life is a bond we share and should act as a unifying force the world over. The philosophers have coined the expression, 'the tragic sense of life' to signify that we, as human beings (supposedly unlike other higher forms of life), have to live with the lurking knowledge that one day we will die. That irrespective of race, creed, or rank we are all in the same boat, all serving a death sentence, all heading towards an unknown destination where this condition we think of as life is going to end. Christians may see this as a 'happy ending' yet those who experience the intellectual adventuring offered by the mind, or are strongly affected by the intensity of feeling that comes with loving and caring, or in encountering some form of the beautiful or sublime, will justify the use of the adjective 'tragic' inasmuch as death puts paid to all such transcendent-like experiences. For there is indeed sadness in the thought that one may have been 'led up the garden path' by a consciousness that has taken one out to wander beyond the world, only to die to the void.

To ponder our common mortality is to take the first step in coming to feel for others, and in so doing embrace the most basic of moral principles: that of empathy. What is likely to follow is an increasing broad-mindedness and tolerance with regard to the way others live their lives, bound up as they are in the traditions of their own time and place. If members of the world community can learn to live together, free of prejudices, then we may be able to justify the human presence on this planet. I am reminded of all this by the line in the English Book of Common Prayer, *'In the midst of life we are in death'*. But this comes at the end of the service for the Burial for the Dead. Better if it were brought to our notice at some earlier stage, at some prior rite of initiation of children into adulthood, for example, when there is a lifetime to think it over and come to fully appreciate its significance.

'Why, what is pomp, rule, reign, but earth and dust?/And live we how we can, yet die we must.' This is one of the last lines spoken by the stricken Earl of Warwick in Shakespeare's *King Henry VI*: a near-to-last breath lament on the folly and ultimate meaninglessness of living simply to pursue power. While more than a hundred and fifty years after Shakespeare, the Scottish poet Robert Burns (1759-96) wrote the well-known ballad, *A Man's a Man for A' That* - a work stressing that wealth, rank, and power do not place one man above another on the scales of human worth and achievement. And what he writes, particularly in the last two verses, tersely expresses the moral truth of this:

> *A prince can make a belted knight,*
>   *A marquis, duke, an' a' that;*
> *But an honest man's aboon his might,*
>   *Gude faith, he maunna f' that!*
> *For a' that, an a' that,*
>   *Their dignities an a' that,*
> *The pith o' sense an' pride o' worth,*
>   *Are higher rank than a' that.*
>
> *Then let us pray that come it may,*
>   *(As come it will for a' that,)*
> *That Sense and Worth o'er a' the earth*
>   *Shall bear the gree, an a' that.*

> For a' that, an a'that,
>   It's coming yet for a' that,
> That man to man, the world o'er,
>   Shall brithers be for a' that.
> (Note: *'bear the gree'* means 'bear off the prize.')

However, I think Robbie Burns would be sadly disappointed were he to return and see how far we are from being *'brithers, for a' that.'*

RELIGION and SPIRITUALITY

'Every religion is good that teaches man to be good,' wrote Thomas Paine in the *Rights of Man* which he wrote between the years 1791 and 1792. And Lucius Annaeus Seneca (the younger), the Spanish born Roman statesman and philosopher living between 4 B.C. and 65 A.D. wrote in *Epistulae ad Lucilium:* 'Do you ask where the Supreme Good dwells? In the soul...

Move over the spread of recorded history, West and East, and note how often a religious note is sounded in the writings and sayings of statesmen, thinkers, priests, poets... in the comments of the great explorers who have adventured over the planet and those of the cosmonauts who have ventured into outer space... in the conclusions reached by some scientists, and by some pioneers in neurological research working to charter the labyrinthine neural channels of the brain. If one were to embark on searching out, and bringing together in one volume, all the extant sayings referring to the way consciousness works intuitively, imaginatively, to convey a sense of life's intrinsic and profound worth, it would result in a substantial book. And in browsing through its pages one would discover how frequently the same train of thought is expressed, spread across the geographical and temporal spectrum - themes employing a similar vocabulary in voicing the necessity to honor such abstract concepts as *truth, right and wrong, good and evil, the sublime and mundane, conscience, hope, love, spirit, soul...* urging one to live with due respect for the *mystery* that pervades the existence of everything in a state of being.

The words 'spirit' and 'spiritual' appear constantly to serve the metaphysical theme running throughout this book, but it is never suggested that they apply specifically to any one religious creed. For it is possible to be spiritual without being religious, or religious without being spiritual.

The general connotation of 'religiousness' is that of adherence to the tenets of a particular theological persuasion. Whereas 'spirituality' connotes a strong *personal* sensibility to conscience, moral issues, the power of love and compassion, the presence of a guiding force...: in other words to values seen as truths that serve a higher purpose than biological imperatives (even survival); that surpass the reality offered by the senses; and that transcend self-interest, self-gratification. But then, you ask, "Can't one be religious and spiritual at the same time?' And the answer would have to be, "Yes: of course, when the spiritual basis of the religion is in accord with the nature of one's personal faith. I would not say that either Shackleton or Worsley were conventionally religious men, but were certainly spiritually grounded in that they were not strangers to mystical happenings. For example, when they were attempting to cross the Allardyce Range on the island of South Georgia, they were both aware of an invisible presence that confirmed Shackleton's decision to make a critical and dangerous descent of a steep, night-shrouded snowfield.*

The mystical nature of this intervention seems to have been accepted almost matter-of-factly as spiritual guidance by these highly courageous men. But then they were individuals incredibly strong in spirit who had lived under tremendously harsh and seemingly hopeless conditions for a couple of years, drawing on strong inner resources of fortitude and an unbreakable will to survive.

Professor Andrew Newberg**, talking about experiences such as that recounted by Shackleton and Worsley, describes them as being *'born in a moment of spiritual connection, as real to the brain as any perception of 'ordinary' physical perception.'* And with this observation he returns us to the enigmatic problem posed so often throughout these pages. Namely, how a physiological system of neurons manages to autonomously create such rarefied mental experiences as moments of *'spiritual connection'* - using the word 'spiritual' in its purest sense to denote mental experiences not attributable to the world of the senses, or necessarily dependent on the doctrines taught by differing denominational and sectarian religions. It eclipses any egocentric spiritual ambition; has no political agenda; and is totally devoid of the, 'My God is better than your God,' kind of nonsense. Rather, I would

---

* An account of Shackleton's journey is given in Chapter 4.
** See the account of his SPECT scanning experiments in Chapter 9.

say it represents an unconscious force in the human psyche described by the poet W.B. Yeats as, 'A Dialogue of Self and Soul.'

It has been given to certain individuals throughout history to think, feel, and act in accordance with the ideals that such a dialogue inspires - ideals and principles that are intuitively felt to convey a surpassing level of truth. One of the truly compelling accounts of the insights that such personal reveries inspire was written by the most powerful (and some historians would say revered) man of his time in the Roman world. The Emperor, Marcus Aurelius (full name Marcus Aurelius Antoninus A.D. 161-180) wrote the stoic philosophical reflections that were later collected and published as *The Meditations of Marcus Aurelius*. He composed them, working in his tent on the Danube, during the seven years he spent campaigning against hostile Germanic tribes on Rome's northern borders. They are writings telling of the moral principles that direct a purposive life and bring a calm and spiritual peace in their wake: of the power of goodness, honesty, compassion, truth, and respect for the sublime nature of those things that imbue human existence with significance and meaning, and which warn of the hopelessness and nihilism induced by evil machinations and actions; by anger, hate, deceit, lack of conscience, and blindness to sacrosanct principles... He advocates a similar way of life as that proposed by Buddha, Confucius, Moses, Christ, Mohammed... yet he knew nothing of their teachings and was certainly not sympathetic to Christians living within the Empire.

Editions of the *Meditations* have been around for close to two thousand years. A new version was produced in Britain recently, and a 'Books on Tape' edition was published at the same time entitled, *The Spiritual Teachings of Marcus Aurelius*. Yet despite the example he set, and the fact that the ideals for which he stands have been in circulation for so long, we seem collectively to be still impervious to his message. In the first years of this new millennium the scale of human depravity and violence on the world scene has reached what can only be described as tragic proportions. Bombings, massacres, sadistic atrocities... bringing suffering and grief to thousands of individuals and families - acts which fly in the face of reason in that they eschew the kind of wise deliberations (assuming that wise counsel is available and desired) that would seek 'a meeting of minds'. Such acts are carried out by those who are totally unaffected by feelings of compassion for the victims - who are, after all, fellow human beings. In view of the millions of lives lost - even in the course of one's own lifetime - it is difficult to be sanguine about the success, or even likely success, of our long journey to formulate the humane and ethical principles by which to live.

Marcus Aurelius may be dismissed scientifically by astrophysicists when he refers to *'One Universe,'* but he enunciates historic and moral imperatives considered to represent spiritual values by philosophers of both East and West - teachers who lived hundreds of years before his own time.

> *One Universe made up of all things; and one God in it all, and one principle of Being, and one Law, one Reason, shared by all thinking creatures, and one Truth.*
>
> Meditations, Bk, vii, sec. 9.

Compare these words of a soldier and Emperor of the second century A.D. who was defending the borders of his country, with those of a top German Nazi General fighting a war of territorial expansion - Karl Rudolf Gerd Von Runstedt - addressing the Reich War Academy in 1943.

> *One of the great mistakes of 1918 was to spare the civil life of the enemy countries, for it is necessary for us Germans to be always at least double the numbers of people of the contiguous countries. We are therefore obliged to destroy at least a third of their inhabitants. The only means is organized underfeeding, which in this case is better than machine guns.*

So here we have examples of two extreme types of human being - made more telling, and more chilling, by the fact that both men are soldiers engaged in war, and that the spiritual man should have lived two thousand years before the godless one. For might we not expect that it would be the other way round? That by 1943 something of the Aurelian spirit might have found its way into European sensibility - at least to the extent that the callousness and evil natures of the Von Runstedts of this world might be tempered by a little humanity? That in the course of two millennia we might have expected to move spiritually and morally forward instead of standing still or even regressing? But the fact is that here we are, well into the first decade of the twenty-first century, and the *'moments of spiritual connection'* that lead to the kind of wisdom Marcus Aurelius describes would seem to be in short supply as the world stumbles from crisis to crisis - moral, political, environmental...

Even so, there are many individuals amongst us who know perfectly well what the Roman Emperor was talking about; have experienced it for themselves. On the other hand there are others who live devoid of any such spiritual connection - whether they attend religious services or not. Such an essential dissimilarity between individual consciousnesses - the fact that some are spiritually-disposed and others totally secular and ego-driven, creates a significant ontological problem in the study of human nature. And one to which we keep returning throughout these chapters.

How is one to account for states of spiritual barrenness - for the complete absence of conscience and compassion that earlier ages would have ascribed to ' lost souls?' The question will become more compelling if I provide a very recent example of the depths to which man can sink, reaching a level of total depravity.* It is taken from Damien Lewis' account of one particular British special forces operation in Sierra Leone's civil war in which Britain and the United Nations intervened. The savagery of the RUFs (Revolutionary United Front) tactics in this small West African country - and particularly the terrible atrocities committed by a guerilla gang calling themselves 'The West Side Boys' - demanded international intervention. Lewis writes: *'The mission was to rescue six British soldiers held hostage by the West Side Boys, a guerilla gang with a pendant for fluorescent wigs and voodoo charms, who stuck enemy heads on poles.'* He goes on to say that, *'The RUF were notorious for the unspeakable 'games' they played with captured villagers, such as 'sex the child': rebels would take wagers on the sex of a pregnant woman's baby, and slice open her belly with a machete to find out who was the winner.'*

Where are the words that can adequately convey the utter despondency that settles over one when reading of this *'game?'* Of all the atrocities committed by human beings and mentioned in these pages, none seem quite so monstrous, cruel, or morally corrupt. The cold-blooded nature of the act itself, involving the fully-conscious 'hands-on' participation of mind and eye, no doubt contributes to the repulsion, to the downright degradation of humanness this conduct represents. And one wonders how the Nazi General Von Runstedt - were he alive today – would react to the sadistic mentality and barbaric behavior of the West Side Boys.

---

* Damien Lewis, *Operation Certain Death* (London: Century Press, 2004).

Might we ponder just how fine is the moral line separating a 'civilized' Nazi general from a self-styled and uncultured 'colonel' of Sierra Leone's RUF? However, the comparison does suggest that conscienceless and amoral attitudes are innate to such men, dispositions of their own nature rather than the effect of conditioning imposed by environmental factors. For despite the background of a morally corrupt Nazi regime, some senior German officers - General Hans Guderian and Field Marshal Erwin Rommel come immediately to mind - displayed as much a sense of right conduct as warfare permits.

We really have no way of knowing when the humanizing sensibility we call morality first became a force in consciousness. Only with the advent of 'writing' - the primitive pictographs representing ideas, or the hieroglyphic symbols used by the ancient Egyptians and others - can we put a likely date to the appearance and affective impact of moral and aesthetic values: to the spirit of compassion and goodness as expressed in the *Maxims of Ptahotep*, for example, written by the third millennium Egyptian pharaoh of that name and which we discussed in Chapter 12. For although expressed in hieroglyphics on a clay tablet, a thousand years before Marcus Aurelius wrote the *Meditations*, they express similar humane ideals. While in another part of the world Siddhartha Gautama - the religious philosopher and teacher who founded Buddhism in India half a century before the Roman Emperor expressed his philosophy of life - describes the *Eightfold* path to wisdom in terms that would have been well understood by Marcus Aurelius: *'Right view, right aim, right speech, right action, right living, right effort, right mindfulness, right contemplation.'* In China, Confucius (551-479 B.C.) was delivering his famous aphorisms on human nature and on the true path to follow in life: short and concise statements of principle that echo the words of Marcus Aurelius and the Buddha. In *Analects* (a collection of Confucius' sayings, one finds the words *'The aim of the superior man is truth.'* Neither should we forget the thousand year influence - from roughly 500-1500 A.D. - of the Rule of Saint Benedict in Europe: emphasizing the physical and spiritual duality of human existence.

As one browses through the published sayings of the most renowned of the world's thinkers and philosophers, the similarity of the humanistic ideals they expound is inescapable - ideals which can justifiably be said to possess metaphysical overtones, introducing values that transcend regional and cultural attitudes of the time. One would wish that more of those who govern and have been responsible for the destiny of peoples and nations over the last few thousand years, had based their rule on the moral truths expressed by such men of vision.

Confucius refers to the man who seeks truth as 'superior' - thereby implying that all men and women are not equal in this regard. But if we agree that such is the case we are admitting that to seek after truth is to be engaged in a very high calling.

'That all men are equal is a proposition to which, at ordinary times, no sane individual has ever given his assent,' wrote the English novelist and essayist Aldous Huxley in *Proper Studies*. Any list that attempted to enumerate even just the essential ways in which we differ from each other, would have to do so under two distinct headings: Psychical and Physical. *Physical* differences are self-evident: no two of us are alike in matters of weight, height, and build; the genetic morphology and mechanical functioning of body and brain. Whereas *psychical* differences are based on the level and range of mental activity that goes beyond serving the practical needs of everyday living, and creates an awareness of human principles and values: differences that can result in someone becoming either a serial killer or a doctor-missionary.

But let us go back to Confucius' statement for a moment and ask just what is truth? Definitions are numerous - some seeking sincerely to establish its credentials; others to make ambiguous and witty allusions that may both clarify or confuse: *'My way of joking is to tell the truth,'* wrote the Irishman George Bernard Shaw - (who received the Nobel Prize for literature in 1925) - in *John Bull's Other Island* II. And then there is Byron, the celebrated English poet of the Romantic movement who died in 1824, writing in *Don Juan: 'And, after all, what is a lie?/'Tis but the truth in masquerade.'* However, throughout history many thinkers, poets and philosophers have sought to reveal the fundamental nature of truth as they see it. Homer, in Book iii of the *Odyssey* writes: *'Urge him with truth to frame his fair replies;/And sure he will; for Wisdom never lies.'* The poet Robert Browning, Byron's fellow countryman who was 12 years old when Byron died, and who is noted for his psychological insights into character and human motivation, expressed his thoughts in an early poem, *Paracelsus (Part i)*, as follows:

*Truth is within ourselves; it takes no rise
From outward things, whatever you may believe.*

*There is an inmost centre in us all,
Where truth abides in fulness.*

The Swiss philosopher Henry Frederic Amiel (1821-1872) - whose *Journal* entries I have found to provide a constant source of wisdom - writes in similar vein to Browning with his entry for 17 December, 1854: *'Truth is the secret of eloquence and of virtue, the basis of moral authority: it is the highest summit of art and life.'* Some years later on January 22nd, 1874, he commented that *'A lively, disinterested, persistent liking for truth is extraordinarily rare.'* On reading this I thought of Pontius Pilate's famous question to Jesus, *'What is truth?'* And of the remarks made by Lancelot Andrewes, Bishop of Winchester, in a famous sermon of 1613 intimating that he saw Pilate's query as one of those philosophic inquiries that, unfortunately, get brushed aside by the press of business. The pertinent sentence in the sermon reads: *'And then some other matter took him in the head, and so up he rose and went his way before he had his answer.'* While Sir Francis Bacon (1561-1626) anticipated Bishop Andrewes on this matter of Pilate's question to Christ by writing the following in his *Essays: Of Truth: 'What is truth? said jesting Pilate; and would not stay for an answer.'*

The Roman Governor was obviously no Confucian *'superior man.'*

Some thirty years ago I was talking to an eminent gynecologist in Atlanta, Georgia, about the marked differences between individuals in both character and physique. He agreed that physiological distinctiveness was determined by a combination of inherited factors coming together from a vast gene pool. But I was surprised when he gave it as his opinion that the psychologic forces shaping our character and destiny were also genetically determined: for at this time the relatively new science of genetics was committed only to the concept of inherited physiological qualities. He told me that he could tell, within a minute or so after delivering a baby into the world, the kind of person it was likely to become. But he went on to say that this observation did not negate the fact that life would work its transformations on any one of them - for good or ill. When I put it to him that in terms of the *nature* versus *nurture* argument he seemed to unreservedly favor the premise that psychological traits, no less than physiological ones are passed on, he said, 'Absolutely... no child comes into this world as a *tabula rasa*: no infant mind can be regarded as a clean slate, blank until impressions are recorded on it by experience.'

'I take goodness and a level of wisdom to be a mark of moral quality... and would say that a strong overall moral endowment may result from some fortunate constellation of 'morally wise' genes in the ancestral gene pool that start to work influencing the formation of character immediately a child is born; and continue to mold him or her positively throughout life, through thick or thin. Then again, it might be that not every one of us inherits an equal quantum of such active 'morally wise' genes. Some - psychopathic personalities, perhaps - may inherit very few, or none at all. And then I have often thought that whatever the quantum of these 'good' genes, there is always the possibility that some or all may be dormant, quiescent... only becoming operative to present consciousness with a sense of right and wrong when life presents its moral challenges. As to how one then responds is, I would think, determined by the level of emotional and intellectual arousal we call conscience that is excited by these specialized genes - by their 'psychic resonance' as it were. One way or the other, the concept of 'morally wise' genes - active or dormant; present or absent - may serve to account for the differences in moral standards and compulsions across the human spectrum. Even so, it should not be forgotten that environmental factors may also modify inherited genetic characteristics. This is the nurture side of the discussion. Negative or psychologically injurious experiences can affect the genes; as can positive and self-fulfilling ones. The French biologist Lamark was onto that in the 1790's. Today they call it epigenetics."

He lit his umpteenth cigarette of the evening, giving me a chance to collect my thoughts. And then I asked him how he responded to the maxim that, *'All men are equal before God.'*

"Well, " he said between puffs, "that may be so in the heavenly realm, but so far as our biological endowment is concerned we are certainly not all created equal; neither do all our little psyches possess the same potential to deliver the goods." He paused for a moment, short bursts of cigarette smoke signalling the intrusion of new thoughts. "I spent six months in India three years ago and had many discussions concerning the age-old

Hindu belief in the process of reincarnation - the means by which souls return to perfect themselves through yet one more lifetime on the character-building treadmill... until they become 'old souls' - which is another way of describing the 'perfected ones.' Not a proposition to dismiss too lightly. Sometimes I have the feeling that I'm delivering quite an old soul."

He stubbed out the remains of his cigarette, walked over to the large picture window that framed a dramatic view of Stone Mountain, and stood silently for a couple of minutes before returning to his chair.

"Let me tell you how a Hindu friend described how you know an old soul. It is in the persistence of the absolute transparency of childhood innocence: that simplicity of outlook, curiosity and wonder, trust in the rightness and goodness of people and events, and lack of guile and duplicity... which is to be found in many children up to the onset of puberty. Qualities that so often fail to be nourished and incorporated in the psyche of the fledgling adult, when rapacious environmental factors have managed to thoroughly expel the child from the grownup: bad parenting, bad neighborhood, bad schools, lack of role models and exemplars... urban ugliness and squalor, the Mall the epitome of civilization. And yet there are those who triumph over such cultural deprivation and become just *good* people, to put it simply - some who never seem to lose the ingenuous manner and outlook of the childhood. *I* know them when I deliver 'em. *You* know them when you meet them - men and women who have retained a wonderful translucency of character: openness of mind, a sense of wonder at being alive, in following their own inner growth throughout the years... calm, gentle, trusting people... whatever misfortune they may have suffered. There's a most attractive wholeness to them. Better to call it 'a retention of innocence,' I would say."

"This is what I mean..." He walked over to a wall lined with books and extracted a beautifully bound crimson volume with numerous slips of paper protruding to mark certain pages. "Here - Byron, canto sixteen from *Don Juan*..."

> *The love of higher things and better days;*
> *The unbounded hope, and heavenly ignorance*
> *Of what is call'd the world, and the world's ways.*

Time for another cigarette.

I was thinking back to the flying Bishop, Herbert Read, Trevor Huddlestone, Dick Guterman... whose lives lent substance to Byron's words. Yet to talk about 'morally-wise' genes seemed to fall short of the mark in conveying the telling presence of these men. And the thought occurred to me that if Byron had heard the doctor talking about 'morally-wise' genes, he might have asked if it would not be more germane to describe them as *'spiritually*-wise.'

"Suppose I was to suggest that it might be more correct to refer to constellations of *'spiritually*-wise' *genes* rather than 'morally-wise' ones...?" I asked my host as I handed back the Byron book.

"I don't really see the difference," he said. "Moral...spiritual awareness...both result from closely related inner sensibilities that are difficult to explain as a phenomenon of our biology: involuntary thoughts and feelings that complement each other in bringing values to mind that speak to a higher order of truth than that offered solely by material realities." He carefully replaced the book, smoothing it carefully into position with the back of his hand so that the spines were all perfectly aligned. "But I'll tell you this: sometimes I think I should interview all the prospective parents who come into my office, and make sure they understand what a miraculous series of events they have set in motion. If they don't seem to have a clue as to what I'm talking about - particularly when it comes to displaying some awareness of the responsibility involved in raising a child - then I feel I should send them somewhere on a course of instruction before agreeing to deliver their baby. You know - it seems an arrogant thing to say - but there are far too many would-be parents who just don't seem to belong to the Working Brain Society. Who don't inspire one with the confidence that they possess the common sense or the seriousness of purpose required to guide a kid through the various stages of infancy, childhood, puberty and adolescence; who fail to comprehend that each of these transitions plays a vital part in the formation of character and good ego. It is difficult to get some of them to understand that apart from looking after the physical well-being of the infant, the more difficult task by far - for the first 15 years at least - is to try and gain some appreciation of the way the child-consciousness is beginning to work. Getting to know the child's particular way of thinking, seeking to understand what triggers what kind of feeling... acquire some insight into the nascent development of the kid's psychology - character, temperament, personality. This is crucial if mother and father are to be in a position to help the fledgling adult make sound judgments and develop worthwhile standards and values. It is not enough to be just loving parents. It is also necessary to work at being amateur psychiatrists and wise counselors. But in my view, the most important thing that parental care and the home environment can do for any boy or girl is to ensure that something of the childlike spirit of wonder - the 'make believe' in which the young mind indulges - persists into adult life. For this initial mental adventuring is unconditioned and free of prejudice - innocent in the full meaning of the word - and serves as the platform from which the youngster is launched into his or her uniquely personal way of seeing things."

Now, writing about this conversation so many years later, I still find myself in complete accord with my doctor friend's views. Yet if he felt the level of parental ignorance and incompetence to be widespread thirty years ago, how would he respond to the situation today in the United States - as well as in other developed countries - where society is doing its best to shorten the childhood years of guilelessness and wonder? 'Beauty Pageants' in America, for instance, that involve little girls as young as three or four, lacquered in lipstick and mascara, and coached to cavort like dwarf sex symbols. Contests where mothers hope to salvage the perceived failure of their own femininity through the substituted success of their daughters: a neurotic situation where the failure of their offspring to win – in any competitive arena - seems to assume the nature of a death sentence. Not to mention what this unnatural competitive posturing does to the normal development of the child who, more than likely, will suffer an early loss of the kind of innocence we have been discussing. Unfortunately, the emphasis on competition and winning prevalent nowadays in so many aspects of a child's life, is based on the quite erroneous view of what constitutes 'success' pervading western culture: the outward trappings of social and economic ascendancy. To apply these standards to a youngster can seriously interfere with his or her own built-in, genetically 'set up', psychological timetable for reaching levels of adulthood – the uniquely personal self-realization that determines both character and destiny. As the American poet Walt Whitman said, *"You can't buck nature..."*

But such enlightened thinking is relatively new. During the Napoleonic

Wars young drummer boys scarcely out of puberty would die alongside adult soldiers. About a hundred and fifty years ago in England children were doing heavy manual labor in coal mines; climbing up sooty chimneys as chimney sweeps; laboring long hours in the cotton and woolen mills. In 1850, over a third of Britain's adult population were illiterate: the State only assumed responsibility for primary education in 1870. And here we are in 2004 with a police presence in many state schools in the United States; gangs of young teenagers terrorizing housing estates and neighborhoods in Britain; and a high illiteracy rate among the juveniles committed to prisons and correction centers. Palestinian youths are in the streets throwing stones at Israeli soldiers. Young boys aged ten or eleven are touting AK 47's in revolutionary armies in Africa and Indonesia, reverting seemingly easily to the savagery of killing. While in large parts of India and the Far East female offspring are considered to be less significant than male and are treated shamelessly as inferior human beings.

The desperate need in so many of the poorer countries is for adequate education of the young. If young minds can be activated by being pushed to think: given the chance to become intrigued by the revelations of science, the achievements of engineering and agricultural know-how, the tremendous significance of environmental issues... resulting in a broadening of personal horizons through reading about history and geography; the development of a personal world of ideas gained from reading, reading, reading... we would have a wiser, saner and more fulfilled world society. And revolutionary gangs like Sierra Leone's West Side Boys would be starving for lack of recruits.

But does not such a proposal represent yet another pipe dream? Even among Western nations, spending on education is not always a priority. In the United States, for example – military expenditure is estimated at 42% of the national budget, compared with 7% for education. (Other published figures show the discrepancy to be even wider than this.)

It is difficult not to be moved when seeing the faces caught on camera of young children – say between three and five years old - in war-ravaged and undeveloped regions of the world; by the childlike openness and directness of their gaze: the trust, expectancy, unselfconscious transparency and simplicity, and absence of guile. The gazelle-like eyes of the healthy ones appear to know no evil; they look dispassionately and clearly at the camera, yet reflect the harshness and apparent hopelessness of their environmental situation; while the terribly haunted eyes of the sick and dying cry out mutely in their pain and fear. We witness the faces of the innocents who, all too soon and too abruptly, are enduring the harsh world of nature's predations (famine and malnutrition, AIDS...), together with that of a harsh and dangerous environment created by the grossly egocentric ambitions of hypocritical political leaders and would-be leaders.

Souls under siege, with little chance that the spark of innocence will continue to ignite the neurons on the path to self-realization.

Older and wiser tribal cultures referred to the rituals that assisted the young on this journey as *rites of passage*. Once childhood was left behind and puberty took over, such rites involved the counseling of the initiate by the elders (rather than by the parents), informing of the psychological and physical changes to be expected during life's developmental transitions. This would be reinforced by training sessions involving isolation and hardship designed to test and train physical endurance and moral toughness of spirit: all calculated to induce the kind of self-reliance and self-knowledge required to move purposefully through life from one transitional level to another. Granted, such rites were intended to condition young males on their life-journey, while the development of young girls was obviously considered less necessary when it came to such training, and so could be left to the counsel of women, with whom they spent most of their time. Could this be because matriarchal wisdom was seen as deriving from woman's closeness to the mysterious origins of life – therefore suggesting that the feminine psyche was innately endowed with far-seeing insights that ensured the vigorous continuance of the tribal 'collective soul,' and so had no need of initiation into the adult skills and philosophical attitudes necessary for the young males who would be ultimately responsible for the survival of the group.

I mention such 'rites of passage' through which the fledgling adults must pass because not too long ago the equivalent 'training' periods in Western societies were well established: particularly in progression through the school system, and in the rules parents set concerning life *en famille*. But in contemporary Western society the distinctions made between the years of infancy, childhood, adolescence and teenager have become blurred, furthering a harmful rush to 'grow up' without benefit of 'growing pains.' Our culture has little regard for nature's timetable when it comes to attaining levels of adulthood; for allowing each stage of growth to become consolidated in the psyche of the developing adult – by which I mean that each level of growth should be sufficiently significant to be retained in the memory – ultimately to promote a natural intelligence: a natural sagacity, a degree of wisdom.

Once again I turn to William Wordsworth, my old ally; let his poem, *The Rainbow*, tell of how childhood innocence and its *'natural piety'* returns to bring meaning in one's later years. As Jung wrote in *Psychological Reflections*, *'...children are most uncannily sensitive...They know intuitively the false from the true.'*

> *My heart leaps up when I behold*
> *A rainbow in the sky;*
> *So it was when my life began;*
> *So it is now I am a man;*
> *So be it when I shall grow old,*
> *Or let me die!*
> *The Child is father of the Man;*
> *And I could wish my days to be*
> *Bound each to each by natural piety.*

An overly romantic view of childhood? Perhaps. But it's one that with the advent of psychology as a medical discipline in the early 20th century, has confirmed the poet's insight.

Note: The names of RAF personnel mentioned in this Chapter – save that of Air Marshall Harris – have been changed for obvious reasons.

15

THE MORAL FIBER of NATIONS: GREECE and ROME

*The glory that was Greece
And the grandeur that was Rome.*
Edgar Allen Poe (1809-1849) in *To Helen*.

Glory or grandeur? Although these two words are often used synonymously, it is obvious that in the quotation above the author is suggesting that a subtle shade of meaning does indeed separate them - a distinction that marks the most significant difference between the civilization of Classical Greece and that of Ancient Rome. The comparison cannot be avoided when mentally reviewing what the Greeks gave to the world through their psychological and philosophical wisdom and the standards of aesthetic excellence they set, vis-à-vis the achievements of the Romans who introduced efficient systems of civic organization, created marvels of architectural engineering, and produced a military machine second to none. Contrasting these legacies, it can be said that Greek civilization possesses a radiant quality: a celebration of fineness in both defining beauty and creating the beautiful... in architecture, sculpture, and verse; a praise and honoring of things human in their drama and knowledge of mans' nature. While Rome's grandeur resides in the monumental scale of its amphitheaters and aqueducts, the pomp and circumstance of its ceremonies, the sumptuousness and lavishness of its appetites... displays of richness and splendor. And if one was wishing to imply that the moral ethos of one culture was more profound than that of the other, then would you not agree that Poe has chosen his words well - that glory has the edge over grandeur in this regard; and that glory is attributed to Greece?

But the fundamental difference between the two cultures can be put more succinctly by stating that the Greek consciousness was more inwardly and intuitively oriented than that of the Romans - their inquiring minds setting the standards for the great thinkers and artists of later western civilizations. They, of all peoples, knew the positive effects of 'right' actions and the harmful results of 'wrong' ones - attitudes which significantly affected the thinking of their great philosophers and which they exported to their colonies throughout the Mediterranean. Rome too had its great poets, philosophers and wise statesmen - many of whom appear in these pages... Cicero, Ovid, Senecca, Horace, and Marcus Aurelius - but their collective influence on the educated citizenry of Rome and empire appears to be less universally efficacious than the authority exercised by Homer, Euripides, Pericles, Solon, Diogenes, Pythagoras, Euclid, Leucippus, Epicurus, Socrates, Plato, Aristotle... on the populace of Athens, capital of ancient Attica in the five centuries preceding the birth of Christ. The glory of Classical Greece - which comes to fruition in the 5th century B.C. and dissipates into an international Hellenism with the death of Alexander the Great in 323 A.D. - was finally overwhelmed by the conquering Roman legions in the middle of the second century B.C. They broke with the political and philosophical legacy of Athenian Greece and the Hellenistic world to which it gave rise; destroyed the civilization of Carthage; levelled temples and houses of worship, pillaged great libraries, and everywhere left Mediterranean culture in ruins. In its place Rome created an empire founded on military law, abounding in monumental structures, and flourishing in commerce.

Consider for a moment the difference between the Olympic Games of Greece and the Colosseum 'Games' of Rome. The Olympic Games - first recorded in 776 B.C. as a festival to honor Zeus, then held every four years at Olympia - were contests in poetry, music, and athletics. The Roman 'Games' - first recorded as public contests between gladiators in Rome in 264 B.C. - were fights to the death between two gladiators or between gladiators and wild animals: gladiators being generally slaves or captives.

The contemporary World Olympic Games follow only the athletic phase of the Greek tradition. I suppose the nearest equivalents to Roman gladiatorial death-fights in modern Western society are professional boxing matches, where the audiences come alive, wild-eyed, screaming and yelling, as a man becomes a punching bag, gore much in evidence. And then there is professional 'wrestling' so-called, bullfighting and cockfighting... and, more recently, the electronic violence and killing simulated in video 'games'.

Bring to mind two visual images of the ancient world. One of spectators sitting on the hillsides of the Greek Stadion - the foot racecourse at Olympia - cheering on their favorite athletes in the 200 meter dash. The other of fifty thousand Roman on their feet in the vast, tiered ellipse of the Colosseum in Rome, roaring their approval of a victorious gladiator's victory, and giving the sign for him to finish off his hapless victim.

> *Of all peoples, the Greeks have best dreamed the dream of life.*
> Goethe: *Spruche in Prosa.*

The Greek 'dream' of life could be described as a grand adventure of the mind, involving all its imaginative and creative power, and producing geniuses who lifted the veil of ignorance covering so many aspects of our own and nature's existence. Today's psychologists and psychiatrists should steep themselves in the dramatic world of the Greek theater if they wish to come *face to face* with the drives, conflicts and tensions which drive the human psyche: such a confrontation may well teach them more about their profession than is to be gained from any academic training. In addition to providing material for the great dramatists, their drive to illuminate the 'why's' and 'wherefore's' of human existence furthered the discipline we know as philosophy, raising it to the sophisticated levels which still provide the basis of contemporary thought concerning these perpetual questions. They also declared that there were certain rules of proportion that determined perfection of form - particularly in architecture where length, breadth, height, and overall area-volume must be carefully designed to ensure the degree of contrast that made the structure visually interesting and dynamic, yet at the same time created an aesthetic balance: thus, bringing mind-satisfying, rhythmic interplays of form and space, to create the perfect edifice. The Parthenon in Athens, built between 447-432 B.C. in the time of Pericles, is the most illustrious example of the formal beauty resulting from the application of these rules of proportion - the formula of the so-called Golden Section utilized by the architects Ictinus and Callicrates. Polyclitus, the famous 5th century B.C. sculptor, instituted a numerical axiom to be followed if the nude female torso was to be sculpturally created beautiful: namely that the distance between the nipples, from the nipples to the navel, and from the navel to the division of the legs, should be uniform. And it is worth remembering that as early as 450 B.C.

Leucippus and Democritus, in searching to discover the intrinsic nature of matter, created the vocabulary of Greek atomism which can be described as the prototype of contemporary atomic theory. While Pythagoras (578?-510?), and Euclid (c.335-275 B.C.) gave to mathematics the form which saw us through well into the nineteenth century.

The creative spirit of Classical Greece certainly inspired the philosopher-statesmen and philosopher-poets of Rome; and what has survived of Roman painting in the form of murals, together with the fine mosaic designs covering floors and walls, is powerfully expressive and heralds the takeover of mood and feeling - and what at times appear to be dream-states when contrasted with the more cerebral Greek aesthetic of planned and balanced formal harmonies. Roman sculptors also excelled in portraiture, producing not only excellent likenesses but revelatory psychological studies of their subjects. However, having achieved this, whenever a full-length figure sculpture was required they would usually select a suitable body-type from a ready stockpile, and simply connect the portrait-head to it. One can imagine the Greek sculptor's condemnation of this practice - of deliberately bringing together the unique with the commonplace, thus denying the finished work any pretense of psychological, aesthetic and formal unity. But generally speaking, and taking account of the exceptions I have mentioned above, the Romans could not match the idealism and profound sense of formal perfection achieved by the Greeks. (And yet, I have to say that in walking through the Roman cities of Pompeii and Herculaneum - buried beneath the volcanic ash of Vesuvius in 79 A.D. and excavated over the last two centuries - the intimate scale and proportional elegance of the villas, and the visual delight offered by their murals and mosaics, suggests that here at least, existed a high degree of artistic discrimination and sensibility.)

But the big picture shows that the Roman *march* through life - (at least until the eventual collapse of the empire when the emperor Constantine moved east to Byzantium in A.D. 330) - was more an external adventure than an inward journey. And the images that persist are indeed those of legions on the march, of the roads that have survived after almost two thousand years of continuous use, the multi-tiered acquaducts carrying water over long distances to the cities, the colossal ampitheaters, and the splendid basilicas and public baths. (In visiting the new Italian scientific station of *Terra Nova* in Antarctica a few years ago, I was both surprised and delighted to see how the old Roman road-building gene had survived. For there was a road - a wide, straight-as-a die road the Italians had cut through the ice and rock-bound wilderness, going from the base to end at the water's edge of a small bay some two-hundred yards beyond the buildings and principal deep water anchorage. Nobody could tell me what it was to be used for. And the raised helicopter-pad had been most carefully levelled and its sides paved with flat, cut and trimmed stones, to create a mosaic-like façade. The usual practice on Antarctic bases was to select a relatively flat end even surface and mark it with four large, yellow-painted rocks at the corners.)

Finally, I should mention that the great architectural achievements of Roman architects and engineers were facilitated by developing the structural possibilities of the round arch - taking it to its limits in spanning and defining space, and in utilizing its load-bearing potential. Also, that it was their discovery and exploitation of concrete's plasticity and cohesive power - and of its strength when set - that brought about the structural innovations responsible for the creation of the great domes and vaults which are marvels of architectural engineering. And in the strange way that symbolic associations develop in one's mind, I have come to regard concrete as the symbol for most things Roman; and the crystalline shining quality of marble as that for most things Greek.

Yet it must also be born in mind that although Athens was the center of Greek culture, the Athenians were adventurous and warlike in their own right. They formed Greek colonies throughout the Mediterranean; and fought most heroically at the mountain pass of Thermopylae to save the State and defeat Xerxes and a large Persian army in 480 B.C. - at the same time deploying their powerful navy to defeat (with the help of a fierce storm) the Persian fleet. But in 404 B.C. Athens (under Pericles) lost the war with Sparta when seeking to protect their relatively democratic form of government from the threat of Sparta's oligarchical system.

It should also be pointed out that many events in the Greek Games demanded feats of endurance and athletic skills that could be turned to good account in times of war - speed of foot in the 200 meters, physical strength and willpower in the marathon, precision of hand and eye in throwing the javelin, hand to hand combat in wrestling... and while the competitive spectacle, together with the specifically Greek philosophy of a *'healthy mind in a healthy body,* provided the basic reason for the Games' existence, the military bonus should not be overlooked. And there was one particularly dangerous and potentially deadly contest between two protagonists: the Pankration - a no-holds-barred wrestling, boxing, strangleholds and pressure locks contest that was only stopped when a contestant signalled his unwillingness or inability to continue. Consequently it was not necessarily a fight to the death. It seems unlikely that there would be any of the enthusiastic roaring for blood that characterized Roman gladiatorial fights; the thrills of delivering judgement as to whether the vanquished should live or die; the display of morbid excitement at seeing the sword go in.

The Greeks of Athens were not sadistically inclined. Neither did they like extremes. They preferred to see a balance in the imaginary scales of destiny and fate. We have had recourse to the pithy comments of Horace, the Roman lyric poet and satirist, several times in these pages. In *The Epistles*, composed 22 -8 B.C., he writes: *Greece, taken captive, captured her savage conqueror, and carried her arts into clownish Latium.* With these words he grants Greece her glory but, significantly, not Rome her grandeur. Would he agree, I wonder, with the premise that the various life-enhancing philosophical, scientific, aesthetic, and poetic sensibilities that made up the Greek view of life, characteristically came to work together to create a many-faceted mental equilibrium: a state of balance in consciousness between differing perceptions, intuitions and passions that allows each their due, yet integrates them to create an amalgamated set of principles. A wide-ranging intelligence system through which they were able to scan the various aspects of 'reality' that present themselves to consciousness. And, in so doing, discern the worth of each in terms of the greater significance and meaning of the whole. For it is only through such a comprehensive awareness that the purpose of our journey - played out in a cosmos of infinite dimensions and untold forces - is able to flicker from time to time before the mind's eye. Such is the nature of the Greek genius - the means by which they arrived at an overall understanding of the psychical balance that, when achieved, illuminates consciousness: albeit with an uncertain and flickering light.

Horace's comments illustrate how the Athenian way of seeing things has been idealized throughout western history. As a boy in school I remember reading *The Greek View of Life* published by the English writer

Lowes Dickinson in 1896, and particularly the lines, *That sunny and frank intelligence... that unique and perfect balance of body and soul, passion and intellect, represent, against the brilliant setting of Athenian life, the highest achievements of the civilization of Greece... With the [end of] Greek civilization, beauty perished from the world.* And in W.B. Yeats' poem, *The Second Coming* (to which I refer at greater length when discussing the concepts of spirit and soul in Chapter 18), there are two lines that capture the significance for western civilization of Greece's downfall:

> *Things fall apart; the centre cannot hold;*
> *Mere anarchy is loosed upon the world...*

The moral fiber of Ancient Greece results from this sensitivity to the rules of proportion, balance and harmony: principles seen to govern both the relatively stable, continuous persistence of things in the material world, as well as regulate the energy forces animating the human psyche, thereby inducing a psychologic stability maintained by the assiduous application of mental checks and balances. Altogether a wise and balanced way of looking at the world and oneself, eliciting as true a mental perspective on human strengths and weaknesses, and nature's complexities, as I think we are capable of achieving: an intelligence of heart and mind that leads one to affirm and preserve life, rather than denigrate and destroy it.

Francis Bacon (1561-1626), Lord Chancellor of England during the reign of James I and the most learned and insightful thinker and statesman of his time, whose discourses on how the revelations of science were going to drastically change man's attitude both to himself and to the mysteries of the universe have proved singularly prophetic - has been aptly described by Loren Eiseley - in my view America's foremost philosopher-scientist of the 20th century - as *The Man Who Saw Through Time*: (the title of his book on Sir Francis.) Certainly, Bacon was a man for all scientific and philosophical seasons, and even a cursory reading of his works reveals how Greek thought inspired the workings of his singular mind. In *The Advancement of Learning* he wrote:

> *Mere power and mere knowledge exalt human nature but do not bless it... We must gather from the whole store of things such as make most for the uses of life.*

How does one *'bless'* life? By gathering *'from the whole store of things...* says Bacon - imbued with the spirit of the Greek passion for breadth of learning - and then following-up with these most significant and wise words, *'such as make most for the uses of life'*. This short and forceful phrase again reveals his affinity with Athenian philosophy, for it subscribes to the Greek view that while nature may be considered mindless, mankind is not. Therefore it is left to human consciousness to understand her ways and act as the steward of her estates, the protector of her interests, in order to assist in the flourishing of life. In this the human role in life is seen as twofold. First to act as the mind of nature. And then to go beyond stewardship of the world and become master of oneself - employ the manifold resources of consciousness to divine one's own purpose in the scheme of things. Follow up on the possibility that in being possessed of a creative vision, and thereby taken into regions of the mind out-of-bounds to the senses, we may be subject to a destiny that lies beyond nature's spheres of influence.

For the Greeks it was a destiny most likely to be realized by attaining the sensitivity that allows for balanced judgements - for conclusions reached when in a state of mental and emotional stability. A balanced and contemplative state of mind that permits of wisdom. And the greatest gift bestowed by wisdom is a sensitivity to beauty and moral issues. (Even so, it should be mentioned that - although knowing it to be an immoral practice - both the Greeks and the Romans kept slaves. Yet better a Greek master than a Roman.)

Certain questions come to mind concerning the way we live our lives today. Are we not firmly in the Roman camp - morally speaking, that is? The high levels of violence provided by the western world's entertainment industry; the attraction of video-games simulating brutality, death and destruction; the numerous incidents of assault and homicide in the cities; and a gun-culture that leads to sniper attacks, shootings in schools and places of work... all reveal the violent undercurrents permeating society. Or can we at least claim that our appetite for sports is Greek-like... bearing in mind the popularity of aggressive body-contact contests - the way in which extreme aggression and the prospects of bloody physical damage will set the fans roaring?

So how many of us, do you think, if given the chance to go back in time, would choose a Roman life with the Colosseum's offerings on hand to fill the weekends? Or how many would take the Greek way of life, preferring to nip down to the Stadion and see some track events, or watch Socrates at wrestling practice?

Brutality and violence are much in evidence in today's world; indeed, so they have been for much of the last century. New horrors created by political and religious extremists occur all too frequently and flash across the television screens of the world - while horrendous scenes from the past persist vividly in the memory of the older generation. And with the global proliferation of the arms trade, the belief in military power as the way to impose ideologies on others, and the sore lack of wise leaders... the spirit of our age does not seem to have much in common with the ethos of Greek humanism.

16

MIRROR, MIRROR, ON THE WALL...
Putting oneself together: the Faces of Ego

We are all framed of flaps and patches and of so shapeless and diverse a contexture that every piece and every moment playeth his part...
Montaigne: *Essays II. xxv.*

INTRODUCTION

A strange image this - a self made up of 'flaps and patches'? Yet it certainly conveys the characteristic workings of the mental patchwork we have been describing throughout these pages. Take, for example, the Chart of Consciousness that launches Chapter 4: even as a relatively fundamental diagram it illustrates the dynamic and complex interplay that can develop between the deductive (sensory) and inductive (intuitive) streams of consciousness.

Montaigne, never guilty of pedantry, refers aphoristically to these differing interactive drives of thoughts and feelings as the 'flaps and patches' contributing to the formation of every individual's distinctive personality and character. And, reading between the lines, his observation implies that if the differing mental experiences are not harnessed to work in unison - but left to jostle independently and anarchically in a consciousness that has not learned to integrate them and build a comprehensive view of self and life - then a person will suffer the consequences of a fragmented and unstable personality. He or she will wander through life like a ship buffeted by wind and current with no one at the helm.

So how can one try and ensure that 'every piece and every moment 'playeth its part'; establish the modulation (ordering, regulating, adjusting) of the 'flaps and patches' in order to become individuated - a unique and whole person? And this so that one can assume command of one's own little ship of state, skilled in navigating the depths and shallows of life. In my view, the most effective way to achieve this kind of control is to practice a reasonable level of introspection. There is nothing 'unhealthy' or morbidly unnatural about this.... as some contemporary psychologists tend to believe. To know periods of introversion (Jung's word), and peruse the nature and pattern of one's thoughts, and the range of one's feelings, is a necessary and normal function of consciousness. It allows one to become critically aware of the psychological factors governing one's overall attitude and behavior.

The importance of such self cross-examination has been mentioned previously in these pages, for it bears so directly on the process of self-knowledge. Indeed, there is really no other way of surveying the *course* of life travelled, and of realizing *how* and *why* one has travelled it thus. One's essential character gradually starts to emerge. Without such reflective insights, the possibility of having a say in the fashioning of one's own character and personality is pretty remote. The alternative – if I read Montaigne correctly – is to remain a person of *'...shapeless and diverse a contexture'*, one suffering a fragmented dysfunctional consciousness capable only of responding impetuously to each random thought or flair-up of feeling.

The opposing view is that the such introspective disclosures does more harm than good - inhibits future mental and behavioral development. 'Let sleeping dogs lie', say some psychologists. Yet the significance of the Delphic Oracle's constantly given advice - sought out as a matter of course in the Classical Age of Greece, 'To Know Thyself" entails precisely the opposite - requiring that one should face up to the psychological complexity of one's own nature and, in so doing, gain the measure of the conscious and unconscious mental forces which are in contention shaping character, temperament, personality. Take charge - as the Greeks would put it - of one's destiny.

Some people are wiser and better than others when it comes to pensively recalling the paths they have taken through life. There are those so taken over by egotistical needs in the pursuit of power and self-aggrandizement, the satisfying of compulsive hedonistic appetites, that the process of introspectively sorting themselves out is devoid of significance or meaning. Concern for psychophysical health cannot compete with obsessive self-indulgence. At the opposite end of the scale stand those innately 'good' men and women in whom ego plays the kind of natural and necessary role which is discussed throughout this chapter: who are kind and unselfish, feel for others and reach out to help fellow human beings, often at the expense of their own personal comfort and prosperity. Individuals who, through an introspective nature, are able to say, *'There but for the Grace of God, go I...'* While somewhere in the middle are to be found the rest of us - caught between the devil and the deep blue sea - inwardly wrestling with our psychological urges, struggling to live a balanced and meaningful life, heading into the future only after reevaluating our responses in the past.

My favorite poet William Wordsworth used to talk about *'Thoughts and feelings recollected in tranquility'* - a markedly expressive way of describing the consciousness that turns to look in on itself. Whatever current psychiatric practice thinks of this process of recollection, it has long been seen as the most effective way to set the mind on the path of self-evaluation. It has been the way adopted over time in many societies by mentors, shamans, elders, priests, parents, and teachers.

I remember well enough when in the throes of searching for an adolescent identity that stubbornly refused to show itself, being exhorted to 'fight the good fight' - advice that made little sense at the time. The suggestion that moral qualities and strength of character are both tested and shaped in the 'fight' to know oneself, only took on meaning when, at the age of sixteen, I read the following lines in Robert Browning's *Bishop Blougram's Apology*:

> *No, when the fight begins within himself,*
> *A man's worth something.*

(The Athenians would have liked Browning: for the mark of the hero in their eyes was the resolve to stand and fight against physical and psychological weaknesses - to overcome fear, hopelessness, helplessness, pessimism... in the face of seemingly impossible odds, triumphing in the end due to the mobilization of spirit and will. Sir Ernest Shackleton - who we met in an earlier chapter - would have been readily admitted to the Greek *pantheon* of illustrious heroes.)

But the proposition that self-discovery is promoted by mulling over the pattern of one's most privately held thoughts and feelings seems to carry little weight nowadays, either in the home or the school. Whereas in my time it was expected that one would learn from one's likes and dislikes, successes and mistakes, adventures and misadventures...as well as from the dastardly acts of villains and the courageous and moral displays of heroes in literature and history, by always imagining oneself in their shoes and wondering how one would respond to the challenges they faced. All too often in my own case I would come to the honest conclusion that I did not possess the stuff of which heroes are made; at best that I might have been able to just about cope in difficult circumstances *under the eye* of a Shackleton in the Antarctic, or an Albert Schweitzer in darkest Africa.

As a boy I was lucky to have a stepfather who constantly anticipated where I stood intellectually and emotionally in my little life, and kept the appropriate books coming. Also, he displayed the uncanny knack of initiating conversations absolutely pertinent to my particular frame of mind at any given moment. Talk and more talk was the order of the day – constant question and answer – fact and fancy equally important: *Alice in Wonderland* enjoying equal time and significance with *The Electrical Life of the Cell*. I was never out of the books for very long. Fairy tales, illustrated natural histories, world gazetteers (from where I first began to make pen and ink copies of the facades of European cathedrals), histories of empires, the biographies of explorers and the writings of famous men - from which those of Marcus Aurelius, Arrian on Alexander the Great, Xenophon, Saint Francis, and the explorer Sir Richard Burton emerged as the favorites... The legendary King Arthur exemplified the ideal knight as the man who righted wrongs; while Merlin proved to the young reader that supernatural forces existed, that metamorphosis from one form to another was indeed possible, that mystery pervaded the cosmos.

In addition, my stepfather was always ready with psychological analyses of both his own friends and mine; always wanted to know how I thought they 'ticked,' as he put it – a practice which instilled in me an enduring curiosity concerning the nature of the inner life of people I met – what was going on in there behind the scenes?

I would lie awake at night (don't we all in those single-digit years?) mind awash with thoughts of how likely it seemed that one would die young, how prevalent was death and personal tragedy in the newspaper reports, and of what it would be like in those final moments. And wonder whether my selfishness, meanness, jealousy, ability to lie, latent viciousness when crossed, cupidity, pride, and masturbatory tendencies... were not already responsible for moving the Finger of Fate - in the form of an early demise - steadily in my direction. Bad things were happening everywhere in those days: there were thousands of hungry people on the march as a result of the Great Depression; the 'forces of evil' - in the form of Germany - were, as my stepfather declared, 'on the march again.' And I felt to be disliked by everyone and lay awake planning how to avoid people and live a solitary life.

There were long periods when such baleful workings of the mind overcame optimism and a friendly self-regard. Yet after living vicariously through delving into the lives of others I knew that life was full of 'ups' and 'downs'; that the very best of us were beset by doubts and fears, given to ungenerous feelings and dislikes - even the men and women I admired the most. But it gave me a great head start to be immersed in books where such a cross section of human types came to life - causing me to recognize those to whom I intuitively felt drawn, and those who aroused an immediate antipathy. Nevertheless there were plenty of times when this serious, youthful idealism and intellectual curiosity took a back seat. Periods when I was totally into lust and self-assertion – seduced by images of sensuality, driven exclusively by Freudian fantasies and desires, amassing a grand collection of legs and breasts which I kept secretly beneath the bedroom carpet; and occasions when I found myself a half-willing voyeuristic fellow traveller with those whose ruthlessness in violent conquest, or in the pursuit of power, appealed to some lingering primitive and sadistic streak held captive in some more primitive part of the brain.

But it was all grist to the mill of growing up - of sorting out of one's *flaps and patches...* coming to the realization that the revelatory mental discoveries determines something called character which, like a fingerprint, is unique to oneself.

"Remember," said G. M. Lyne, my Classics master at school, "*that the Ancient Greeks knew all about walking the tightrope between the heights of nobility and the depths of depravity. And as we all stand on the shoulders of those who've gone before, you'll do yourself some good by getting to know them. Then you'll have a firm foundation on which to build your own young lives.*" And I followed his advice by reading avidly everything pertaining to the Classical Age of Greece that came my way. Yet for anyone seeking to know and harness the mental dynamics of their own consciousness in a Montaigne-like way, one formidable hurdle presents itself - the wild card known as the unconscious. This 'underground' source of dreams and dream-like flights of fancy and inventive reverie - hidden mental themes and persuasions that constitute the stuff of the deepest seams of one' psychic disposition - operate independently of normal conscious processes. As we have seen they come to mind most readily when sensory and rational processes are off-duty, so to speak, and consequently possess the element of surprise together with a high degree of significance. The most persuasive and influential of them are charged with an air of mystery, for they relate to the presence within the psyche of the two most profound archetypal forces that lend significance to life: the one carrying indistinct yet compelling mental reminders of the darkest and most germane secrets of nature to which we were once much closer; the other bringing to consciousness at some point in our evolution the image of transcendent, so-called spiritual values to... introducing the *human* into being. An element I believe we have to discover if we are to live with purpose. Over two thousand years ago both Socrates and Plato brought the rational powers of the Greek mind to accept - as a reasonable proposition - that so-called irrational factors are present in the whole scheme of things. They envisaged, for example, the existence of an 'occult' self as part of the overall psyche, potentially divine, and bringing special extrasensory insights to mind: a psychological phenomenon not unlike that characterized by Freud and Jung in our own time as 'the unconscious'.

It was Carl Jung's publication *Psychology of the Unconscious* in 1912 that helped shaped contemporary awareness of this subliminal imagistic world. The picture he presents of mental processes over which we have no control, intruding anarchically into the everyday life of consciousness and circuitously influencing attitude and behavior, adds to the difficulties in attempting to reconcile Montaigne's diverse 'flaps and patches'. For to be faced by conjectures that arise involuntarily and take one beyond familiar trains of thought, simply introduces more, and less easily identifiable, aspects of character and motivation which must fit into any overall picture of oneself already under way. Consequently, the job of bridging the gap between the workings of a regular consciousness and the unbidden incursions of the unconscious demands a great deal of perseverance, and takes a great deal of time. Most likely, nothing short of a lifetime.

> *The uttered part of a man's life, let us always repeat, bears to the unuttered, unconscious part a small unknown proportion. He himself never knows it, much less do others.*
> Thomas Carlyle: *Sir Walter Scott*

Thomas Carlyle, Scottish essayist and historian was born in 1795 and died in 1881. In his day the concept of the unconscious and the significant part it plays in shaping the life of the individual was hardly a common topic of conversation. Yet in the passage above he aptly describes what has been called in modern times 'the great divide' in human consciousness.

CONSCIOUSNESS AS A PATCHWORK QUILT

Quilts made by sewing and stitching bits and pieces of fabric onto a coverlet have, over the years, come to assume the status of works of fine art. It is a slow and laborious process, involving the assembly of dozens of fragments of cloth gathered from many sources and painstakingly juxtaposed to create the most satisfying result. But what is 'satisfying' in the context of a quilt on a bed? Those who follow the craft say it should please in the way that a stitched and sewn tapestry hanging like an extra-large 'painting' on a wall regales the eye. And I would say that this is certainly the intention of those who spend months creating these visual extravaganzas for the bedroom. These artists must be constantly collecting snippets, large and small, of stitchable materials comprising varying textures, shapes, and colors - some displaying recognizable motifs, others purely abstract patterns. I would think it necessary to cover a large area of floor with them, in order to facilitate the selection of a particular piece for its special place in the work. In making these decisions the artist's eye and aesthetic judgment must work together, deciding the evolving characteristics of the design as it gradually takes shape - overseeing the visual rhythms created as differing sizes and patterns of cloth are stitched to fit together; the vibrancy and compatibility of their colors and textures; and the expressive and symbolic links revealed between the differing representational or abstract pictorial qualities presented by each piece. All of which being aesthetic achievements that the great French painter Paul Cézanne regarded as vital in the creation of any significant work of art. He gave the word *moduler* to the process of organizing the forms, spaces, and colors (the basic elements in pictorial design) in such a way that they harmoniously interrelate, rather than compete or 'fight' with each other... thus imparting to the work a balanced, visual wholeness - the kind of structural homogeneity within the cone of vision that our visual perception of the world presents. The strikingly successful patchwork quilt is therefore a work of art in which each individual piece retains its own visual identity, yet plays a vital part in contributing to the uniqueness of the design as a whole. Remove any one part from such a complete work and it appears to fall apart, its aesthetic impact substantially diminished.

Not all such quilts achieve this level of aesthetic success. It takes an artist capable of feats of patient and creative modulation to bring it off.

And so it is with bringing together the *flaps and patches* of consciousness to fashion and become aware of oneself as a total entity - a complete 'patchwork quilt' of a self.... formed by mustering, recognizing and evaluating those parts of one's psychological fabric available for inner scrutiny. Which is akin to creating a work of art - only this time with the tapestry-like, psychical unity of oneself as the result. It is a process which can go on until the last minutes of one's existence. I think of Michelangelo, at the age of 88 and just a few weeks before his death, sending the marble flying as he worked away on the Rondanini Pieta... declaring that he was not ready to die, pronouncing, *"For I am still learning..."*

> *Men can starve from a lack of self-realization as much as they can from a lack of bread.*
> Richard Wright, Native Son

For we make our own psychic weather, like mountains make their own elemental systems. Yet unlike the mountains we have the ability to be our own meteorological psychologists, able to present a united front to the heavy mental weather we experience from time to time.

MIRROR, MIRROR, ON THE WALL...

Nothing can be more disconcerting than staring at oneself in the bathroom mirror – not in the routine way one does when shaving or applying makeup – but quite involuntarily, under the sway of some irresistible compulsion.

What the Hell are the Neurons Up To?

Twice only, in the course of the last forty years, have I found myself in this situation, trapped for two or three minutes by the unfamiliar image in the glass – a long time to spend in such disconcerting confrontations, particularly when unable to find the will to pull away; return to the everyday world. I first experienced the phenomenon one night when, after brushing my teeth and turning towards the bathroom door, I found myself unaccountably drawn back to the mirror where the old familiar face returned my gaze. There I was, as usual. Yet only briefly. The image was already beginning to transform itself – outlines of features wavering causing definition to be lost, before settling down to present a subtly distorted version of myself – mocking, scornful, unpleasantly sneering. Not the face of a man one could like or trust. And I knew instinctively that it must be stared down – that there must be no looking away, no obvious consternation or apprehension on my part. The apparition gave ground slowly, fading to the contours of the old familiar face – yet one barely recognizable as myself, for it was translucent, a waxen mask, smoothed of all lines, wrinkles and sags; strangely detached, gazing into a distance far beyond where I was standing, unaware of my presence. It was a reflection with which I suddenly felt more ill at ease than I did with its predecessor. For it was a lie: there was no way I could consider myself to be related to, or worthy of, this calm and inscrutable visage. But this discomfort was nothing compared to the shock which followed. Abruptly – and without my being able to perceive how the transition was effected – I was confronted by a face so alien and of so menacing a presence that I felt a quick stab of fear. It was a countenance given power by the bones: assertive and sharp-edged bones, avian, suggesting a creature of sky not earth; demonic in one sense yet ethereal in another. But somewhere in this phantom image my likeness lingered. Above all it was the darkling eyes which set the scalp prickling. Brilliant, without a hint of gentleness or compassion – they bored through me, relentless and judging.

I knew at this point that it was essential to beat a retreat – the image itself showed no signs of fading. Yet limbs were disinclined to obey orders. Only the sound of my voice, telling me loudly to 'get the hell out of it' broke the spell, straightened me up to turn away and leave the room without risking a backward glance.

Twenty years were to pass before I was possessed in a similar way. Yet on this second occasion the faces in the glass were relatively benign. The visitation was still unnerving, but the previous cynicism and judgmental hostility were less evident. I felt to be more on equal terms with the mirror-presences, and as it seemed that peace between us was established, I did not find it difficult to turn away and go to bed.

An old and wise friend told me that the phenomena I have described, while not widely experienced are, however, not unknown; that they can even be produced voluntarily – initiated by staring intently at one's natural reflection, and continuing to gaze with patience and concentration until the 'other', in one guise or another, appears in the glass. You might want to try it – discover how many of you there are. It may turn out to be less of an ordeal if you, rather than 'they,' assume dominance when standing before the mirror. However, if one becomes fixated on the appearing images by reason of some compelling urge to stand and stare – it would seem that a sudden up rush of the unconscious is responsible. In this event, an encounter with masks of the psyche is no bad thing, for it can signal schizophrenic imbalances between critical aspects of consciousness – an ineffective integration of 'flaps and patches.' Like the significant dreams of sleep, the mirror 'dream' can serve to arouse awareness of serious problems in coming to know oneself and comprehend the structure of one's character.

Three outstanding men of the 20th century have written of the need to travel the inward journey - to experience what the poet Edwin Arlington Robinson described as, *'The sight within that will never deceive'*. To which I would simply add that one should especially value those archetypal images which possess a transcendent quality: any or all of the many forms of the *dei imago* such as the transfiguring power of love, the truth of perfection or beauty... all those heart-warming insights which can stop us in our tracks, stop the clock, render speech inadequate; cause us to *'move out of our senses'*, as Plato said. While in *The World as I See It* (1934), Albert Einstein wrote:

> *...everybody has certain ideals which determine the direction of his endeavours and his judgments. In this sense I have never looked upon ease and happiness as ends in themselves - such an ethical basis I call more proper for a herd of swine. The ideals which have lighted me on my way and time and time again given me new courage to face life cheerfully, have been Truth, Goodness and Beauty.*

From the American writer and philosopher Lewis Mumford - in his, *Condition of Man* (1944) - we have:

> *A day spent without the sight or sound of beauty, the contemplation of mystery, or the search for truth and perfection, is a poverty-stricken day; and a succession of such days is fatal to human life.*

While Carl Jung - considered by many to be one of the most insightful thinkers of our time - wrote in *Memories, Dreams, Reflections* (1961):

> *The decisive question for man is: Is he related to something infinite or not? That is the telling question of life.*

There are times while writing this book when I think how utterly anachronistic are the concepts and experiences it embraces. Life in the last thirty years of the 20th century has become increasingly outgoing, extroverted (Jung's terminology)…. and the sort of inner experiences we are constantly advocating here will no doubt seem irrelevant to many in today's consumer society. Yet if the American poet Walt Whitman was right when he suggested that you cannot buck the given order of things - be they within the natural order or the human order - without risking unforeseen consequences, then we are heading into rough water so far as maintaining the interior world of our consciousness is concerned. Personally, I can think of no worse fate than to become a totally outgoing person with no private agenda to occupy the mind and, consequently, having but a diminished sense of personal identity.

In an affluent and well-nourished society where ample food and creature comforts are taken to be vital constituents of the 'good life', might one not expect there to be some compensatory hunger for the kind of insight that hints at the presence of what we call - parrot-like and conventionally - the soul? For although the expression, 'body and soul' is still habitually used, I wonder if today's prosperous citizens in the world's developed nations would not find it more agreeable to suffer a lack of nourishment for the soul, rather than be deprived of food for the body? Which would be considered the greater hunger? Robert Lacey, co-author of the book, *The Year 1000*, indicates that in Britain a thousand years ago the response to such a question might have been exactly the opposite. In a London *Sunday Times* article about the book, Lacey writes:

> In August 1000, the importance that people attached to the loaf-mass showed the difficulty they had in meeting their lesser hunger - while the greater one was satisfied almost without trying, through their unquestioning religious faith.
>
> For us today, the opposite is true. Few Britons will suffer tomorrow for lack of food. But as we relax this bank holiday, how many of us could say that our greater hunger has been satisfied?

A QUESTION OF IDENTITY

I have spent a great deal of time wandering in and around Europe's mediaeval Gothic churches and cathedrals and the images they leave in memory are always coming to mind, reminders of the fact that their soaring magnificence served to sustain a religious faith for a largely illiterate populace. More than once throughout these pages I have used the architectural features of these wonderful structures to symbolize and illuminate certain aspects of human consciousness - for in my view these great churches are the most impressive and tangible results of man' more profound aspirations.

In talking here about 'putting oneself together' it becomes clear that the basic task is one of reconciling the tangible reality of one's physical characteristics as readily revealed in the mirror, with the invisible and *intangible* psychological traits of the inner 'I' - a process requiring that one engages in periods of disciplined introspection if a level of self-knowledge and individual wholeness is to be achieved. It is an exercise during which the two selves constantly surprise the heck out of each other.

Gothic ecclesiastical architecture arouses a similar kind of surprise between outer and inner realities, and just as dramatic in its own way. For when one leaves the light of day after being dazzled by the mass and sheer physicality of the stone exterior, and steps into a seeming infinity of space and refracted light... the impact of weathered façades enlivened with sculpture and heavily recessed doorways, of soaring buttressed walls pierced with delicately-spoked 'wheel' windows and high-reaching stained glass windows, is surpassed. One wonders how on earth the skeletal stone-mass, load-bearing structure of the exterior can be built to reach so high.... and still manage to support the stone-vaulted roof inside that spans so broad an area of nave: resulting in the creation of a mysterious and ethereal architectural cosmos in which the clock has no place, mind and feelings take wings, gravity's hold is challenged.

But wander about for a while to sort it all out. Survey the edifice more intently on the outside, starting at the west front and walking as far around as ground access permits - and the engineering principles that permit such a vertically dominant framework of stone and glass to support such a high stone-ribbed roof.... readily reveal themselves. You will notice that the walls of the nave alone, rising high above the aisle roofs, are not bearing the heavy load of the roof. Instead, they are being bypassed, the weight carried to the ground through the heavy mass of the crucially positioned external stone buttresses - whether these are simply embedded into the walls or are free-standing and throwing out flying-arches to reinforce key stress-points high on the nave wall: a wall which, incidentally, by the end of the 15th century, is penetrated (and thus weakened) by large areas of glass in the form of high clerestory windows (See photograph of Bourges, p. 165).

It was the genius of the mediaeval builders that enabled them to construct such an architectural paradox as the Gothic cathedral. For the substantial massing of stone in soaring elevations and towers - a physical statement that takes eye and mind by storm - is sufficient an aesthetic experience in itself. But then the interior element that dominates is just the opposite, totally insubstantial, space and air that take the breath away and sets the spirit loose to fly close to the vault. Saint Bernard (1091-1153), Abbot of Clairvaux, whose Cistercian order founded over 62 monastic houses, wrote that *truth is to be found in the tangible...* meaning that awe, wonder, and a sense of the marvellous can be experienced in the presence of certain phenomena, physical things, organic or inorganic, that inspire by their beauty, structural complexity, and fitness for purpose - be they ants or cathedrals. The late Romanesque or early Gothic great churches built by the Cistercians would surely have been amongst those *tangible* things of which the Abbot speaks, conveying a sense of things sacred that move one from the reality of the physical world into a mental realm of pure feeling: a transcendent state of expanded personal awareness not hedged around by time or environment. The effect of visually moving from the solidity of stone surfaces to the 'nothingness' of soaring space creates a sense of structural and aesthetic wholeness - a unity resulting from the expressive, symbiotic integration of form *and* space. For space is the unique element with which architects work, and when a causal relationship between its 'shape' and that of the enclosing fabric is clearly perceived, then a vision of architectural completeness results: achieved by the complementarity of the material to the immaterial. And in my view, nobody has exploited this union so significantly for spiritual ends as the stonemasons of the Middle Ages - the Master Mason being, in fact, the 'architect' of the time.

This commentary on the Gothic is not meant as a digression, but rather to serve as a symbolic representation of our own physical outer edifice of flesh and bone, and inner immateriality of spirit. In terms of the-face-in-the-mirror analogy, the external physiognomy is immediately apparent. But look in the eyes - frequently said to be *'... the windows of the soul'* and there is a chance the inner cathedral will welcome you in.

However, there are less contrived ways to experience the inner mysteries of oneself. The autonomous mental faculty sometimes described as the workings of the 'mind's eye', can have you wandering 'in pensive mood', as Wordsworth identified the source of his poetic insights. Mozart wrote of going to the 'ragbag of memory', in order to give his musical inner life tangible form on paper. And certainly one may encounter a 'shadow self' at times of going a mental 'walkabout'..... images emerging from the shadowed reaches of consciousness like bats fluttering beneath a high Gothic vault. In my youth it was commonly said of someone who appeared unusually detached and solitary that they were too much 'into themselves' or, more disparagingly, that they 'had bats in the belfry'. It was a derogatory statement then, but nowadays if it were said of myself, I'd take it as a compliment.

I wonder whether Montaigne would attach much importance to this kind of introspection? For although he states in his *Essays* subtitled *'To the Reader'* that, *'The greatest thing in the world is to know how to belong to oneself...'* he also writes in *Essays* I.x: *'Where I seek myself, I find not myself: and I find myself more by chance than by the search of my own judgment.'* Yet he might agree that 'chance' might more readily reveal a significant thought or sentiment when one is wandering through the inner landscape of reflection, than at those times when the senses are very much in control, both feet firmly on the ground.

We have made frequent references to personality and character throughout these pages, concentrating particularly on character ('for good or ill') in Chapter 12. In *Troilus and Cressida*, Shakespeare distinguishes the 'good' character as that illuminated by *'manhood, learning, gentleness, virtue... / the spice and salt that seasons a man...'* But the issue of personality merits discussion in its own right. For although personality and character are generally seen as synonymous terms, there are subtle distinctions to be made between them. Personality is often more readily associated with an individual's external appearance and style - with the persona, the social facade or 'front' he or she presents; which can be 'read' when in their company and make an immediate impression. Whereas it is said that to know a person's essential nature it is necessary to 'plumb the *depths*' of their character: take time to discover the underlying and habitual ways they think, feel, and behave in both good times and bad. On the other hand, we tend to make judgments as to someone's personality after a brief encounter: perceptions based on their manner - their overall demeanor whether it be open or standoffish, cheerful, sullen, positive, pessimistic, patient, bland, flamboyant... and so on. Yet we don't necessarily go on to make a *moral* judgment as to character, although indeed there are times when we do so because an individual's personality is so transparently wholesome or, equally obviously, somewhat shifty. No: forming an impression of an individual's personality does not necessarily guarantee that 'what you see is what you get'. It is all too possible to be hoodwinked by taking personality as a yardstick of character. Many of us have found to our cost that an engaging personality is no assurance of trustworthiness, or that an aloof one a sign of unreliability.

The physical body can be mapped, in whole or in part. Even the physiological structure of the brain can be mapped - at least its more surface characteristics. Yet although this helps the neuroscientist to navigate its terrain and understand its mechanistic workings, the operational means by which the brain conjures up, and transmits to consciousness - chemically and electrically - the more abstract thoughts and feelings that determine the psychologic nature of character.... remains something of a mystery; perhaps impossible, to map.

'There is a great deal of unmapped country within us', wrote the 19th century English novelist George Eliot in *Daniel Deronda* over a hundred years ago - and if one were to take her statement as referring to character-country, then it is just as relevant today. Even though we know considerably more

nowadays about the brain's electrical and chemical life, the presence in consciousness of judgmental moral value systems informing of principles as opposed to facts - is still as much an enigma for philosophers as for scientists. The old nature versus nurture argument crops up again here. Yet while agreeing that environmental circumstances play their part, I would say that innate psychologic propensities play the greater role - that mental energy-forces emanating from the *'unmapped country...'* in the form of intuitive feelings and thoughts, bring us to require a certain measure of self-respect - and, further, to hold life itself in some esteem: insinuating that we do no harm to ourselves or to it. When these forces are strong, they make for strength of character and firmness of will; when weak, for deficiencies in character and feebleness of resolve. And this is not to gainsay the possibility that more *metaphysical* forces such as strength or weakness of spirit - or the inability to break through to consciousness of that other mysterious power we traditionally have called the soul - might also play a part in the character stakes.

At this point it is almost with a sigh of relief that I turn once again to the Greeks for an uncomplicated and lucid view of how we should regard the mental phenomenon of human character. The three major determinants of character in Aristotelian terms are *ethos, pathos,* and *logos. Ethos* represents moral strength, the source of one's credibility and therefore ability to persuade; *pathos* is the capacity to feel and so touch the emotions of others; while *logos* denotes the ability to give good reasons for a course of action and stimulate the intellect of others. Confucius, the Chinese sage and philosopher who lived between 551 and 479 B.C. - some hundred years or so before Aristotle - declared that *'Wisdom, compassion and courage - these are the three universally recognized moral qualities of men.'* They are three qualities which correspond pretty closely to Aristotle's thoughts on the principal constituents of character. And now, some two and a half thousand years later, we find little to add to their conclusions. Do we not automatically size up a person's worth according to standards similar to those set down by Confucius and Aristotle? And while first impressions may sometimes prove to be right, it usually takes time and events to fully reveal the mettle of a person's character.

I think of the remarkable heroism and self-sacrifice of the police and firemen in the devastating terrorist attack on New York on September

11, 2001 - individuals who most likely thought of themselves as average members of society possessing, like most of us, their good and bad sides, '... *framed of flaps and patches... shapeless and diverse...*' as Montaigne would say. And yet when those firemen went up that crumbling tower in the hope of saving some of those trapped at the higher levels, most of them would surely have been experiencing some trepidation. And yet not one man dropped out. No doubt their training helped them to overcome fear and anxiety; yet in listening to some of their colleagues later who had not been assigned to that dangerous operation, it was impossible not to be moved by what they had to say about their commitment to saving the lives of others, even at the risk of losing their own. Whether you describe such altruism as courage or compassion, the significant fact is that in the last resort these men found an inner strength - a resource that removed them from concern for self, took them beyond the normal moral persuasions that pertain in daily life. This is a remarkable human quality. And yet there are plenty of men in the world who seem lost to any higher authority of this sort - who feel nothing, can kill without mercy or compassion: men of negative character in whose dead eyes nothing of spirit shows; are totally unknowing, even of themselves. Yet these individuals also have a brain in place, neurons, chemistry, electricity....

All brains are not the same. Great variations occur in the folded and convoluted structure of the cortex - individual differences which can become even more marked when one considers the trillions of neural interchanges possible: a simply astronomical number which surely allows for the possibility that, in some cases, the vital neural interconnections responsible for the moral impulse are just not in place. And one might conclude that, as a result, the light of conscience or the spirit of compassion is never kindled, or may have flickered briefly and gone out. And if the soul - as a concept of our human essence - requires the services of the brain in order to affect consciousness, then it too may be incommunicado.

Those who hold to the doctrines of scientific materialism - who see all of human mental experience as resulting solely from the operation of biological, physical systems - will give little, if any, credence to the metaphysical drift of the speculations we present here. If faced with the following four lines from Wordsworth's *Expostulation and Reply,* they would probably smile indulgently at such poetic fancy:

Nor less I deem that there are Powers
Which of themselves our minds impress;
That we can feed this mind of ours
In a wise passiveness.

THE FLAPS and PATCHES OF EGO as Constituents of Character and Self.

Nowadays, to say that someone 'has an ego' is seen as a derogatory comment - implying negative aspects of character - excessive pride, self-centeredness, self-assertiveness, unbridled self-gratification, lack of concern for others... Yet in the psychoanalytic practices initiated by Freud and Jung at the beginning of the twentieth-century, the term 'ego' had a much broader significance for them, being used to denote the *whole* self, complete in all its characteristics: the personal manifestation of physical appearance and overall mentality unique to each one of us. They worked on the assumption that at some point in the evolution of consciousness we became *self* conscious, aware of our individuality, our personal identity, and were able to say,' Ah... here *I* am... everything else out there is *not* me.' Given the common tendency nowadays to think only of ego as the villain in the patchwork quilt of character, this is an important point to make. It is also a necessary clarification of ego if the discussion in the following pages is to make much sense. And, equally importantly, it returns the word to its classical roots - to what it originally signified when represented by the capitalized 'I'.

A significant linguistic clue tells of the historic connotation of the term 'I' - surely the most used pronoun in any language. Appearing in Old and Middle English, 'I' is seen as deriving from the Greek *Ego'*, the Latin *ego*, and the Sanskrit *aham* - and the general meaning intended was, '*My* presence here'. So the relationship between the English pronoun 'I' and the Greek noun *Ego'* is one telling of presence - the appearance of a complete, one-of-a-kind-*self* comprising body, mind, and (the Greeks would add), spirit. Consequently, I will use the capitalized Greek term *Ego'* throughout the remainder of the Chapter when referring to the all-inclusive 'I".

The people are a many-headed beast, wrote the 18th century English satirist and poet, Alexander Pope in *Imitations of Horace*: which is just another way of saying that we are an unpredictable and inconsistent lot. In terms of *character* we are capable of running the gamut of behaviour from the praiseworthy to the contemptible. We move between generosity and selfishness, are reliable or untrustworthy, prosaic or imaginative, intelligent or witless, murderous or benign, honest or dishonest... - all varied ways in which human beings respond to life, and which are as much in evidence today as they were in Pope's time. Freud and Jung, in discussing *Ego'*, draw attention to the manifestation of such diverse psychological traits in our psychological makeup - the most significant groupings being those I refer to here as T*he Three Faces of Ego'*.

Some years ago when corresponding with the poet and novelist Robert Graves, I used the word 'psyche' in one of my letters. In his reply he said, *'I have always thought that that is a very peculiar way to spell fish'* - tongue in cheek, of course - for as a Classical scholar he knew perfectly well that in the Greek vocabulary the word 'psyche' denoted the vitalizing 'breath of life' induced by the soul and its active agent, spirit.... together with the intelligence provided by mind and brain concerning the nature of the outside world. However, Greek philosophers and playwrights in Socrates' time would no doubt regard this definition as needing qualification.... by introducing the vital role played by each person's *character* in the makeup of their individual psyche. For they were very aware of how psychological attitudes and behavioral responses to life - built up over the years and relatively unique to each person - come to constitute an integral part of each personal *Ego'*.

It is said that no two sets of human fingerprints are alike. Given the 6 billion-plus people in the world such incredible physiological diversity is hard to believe. Yet, similarly, it can also be said that no two persons are psychologically alike - each human psyche being individually structured and conditioned.... functioning in its own particular way. I doubt if anyone has understood this better than the dramatists of Ancient Greece: the tragedies and comedies of Aeschylus, Sophocles, Euripides... which tell of the internal mental dynamics that can create such psychical variety. Shakespeare was every bit their equal in expressing profound insights concerning the potent mix of mental drives responsible for human nature and motivation.

Just to pick up a newspaper is to reveal a bewildering variety of mental attitudes and responses to even local issues, never mind those affecting the world at large - extremes of opinion and behaviour that can cause the reader to shake his or her head in disbelief: news items that would cause my old friend Livio Valentini, the renowned Italian painter, to give one a droll look of disbelief and mutter 'Grande Confusione' as he put the morning paper into the wastepaper basket.

Yet I think that for the great Greek dramatists - and for some contemporary psychologists - the word 'confusion' would be regarded as altogether too mild a term to use in this context: They would see the interplay of psychological forces and events as a conflict: a life-contest for each one of us that must be knowingly entered upon, pitting one's resources of mind and will against - as Hamlet put it - *'the slings and arrows of outrageous fortune'* and, in so doing, discovering the nature of one's own strengths and weaknesses in both confronting and shaping a personal destiny. Life's journey as the anvil on which every man's character and worth is forged: strengths and flaws playing themselves out to result in either auspicious or calamitous outcomes. A truly human peregrination demanding that we engage in our own fate by venturing with some degree of courage into the mysterious depths of the psyche, putting the body to the test and engaging the mettle of spirit - the real heroes being those among us who, while seemingly possessed of but average self-confidence and strength of character, manage to overcome extreme difficulties by discovering profound powers of *Ego'* hitherto unsuspected.

> *Few of us can be saints; few of us are total monsters... It is one thing to be "realistic", as many are fond of saying, about human nature. It is another thing entirely to let that consideration set limits to our spiritual aspirations or to precipitate us into cynicism and despair. We are protean in many things, and stand between extremes.*
> Loren Eiseley: 'The Inner Galaxy' from
> *The Unexpected Universe*

Eiseley's 'inner galaxy' of mental forces conjures images of a veritable firmament of neurons, their trillions of chemical and electrical intercommunications creating a microcosmic web - a universe in itself of trailing, scintillating, axon links - a unique neural cosmos responsible for the form taken by every individual *Ego'*. In referring to 'saints' and 'monsters', Eiseley distinguishes the two extreme polarities of human character and behavior: that of saintliness which - in terms of the three *Faces of Ego'* - reveals the overwhelming influence of *supra*-ego. And of outright wickedness when the machinations of *gross*-ego are in control.

The mid-position is that induced by good-ego *which* strives to achieve a balance between the pull of the other two extremes - and one which I think most of us naturally try to achieve. It is the central psychological plank of Selfhood, particularly in that it brings to consciousness the recognition of our own specific existence (presence).... as distinct from the rest of the world. Few of us manage to cloak ourselves in the radiant patchwork quilt of sainthood (as Montaigne might put it) or, indeed, are naturally at ease wearing the dark cloak of the evildoer.

Good-ego constitutes the psyche's central influence stabilizing mental life. Among the countless images and impulses that parade through the stream of consciousness it manifests as a steadying authority, promoting the kinds of thought and feeling which lead one to develop a balanced outlook on existence and respect for oneself. It brings reason and intuition to work together in appraising the worth of opposing and paradoxical aspects of life's offerings - of determining, in Erich Fromm's words, whether such and such an attitude or action reveals *'a reverence for life... for enhancing life'*; or *'stifles life, narrows it down, cuts it to pieces...'* It works hard to persuade one to take the former course. All in all, the moral stability and self-respect induced by the face of *good*-ego plays a major role in forming a rounded, identifiable and companiable 'I'. A Self finding purpose in life through simply acknowledging feelings pertaining to what is right and what is wrong... doing the best it can to serve conscience and do no harm to the common good.

> *In popular language we say that the (good) ego stands for reason and sanity, in contrast to the id (gross-ego) which stands for untamed passions.*
> Freud: Lectures in Psychoanalysis (1933).
> (The words in brackets are mine.)

It is important to like yourself. A great deal of trouble in this world is caused when people dislike themselves - usually when the positive and supportive powers of *good*-ego have little chance to develop in the face of constant, narrow and negative attitudes to life that inhibit a central sense of wellbeing. It is difficult to think affirmatively about one's life or that of others when disaffected with oneself and life in general. Too low a self-esteem is just as dangerous for the wellbeing of society as too high a self-regard. Whether a deficiency of *good*-ego can be considered congenital or to result from negative life-experiences, its absence can lead to self-destruction of one form or another. (A personal observation here: it seems that in growing older one may experience strong groundswells of *good*-ego's promptings, resulting in a certain serenity. The contest with the outside world becomes less important as this maturing 'I' becomes more autonomous and sufficient-unto-itself - driven by a free-roving spirit that hints of more mysteries to come.)

The belief that the face of *good-ego* represents a psychologic force capable of overcoming the instinctive urge to act aggressively and selfishly in order to meet with success and, more primitively, to simply survive - has been around quite a long time. Confucius, Chinese sage and philosopher wrote: *There is only one way for a man to be true to himself. If he does not know what is good, a man cannot be true to himself.* Lucius Annaeus Seneca (the younger), the Spanish-born Roman statesman and philosopher living between 4 B.C. and 65 A.D., wrote in *Epistolae Morales*: *In every good man a god doth dwell.* Some one thousand and five hundred years later, the French essayist Montaigne declared: *Confidence in the goodness of another is good proof of one's own goodness.*

One would wish to start life automatically aware of *good*-ego's guiding influence, thus setting off on the right foot from the start. But then there would be no formative process; no psychical growth without the pain and scars that come with trial and error. (Diamonds, after all, are only created by the heat and pressure of geological forces.)

> *They say best men are molded out*
> *of faults;*
> *And, for the most, become much more*
> *the better*
> *For being a little bad.*
> Shakespeare: *Measure for Measure V.i.*

In contrast, *gross*-ego presents itself as an unremitting and insidious drive to attain personal power, satisfy every desire and whim whatever the cost to others, have little or no regard for the sanctity of life, and be driven by excessive pride. So we see the 'I' rampant - out to sabotage or destroy that which appears to be productive, purposive, praiseworthy.... or that conforms to some universal moral code: *'Stifling life, narrowing it down, cutting it to pieces...* to quote Erich Fromm once more. Yet it is due to the presence of the psychological tendencies identified as *'gross'* that we are pushed to wrestle with ourselves... try and become fair-minded, just, and 'decent' men or women attempting to understand and feel for the rights and needs of others. ('Decent' is defined in most dictionaries as 'fairly good but not excellent!) *Gross*, on the other hand is a lumpen kind of word, slow and dull sounding. The Oxford English Dictionary defines it as a term 'of conspicuous and dominant magnitude in its early use; connoting inferior, repulsive, coarse, lacking in quality and perception...'

We have talked previously about the opposing forces of good and evil, pointing out that nature herself knows nothing of such judgmental terms and stressing that the concepts are ours - resulting in words that define those human ways of life which are thought to serve the wellbeing of mankind and the world, and those behavioral patterns that work against life in all its forms. Many profound books have been written to try and shed light on this enigma. Voltaire, the French philosopher whose influence throughout Europe was unparalleled in the 18th Century, considered it a question which could never be answered. Yet here we are, philosophers and psychologists still attempting to understand these opposing mentalities. A later countryman of Voltaire's, the poet Charles Baudelaire (1821-1867), expressed the dilemma in *Mon Coeur mis a nu (My Heart stands exposed), in* the following words: *There is in every man at all times two simultaneous tendencies, one towards God, the other towards Satan.*

We talked at length in Chapter 12 about this dichotomy in the human psyche, employing phrases such as ' the character-*shriveling* effect' of mean, selfish and vicious conduct, and the 'character-*enhancing* effect' of virtue and kindness - directly opposing forms of behavior which the early 20th century practitioners of the new 'science' of psychiatry attributed to the 'good' and 'bad' aspects of *Ego*. Being brought up in the Christian tradition inclines one to think that a natural tendency towards goodness prevails in most human beings - that it represents a dominant norm in our psychological makeup. While our dark and destructive tendencies are due to some inherent flaw in our spiritual growth. On the other hand it is also important to consider the theory that *gross*-ego is a psychologic force in its own right - a pervasive human drive requiring its own particular explanation.

I wonder if Voltaire - were he to be wandering around spectrally somewhere, privy to the theories of contemporary neurologists, psychologists and philosophers concerning this issue - - would think that we are shedding some light on the problem? What would he think, for example, of the suggestion that *gross*-ego exemplifies the persistence in consciousness of the fiercely competitive primal and instinctual urges to aggression serving the survival drives originating in the early brain.... but which have been aided and abetted for the last few hundred thousand years by the evolved higher brain's imaginative ability to better serve such self-serving, aggressive urges? Or that this *'grossness'* results from neurological problems in the form of malfunctioning neurons and their interconnections, creating a situation where the more natural mental influences of *good*-ego are inhibited or blocked altogether? Yet again, one can take a more metaphysical approach, and blame the worst excesses of compassionless and a-moral self-interest on the weakness, even absence, of any *spiritual* force influencing the psyche of those whom Loren Eiseley describes as 'monsters'. Finally comes the question as to whether the worst manifestations of *gross*-ego can be seen as the work of evil forces operating independently of either neurological or metaphysical spheres of influence: *'...this dark and disturbing spirit'*, as Carl Jung described it such a possibility. In *Aion* he wrote:

> *Today as never before it is important that human beings should not overlook the very real danger of the evil lurking within them. It is unfortunately only too real, which is why psychology must insist on the reality of evil and must reject any definition that regards it as insignificant or actually non-existent.*

Yet however one attempts to account for *gross*-ego's presence and effect, perhaps the most obvious thing to say is that without its presence we would not be presented with moral choices at all. The challenge of 'the moral' would not be with us if amoral possibilities were unknown. The emotional and intellectual tensions that determine the strength of the faculty we call conscience would not weigh so heavily on human consciousness. Dick Guterman, a wartime colleague of mine - of whom you will have read in Chapter 14 - used to refer cynically to those who remain indifferent to either choice as the *'morally neutered members of mankind'*. The British novelist Anthony Burgess, was more explicit in describing them as, *'soulless, a mere machine, or a mere emptiness, if you like...'* But Burgess' contemporary, that great writer Iris Murdoch who examined the psychology of her characters with such insight, was less judgemental in terms of their moral frailty. Here she is, talking about the fine line we all must tread in the contest between *good*-ego and *gross*-ego:

> *Plato is king. Plato established the notion that human life is about the battle between good and evil. A.N. Whitehead said that all Western philosophy is footnotes to Plato. He set up the first great philosophical picture of the human soul and of this mysterious business we're all involved in, and when I first started studying all Plato's dialogues, I was absolutely enchanted and taken over by this extraordinary mind... If one's looking for philosophical pictures, I would follow one which makes it very clear that human beings live on a line between good and evil, and every moment of one's life is involved in movement upon this line, in one's thoughts as well as in the things one does.**

If we are to live every moment of life aware of the good and evil sides of the line we walk, as Iris Murdoch suggests, it will have to be a more judicious, and certainly a more solitary walk, than most of us are inclined to take. 'Solitary' is the key word here. For when part of a group - and more so when in a crowd - one's most personal and privately held values can be pushed to the sidelines, even supplanted, by the collective mood and point of view of others. As Sigmund Freud points out in *Group Psychology and the Analysis of the Ego (1921)...'all their individual inhibitions fall away and all the cruel and destructive instincts, which lie dormant in individuals as relics of a primitive epoch are stirred up to find free gratification.'* (I would enlarge on Freud's statement by saying that not only do *'individual inhibitions fall away'* but that individual characteristics in general are swamped: the 'self' is lost when part of the crowd.)

One has only to view the films of Hitler's Nuremberg rallies to see how thousands of people can be moved to feel and act as one, and come to both tolerate and indulge 'cruel, brutal, and destructive instincts'. When the passion of the crowd rules it is difficult for individuals to maintain a personal sense of identity. And it is on such situations that *gross*-ego finds

---

* Iris Murdoch, *The Fire and the Sun* (Viking Penguin Inc., New York 1991).

it all too easy to take over. All the more important, therefore, to remember the examples of human beings like Hans and Sophie Scholl in Germany who could not be swayed by the collective acceptance of Nazi politics and values. Their strength of character in remaining true to their own principles and insights, and the untimely death they suffered for their moral and spiritual constancy, bears witness to two lives which reflect not only the influence of *good*-ego but, in addition, the inspiring and surpassing force of *supra*-ego.

Supra-ego - the third face of *Ego'* - is the most metaphysical of *Ego's* three countenances in that it inspires a high level of altruism - a state of being indicative of values more spiritual than mundane. One might think of such qualities as representing the ultimate level of *good*-ego - one step before *supra*-ego finally takes one beyond the days of the Natural Order - inducing a way of thinking and feeling that conclusively lifts us above material considerations, beyond the constant preoccupation with ourselves. *Supra*-ego induces a state of mind in which self-love is less persuasive than the powerful feelings by which we come to identify with others: a kind of loving driven by compassion and caring - a commitment to people, principles, causes.... the very earth itself... that rings true. When subject to the influence of *supra*-ego we encounter that profound nature of humanness which takes us beyond a self conscious preoccupation with our own existence. And - whether we realize it or not - results in rendering us more acutely aware of the mystery surrounding the existence of the cosmos and our presence in it. *Supra*-ego is responsible for an awareness of spiritual truths that induce a reverence for all life, and a victory over the 'psychologies' that conceive of religion in a *'death to the infidel'* kind of way.

When Frank Wild, Shackleton's redoubtable second-in-command, talked about the *'little voices'* which he heard when in the vast wastes of Antarctic silence, he was referring to the dialogue initiated by the innermost world of the self - to the kind of converse traditionally attributed to the breakthrough of intuitively-held understandings of truths relating to what are sometimes called *The Eternal Verities'*. Absolute solitude, total silence and a complete separation from civilization, are marvellously conducive for *supra*-ego to speak from the heart. After Shackleton's death on their last voyage together - (Shackleton is buried at Grytviken on South Georgia) - Wild took over as leader of the expedition which became severely limited in scope due to the northerly presence of impenetrable ice. His life had no direction after that: he was only able to 'find himself' when in the Antarctic wilderness - an environment that inspired trust and mutual understanding between men; and quickened the spirit to evoke fearlessness and courage... without concern for personal safety or material gain.

One thinks of the men on the Titanic who went down with the ship in order to give women and children their chance in the lifeboats. And I remember watching - on the television news - an anonymous passenger standing on the tail section of a doomed aircraft that had come down in the ice-strewn Potomac river, and who was obviously insisting that a flight attendant take the only available seat in a small boat, before he himself disappeared with the remains of the fuselage a few seconds later. I have read of African and European doctors working over victims of the dreaded E-bola virus, knowing full well the danger involved in such close contact with the patient; knowing how many of their colleagues had died in similar circumstances. A few days ago a man here drowned in a swollen river attempting to rescue his dog. Members of the police forces, firefighters, air and sea rescue personnel, constantly put their lives on the line in attempting to save others - witness the hundreds of police and firemen of New York City who lost their lives in the terrorist attack on the World Trade Center on 11 September, 2001. I will not easily forget the television news footage showing firemen moving up into the inferno above them. What face of *Ego*' was triumphant here?

It is important to realize that in a world driven by violence and harsh social and economic inequalities, there are 'Frank Wilds' around who are driven - without benefit of Antarctic-like solitude - to perform greathearted acts of courage and compassion all over the world stage. Such heroes (for that is how I would describe them) help one to maintain a slender measure of belief in our worth as human beings, for their lives confirm that such altruistic manifestations of *supra*-ego mark the transcendent heights to which man can aspire. *'The Super-Ego is the highest mental evolution attainable by man.... The feeling of conscience depends* altogether *on the development of the Super-Ego.'* So writes the American psychiatrist A.A. Brill (1874-1948) in *Psychoanalysis: Its Theories and Practical Application*. Born in Austria, he was the first to translate most of the major works of Freud and Jung; and his views as a psychiatrist are in accord with much of what we are talking about in this chapter. (Nevertheless, I should point out that the prefix '*Super*' possesses slightly different connotations from those implied by my use of the prefix '*Supra*'. The first can be seen as referring to quantitative factors signifying 'higher or greater in quantity or volume'; whereas '*supra*' is more qualitative, indicating 'above or beyond', thus suggesting that a transcendent quality pertains.)

The psychological impact of *Supra*-ego can take over consciousness in a way similar to that experienced when one becomes subject to the sudden onset of inspiration - that breakthrough to moments of vision when a surpassing sensibility takes over: an awareness of clock-time slips away, the tangible reality of the outside world dims. The resulting loss of the everyday Self produces a nirvana-like state of mind. I have found that to be adrift on a sea of music is one of the surest ways to be overtaken by that *force majeure*, mysterious and irresistible, which surrounds one with an air of the wondrous... I remember what was said of Leonard Bernstein after his death: that *'he was a musical soul which just ignited'*. And I would think that anyone who saw him conducting Mahler's Fourth Symphony with the Vienna Philharmonic would surely agree that there were moments when he seemed almost transfigured - that this was a man who, through music, knew something of the Infinite. Here is a note I wrote to myself at the time: *A man in a true state of grace - uplifted, devoid of vanity or self-consciousness, living a truth which lies beyond the world: remember Nietzsche in* The Birth of Tragedy *talking about... the genius in the act of creation merging with the primal architect of the cosmos.*

Many years ago I made drawings of famous British conductors for the B.B.C's weekly journal, The Radio Times. I particularly remember Sir John Barbirolli - with whom I spent many hours on concert tours with the Halle Orchestra - talking with him between rehearsals, and making quick 'action' sketches in pen and ink sketches during performances which later became the basis of a formal portrait. He was a master in interpreting a composer's musical intentions; a superb cellist in his own right, who seemed to have but little need of food or drink; who when not performing trios or quartets with members of his own orchestra between concerts, always had his head buried in some score or another. He was always kind, gentle, courteous, and seemed to have no sense of his own importance and genius. The orchestra loved him, and he drew from them the most sonorously moving intonations: it was said that no other string section in any European orchestra produced such groundswells and surges of transcendent sound. Like Bernstein, Barbirolli was perpetually transported by the music, lost to his surroundings, yet always in control as he became the medium through which orchestra and composer expressed themselves. Laurence Turner, the leader of the Halle, said to me once that the musical authority exercised by Sir John was totally mesmeric because the players were responding to the power of a great musical soul, not simply to a conductor's bodily gestures. Or, as I would put it in my pedantic kind of way, they were inspired by being in the presence of the most profound face of the psyche - that of *supra*-ego.

> *We moderns are faced with the necessity of rediscovering the life of the spirit; we must experience it anew for ourselves. It is the only way in which we can break the spell that binds us to the cycle of biological events.*
>
> C.G.Jung: *Psychological Reflections*

Jung made this statement after surveying a twentieth century world in which the horrific slaughter of World War I was followed in the Thirties by Stalin's uprooting of Russian peasants, causing the death of millions; by Hitler's triggering of World War II resulting in many millions of military and civilian casualties on both sides - numbers swollen by his policy of extermination directed at any Jew who came under the authority of the Third Reich. And then came the atrocities practised by the Japanese (as Germany's ally) first in China and then in their ruthless campaigns in the Far East. Jung saw how terribly warped were the psyches of the leaders, and of those who agreed with their policies, carried out their orders. And how easily and quickly the influences of *supra*-ego and *good*-ego could be eclipsed by the conscienceless and destructive rise of *gross*-ego. Indeed, the bases of Jung's philosophy and psychiatric expertise lies in his belief that human beings are subject to three intrinsic pairs of opposing mental forces: body and spirit; good and evil; consciousness and the unconscious. His great healing powers - as medical doctor and analyst - resulted from his ability to bring the patient to recognize and face the countermanding nature of such forces and, in so doing, reconcile the physical and spiritual aspects of their lives, and not repress the revelatory mental flights of *supra*-

ego when they moved into consciousness. Jung believed that bodily and spiritual health results in living beyond biology with one metaphysical foot, as it were, groping for a hold beyond time and space. If such is indeed the nature of the human condition - and intuitively I have always felt that Jung is right in this - we once again are brought to face the curious anomaly we have mentioned before in these pages: the extraordinary prospect of a consciousness able to reveal the existential and physical realities of the world in which we live, yet provide intimations of *meta*-physical ultimate realities lying beyond the reach of the natural sciences: a two-track intelligence system served by a corporeal brain employing chemical and electrical neurotransmitters to achieve to achieve such a feat.

No smoke and mirrors here.

Can either one of these modes of awarenesses be considered more important than the other? Well: it is certainly the case that philosophers, poets, shamans, priests…have emphasized that to live a full and complete life it is necessary to 'feel' and acknowledge the 'hereness' of the meta-physical dimension of being; advising that otherwise we inevitably reach a point where, despite the gratification of all physical and material desires, a psychical emptiness takes possession. In which case we would likely agree with Lucian - considered the most brilliant wit and Greek man of letters living under the Romans about 160 A.D. - that, *'The wealth of the soul is the only true wealth.'*

The late Bernard Levin, a celebrated and insightful columnist and book reviewer for the London Times and Sunday Times wrote in one his column anthologies about…. *'the gnawing concern that ultimate reality lies elsewhere, glimpsed out of the corner of the eye, sensed just beyond the light cast by the camp fire, heard in the slow movement of a Mozart quartet, seen in the eyes of Rembrandt's last self-portraits, felt in the sudden stab of discovery in reading or seeing a Shakespeare play thought (previously) familiar in every line.**

There are those in today's world who would greet Lucian's and Levine's statements with, at best, a blank look; at worst with a condescending and pitying smile, dismissing such a belief as a superstition which has surely had its day. The Nietzschean idea that God is dead - which was taken up again in the Fifties and Sixties following the Second World War - still

---

* Quoted in a tribute to Bernard Levin by Arianna Stassinopoulos-Huffington in the London Sunday Times, August 22nd, 2004.

resonates in contemporary society as science and technology support a secular view of life, despite the new mysteries they uncover concerning the origin of energies of the cosmos and its incredibly complex patterns of energy manifested over time; giving rise to a philosophy of 'consumerism' that is seen as the way to go in order to achieve success and happiness. But the eminent German-born writer and psychologist Erich Fromm (1900-80) turns the issue around by asking who is really dead - God or us?

> *Theologians and philosophers have been saying for a century that God is dead, but what we confront is the possibility that man is dead, transformed into a thing, a producer, a consumer, an idolater, or other things.*
>
>   In an address to the American Orthopsychiatric
>   Association, April 16, 1966

As we move on into the Third Millennium we should ask ourselves whether or not we have become *'transformed'* in the ways that Fromm suggested over forty years ago. Have we really become a totally secular society in which the concept of some Supreme Spiritual Authority has no real place - is merely a traditional belief to which we pay lip service? A society which has become delighted with technologies, yearning for aggrandizement, living for the sensations of the moment? A society in which the concept of the *three* faces of *Ego*' has no place because the psychological values represented by the other *two* - *supra*-ego and *good*-ego - are in decline? If so I would say with Fromm that we have become, or will ultimately become but 'things' - less than human, in that the influence of soul and spirit will no longer be felt in our lives. Once again I must turn to Loren Eiseley to help me out. Here is a passage from his *The Firmament of Time*:

> *We have come a long road up from the darkness, and it may well be - so brief, even so, is the human story - that viewed in the light of history, we are still uncouth barbarians. We are potential love animals, wrenching and floundering in our larval envelopes, trying to fling off the bestial paths. Like children or savages, we have delighted ourselves with technics. We have thought they alone might free us. As I remarked before, once launched on this road there is no retreat. The whirlpool can be conquered, but only be placing it in proper perspective. As it grows, we must learn to cultivate that which must never be permitted to enter the maelstrom - ourselves. We must never accept utility as the*

sole reason for education. If all knowledge is of the outside, if none is turned inward, if self-awareness fades into the blind acquiescence of the mass man, then the personal responsibility by which democracy lives will fade also.

Schoolrooms are not and should not be the place where man learns only scientific techniques. They are the place where selfhood, what has been called 'the supreme instrument of knowledge,' is created...

But one significant safety net remains in position to ensure that we do not degenerate into mere 'things'. As long as the arts are able to flourish as expressions of our inner meditative and creative capabilities, then we will not be totally seduced by the senses, able to live only in the present moment. For it is due to the creative expression of the imagination - be it in music, poetry, painting, literature and drama.... - that both *supra*-ego and *good*-ego enable the mind to venture into the past, and beyond the present moment, to provoke intimations of states of Being that lie beyond nature's laws.

EGO and its 'CONTRARIES'

Without Contraries is no progression. Attraction and Repulsion, Reason and Energy, Love and Hate, are necessary to Human existence.
      The Marriage of Heaven and Hell: William Blake
                                        (1757-1827)

I think it likely that the visionary English poet and artist William Blake would agree that the three faces of *Ego'* are responsible for our most vital kind of *'progression'* - that which is born out of their dynamic interplay in consciousness. For what would we know of the quality and nature of love if we knew nothing of its opposite, hate? Of hope if we knew not of despair...? There is nothing static about the average person's mental life: rather it can be described as one of opposing cognitive and affective (feeling) states vying with each other for expression, and setting us wandering in a wide-ranging psychological landscape of mental depths and heights. Each day we pass through zones of courage and fear, optimism and pessimism, self-reliance and self-doubt, happiness and sadness... some days coming to a dead-end, others allowing glimpses of a likely purposive future. We move according to conscious or unconscious direction, affected by reason or intuition: the body attuned to the physics of matter, time and space - the soul rambling along in its own metaphysical countryside. Such is William Blake's uncertain, paradoxical, but most human chronicle - the *Marriage of Heaven and Hell:* the Progressive Way of the *'Contraries'* as experienced during the brief span of a lifetime.

Within the reaches of the Universe a system of contrary forces drive the physical structures as the solar system: stars, planets and moons move according to a dynamic system of contrariness ordered by gravitational, nuclear, and electromagnetic forces. There is no crashing around in anarchic disorder - as yet, anyway - which would be the case were one or other of these forces to strengthen or weaken by even the smallest margins to break the overall pattern of existence dependent force countering force. Blake's *'Attractions'* and *'Repulsions'* would seem to apply throughout the whole scheme of things - cosmic and human. For the forceful interplay between the faces of *Ego'* - like the forces governing the universe - must also be brought to work within contained limits to promote a dynamic mental life that does not fall apart in runaway mental anarchy.

Even so, within this dynamic equilibrium of which we have been talking change occurs - change, not disintegration. In the cosmos stars die and are born; the universe expands. We ourselves die biologically - and perhaps are transformed as the soul flies away. The French philosopher Henri Bergson (1859-1941) and 1927 Nobel Prize winner, wrote: *'...for a conscious being, to exist is to change, to change is to mature, to mature is to go on creating oneself endlessly'*.

And so if you were to ask me what is the purpose of the struggle with the faces of *Ego'*, I would say in order to have some control of the processes of personal change: to be responsible for *Creating a Self.*

The supreme example of all *'Contraries'*- and which even William Blake was unlikely to have dreamed up - is that of matter and antimatter. The notion of antimatter is an outgrowth of the discovery of antiparticles, for it is well established experimentally that any particle has a corresponding antiparticle with equal mass but opposite electrical charge. Quantum theory predicts that all interactions between antiparticles are essentially identical with interactions between particles. This implies the potential existence of antinuclei made of antiprotons as well as antiatoms made of antinuclei and positrons or, in brief, of antimatter. It is tempting to assume that matter and antimatter (or particle-antiparticle symmetry) is present in the Universe, but attempts to observe this have not been particularly successful.

> *In any case, antimatter remains one of the most baffling problems in cosmology: if it is present, how is it possible to remedy the difficulties found in the theories? If it is not present, how strange it is that the Universe does not have the particle-antiparticle symmetry which characterizes the laws of physics?*
> Encyclopedia of Physics: Edited by Rita G. Lerner & George I Trigg. (Addison-Wesley Publishing Company 1981, p.37)

17

MIND and BRAIN

What is mind?
No matter.
What is matter?
Never mind.
Thomas Hewitt Key: epigram in *Punch* (1855)

Just what is 'mind'? The word is in constant use today, no less so than in the early years of the First Century A.D. when Seneca (the younger) wrote *'On Tranquility of Mind'*. Yet if you wish to discover the full implications of the term, and search through the indexes of books on psychology and philosophy, you will be hard pressed to find 'mind' listed. The root of the word is to be found in the Greek *menos,* meaning 'spirit' or 'force'; and in the Latin *mens,* signifying 'mind' as the ability to gather perceptions, utilize memory, and express thought and opinion. Thus we talk about 'speaking your mind', 'losing your mind', 'bringing to mind', 'being in one's right mind', 'giving someone a piece of one's mind'... and so on and so forth. Yet the definitive meaning of the word itself remains elusive, shot through with ambiguities; leaving it open to many interpretations and applicable to many levels of consciousness.

However, Thomas Hewitt Key's well-known, witty, and clever linguistic twists, are very positive about what mind is *not*: it is not a *material* entity. In which case the word 'mind' should not be used as a synonym for 'brain' - an organ which, after all, has been described as the most complex material instrument ever invented by the universe. The eminent Canadian neurosurgeon Wilder Penfield who died in 1976 - (to whom we refer later in these pages) - may well have been amusedly thinking of Key's epigrammatic verse when he asserted, a hundred years later, that 'mind can only be thought of as a *non-temporal, non-spatial entity.*' The idea of mind as an abstract force manifesting via the amazingly intricate physiology of the brain is evident in some of the historic records of early Egyptian religious thought and philosophy - attitudes and concepts that were to pervade later

Mediterranean cultures. Among the early Greek philosophers, Socrates stressed the primary importance of Mind's imaginative powers - a philosophical position which his pupil Plato was to develop in suggesting that such ideas and beliefs are delivered to full consciousness *via* the cellular marvel of the brain and - although abstract - are just as 'real' as the brain transmitting them. Western culture, cradled in the Mediterranean region came, in large part, to affirm this view - accepting the proposition that the mind is a 'reservoir' of psychological forces delivering abstract knowledge in a way that cannot be completely biologically explained; allowing sensory perceptions to be instantaneously analyzed and their particular values ascertained.... while the most intelligent way to deal with them is also simultaneously made available to consciousness. Finally, over the course of a lifetime, individual character and behavior become formed as a result of the mind's overseeing capabilities.

Dr. Penfield's response to the question 'What is mind' - saying that it represents a *'non-temporal, non-spatial entity'* - is the most provocative I have come across - conveying the arcane nature of this mental force. For the implications of his words are (a), that the mind can move in and out of linear time; and (b), that it is not a biological organ occupying body-space.

MI 5 is the British Government's 'Top Secret' Intelligence Agency responsible for home security. As such it maintains a low profile; theoretically is not part of the body-politic, and operates according to its own rules and principles. Independent and clandestine, it gathers and analyzes intelligence not available though the normal channels of day-to-day open inquiry and then - when totally comprehending the essence of the problem under purview - initiates the solution. In these respects the Agency is somewhat analogous to Mind. While, in contrast, the conventional, existential problems of government concerning the daily, factual round of life, are sorted out, and pragmatic solutions suggested, by highly visible Administrative Departments - bureaux which, collectively, can be seen as the Brain of Government.

In the chapter describing the Chart of Consciousness I referred to that memorable statement made by Réne Descartes' which has echoed down the centuries: *"I think, therefore I am."* I use it here again - and say more about Descartes himself - in order to help illuminate the course of the present discussion. As we have seen, Descartes had no doubts about the primacy of thought. He did not say, for example, "I *see* myself in the mirror, therefore I am." A man possessed of an extraordinary, inquiring mind, he was engaged in a constant intellectual search to explain the idea of an infinite God. On November 10, 1619, the importance of synthesizing all aspects of thought - objective and imaginative - into one comprehensive way of thinking, was revealed to him in a dream. It led him to take a mechanistic view of physical phenomena, in that it enabled him to explain all forms of matter as systems of very small particles interacting automatically... concluding that animals are machine-like in their existence. Man however, while also a machine-like animal, was separated from the mechanistic world of other creatures due to his ability to *think*... By which he meant to analyze the material nature of phenomena as presented by the senses in order to understand cause and effect. Yet to think *logically* - insofar as Descartes was concerned - it was also necessary to engage our intuitive ability to think imaginatively, creatively; and so come to increase our understanding of both how and why things are the way they are. And as analysis was a function of brain. while intuition was one of mind, he saw this complementarity to represent a particularly human synthesis of mind and brain: a mental system of awareness by which a Divine Authority reveals both material and spiritual truths to man. Descartes eventually came to be considered the father of modern rationalism But it is significant that he saw a *logical* progression of thought as one not only leading to understanding the physical realities of man and the world, but was also capable of lending credence to metaphysical speculation. Brain and mind in partnership.

He was a firm adherent to the Catholic faith, caught up in that terrible and tragic time at the beginning of the seventeenth century when Catholic fought Protestant in what has become known as the Thirty Years War - a conflict marked by brutal atrocities. It was a demoralizing period of general social unrest: of uncertainty concerning established religious and political values; of the value of human life. And Descartes found himself, like many others, uncertain about the state of civilized life in general, wondering how it could be considered significant from any point of view, either existentially or spiritually. It was a fearful doubt about human worth, yet ultimately he concluded that the certainty of his own 'realness' lay in the fact that he was a *thinking* being - possessed of a mind which reigned over the senses, fed the imagination, served God, and thus nourished the soul.

Mind over Matter is a familiar phrase, still implying what it has historically always implied: namely, the ability of the mind's psychologic powers to direct the brain down pathways that inspire hope in times of despair, or bring about practical solutions to apparently insurmountable existential problems. (The reader may recall my earlier accounts of Sir Ernest Shackleton's *Endurance* Expedition in the Antarctic': that most impressive example of 'mind over matter' - a combination of logical reasoning and intuitive intelligence in the face of potentially disastrous situations - a Cartesian display of fortitude and resources of mind.)

Mens agitat molem - (Mind moves matter) - wrote the Roman poet Virgil in *The Aeneid* in 19 B.C. While more than three hundred years before Virgil, Democritus of Abdera (460 B.C?-370 B.C.?) took a similarly dualistic view, suggesting that the physical body is animated by a soul-mind force in the form of fire - fire as energy - having the mythic connotation of supernatural power stolen from heaven for the benefit of mankind. He also developed the atomistic theory of Leucippus, holding that the universe is composed of atoms moving around in space and forming themselves into material bodies. And at yet another level, Democritus may be regarded as the originator of the principles of psychosomatic medicine - the understanding of which is consistently gaining ground in contemporary medical practice: the belief that the state of the *psyche* (mind and spirit) affects the life of the *soma*, (the cellular body).

In today's health-conscious society we are told that 'stress' is the villain in the piece responsible for many of our more serious ailments - high blood pressure, heart attack, stroke, premature ageing, a weak immune system and all the related problems which ensue. And here is Democritus, almost 2500 years ago saying, in effect, that stress must be avoided - stress being the very opposite of mental balance and equanimity. (After all, in the heyday of Greek civilization the slogan 'a healthy mind in a healthy body' - lay behind the extensive use of the gymnasium and the institution of the Games.) Remaining in the Classical world for a moment, it is interesting to note that the Latin term *strictus* - from which the term 'stress' is derived - today signifies a straining force affecting our whole system, physical and mental. The word 'equanimity' also derives from a Latin root - *equanimitas*, combining aequus meaning 'even', with *animus* signifying 'mind': thus expressing the notion of 'even mindedness' - a term which nowadays implies a balance achieved between mind and brain. And there are several commonly used synonyms which I feel would satisfy - if not embellish - Democratus' views on mental equanimity. Words such as 'composure', and 'serenity'; and others less obvious such as 'nonchalance' - implying a cool detachment from situations that could disturb one emotionally; and 'sang-froid' - signifying coolness and presence of mind when life's dramas threaten.

Which brings us back to the central question. Is the physical brain - in and of itself - through the rapid exchange of electrical and chemical signals via trillions of neural interconnections, responsible for creating the imaginative and evaluating faculties of mind? Or, are we to consider (as Wilder Penfield suggests) that the way to achieve 'mind-intelligence' is not necessarily attributable to the neurological activity of the brain alone? That it emanates from an as yet unidentifiable, metaphysical energy-force - one capable of inducing the brain and consciousness to travel in realms of knowledge that lie beyond those offered by the senses and intellect alone?

Well, as the reader knows, we have consistently taken the Penfield view in attempting to account for the highest forms of creative and imaginative experience. And have gone further in proposing that heights of revelatory thought and feeling are achieved when soul and spirit are unconsciously empowered to act as catalysts, and bring insights to mind not generally available to the brain's day-to-day electro-chemical activity. Which lends one to wondering if - with the advent of the cerebral cortex in the last stage of the brain's evolution - we came into possession of the ultimate 'cellular wireless': new 'brain-technology' empowering us to receive electrical impulses capable of engendering the abstract imagery of a transcendent imagination - metaphysical transmissions which cannot, as yet, be linked to any known systems of radiant energy. The theory would be that only when we became capable of receiving such signals did we come to experience the deeper levels of awareness we associate with the creative powers of mind. In which case it follows that before the advent of the fully developed cerebra with their convoluted layers of cortex, our predecessors were relatively 'mind-less', living the simpler, more instinctual form of existence demanded by a pre-cortical brain.

It is often a frustratingly speculative and basically academic exercise to try and plumb the depths of a process as complex as that of consciousness - especially by the analytical and reductive method of naming the 'parts', and then theorizing about the probable ways they work together. Yet there are times when, prosaic as such elucidation may be, it can at least help us to recognize the profound - and indeed mysterious nature of our human potential.

However, when a moment of heightened visual perception is described by a poet, visionary, and priest of Thomas Merton's stature, the limitations of philosophic and quasi-scientific theories seeking to explain the puzzling nature of this very humanness become immediately obvious. For it is the poet's language which conveys that magic combination of mood and enlightenment occurring when the mind - following up on the objective knowledge provided by the brain's perceptual powers - introduces its own illuminating dreams and reflections..... as Carl Jung was wont to call them. Merton writes about coming upon three deer - a stag and two does - in a field, and his words create a moving and illuminating account; reveal how the sighting affected him.

> *The thing that struck me most - when you look at them directly and in movement, you see what the primitive cave painters saw. Something you never see in a photograph. It is most awe-inspiring. The* muntu *or the 'spirit' is shown in the running of the deer. The 'deerness' that sums up everything and is sacred and marvelous.*
>
> *A contemplative intuition, yet this is perfectly ordinary, everyday seeing - what everybody ought to see all the time. The deer reveals to me something essential, not only in itself, but also in myself. Something beyond the trivialities of my everyday being, my individual existence. Something profound. The face of that which is in the deer and in myself.**

This passage is an extract from Thomas Merton's *Vow of Conversation*, an unpublished journal of 1964-5, and it came to my attention when reading Peter France's excellent book Hermits.

---

* Peter France, *Hermits: Insights of Solitude* (St. Martin's Press; New York, 1966), p.191

Thomas Merton, born in France, in the Eastern Pyrenees, on January 31, 1915, was one of the most influential Christian apologists of the twentieth century. Hermit, mystic, contemplative, philosopher, ministering priest.... he was all of these things. He decided to become a Catholic in 1938 and joined the Trappist Order in their monastery at Gethsemane, Kentucky, in December 1941. He died in 1968. Throughout his life as a monk he came to believe that the contemplative life of solitude was the royal way to the true and essential Self; and that in finding this, one found God. In the course of struggling to live as a hermit within a monastic community he published more than 300 articles and some thirty-seven books, including a best selling autobiography.

Throughout these pages we have talked at some length about the fully developed structure of the brain so far as we know it today, describing it as a neural marvel, unique in the universe, and unmatched by anything man has ever made: the formulator of all human experience. And while it may be said that brain is not mind, we would, nevertheless, know nothing of mind without it. Also, I find it worth noting that the important English philosopher John Locke (1632-1704), who established the principles of modern empiricism and the theory that only sense experience can be considered the source of knowledge; believing that the notion of intuitively generated and inspired ideas was not tenable - seemingly vacates his empirical position when he writes as follows about ideas that arrive independently, 'out of the blue', as it were:

> *The thoughts that come often unsought, and, as it were, drop into the mind, are commonly the most valuable of any we have.*
> Letter to Samuel Bold (1699)

These words lend support to Penfield's view that the mind is neither a temporal nor a spatial entity. To talk about thoughts coming *'unsought'* rather than resulting from a particular objective experience, and that *'drop'* into the *'mind'* rather than into the *'brain'*, is hardly compatible with the maxims of the empirical philosophy Locke himself espoused. He addresses his correspondent in almost Platonic vein, reminiscent of Plato's statement that...*'To be inspired is to be out of one's senses'*. And his words also have a Wordsworthian ring to them: telling of the autonomy of a mind that breaks through and both usurps and enhances the objective functions of consciousness. In the following verse Wordsworth's description of a moment of visionary, poetic insight, suggests how the brain would find itself taken over on certain occasions, pursuing thoughts and moods that go beyond the descriptive details expected of a straightforward observer. And in so doing becomes a participant in the lyrical process of apperceiving levels of truth and significance the poet discovers in particular encounters with the world and life.

> *My brain*
> *Worked with a dim and undetermined sense*
> *Of unknown modes of being.*
> The Prelude I. xxxix

Throughout our recorded history it is constantly being asserted that all minds are not equal. One unknown author of a treatise entitled *Minds* writes that *'Great minds discuss ideas, average minds discuss events, small minds discuss people.'* And no doubt we all have commented from time to time on minds we consider to be great, and on others less so. We generally define great *minds* as those making the most extraordinary discoveries in science; composing the most moving musical scores; expressing and shaping life's revelations through literature, music, poetry, sculpture, painting, architecture, philosophy, theater.... Yet, in so doing, should we not be acknowledging the vital part played also by great *brains* in bringing about such achievements? Also - as we have discussed - the role played by the vitalizing and insightful powers of spirit should not be overlooked in discussing the Great Mind Theory. We are talking about a triumvirate of mental forces here. And we should ask what role the trio play in inducing surpassing levels of moral and physical courage; enabling equanimity and stoicism to take over in the face of disaster, pain and suffering; and in bringing some of us to muster qualities of leadership when threatening and unfamiliar situations arise, requiring high personal morale, and intelligent adaptation (tactical brilliance), to meet and overcome them.

The supposition that the indwelling presence of an inspiring and vitalizing spirit-force - seen as the means by which the advanced soul can influence the mind - is rooted in many cultures and has prevailed throughout the centuries. Aristotle, foremost among Greek scientists and philosophers wrote, *Reason is a light that God has kindled in the soul.* Cicero, Roman statesman, orator and author, declared *He found a sort of truth for the soul in cultivating his mind.* Early in the 19th century the French philosopher and essayist Joseph Joubert wrote in *Pensées...The mind is the atmosphere of the soul.* And more recently in 1953 Carl Jung wrote in his *Psychological Reflections: Although common prejudice still believes that the chief foundation of our knowledge comes from outside.... every student of ancient natural history and natural philosophy knows how much of the soul is projected into what is unknown of the outside phenomenon...* And in lighter vein J. R. Lowell (1819-1891), the American poet and essayist wrote, in *A Fable for Critics: Most brains reflect but the crown of a hat...* followed by *...The defect in his brain was just absence of mind.*

When I was wandering around wondering how to bring this chapter to an end, I was unexpectedly assailed by a most graphic, visual recollection of a moment when a great orchestra and their conductor were about to perform the first work of an evening concert. There they were, the full complement of the Hallé Orchestra, spread out widely across the stage in semicircular rows, the eyes of every musician fixed on the conductor, Sir John Barbirolli, waiting for the movement of hand and arm that would turn the written score of Tchaikovsky's dramatic and tortured Fourth Symphony into waves of sound.

And then I had it. Here it was. A symbolic illustration of how brain, mind, spirit and soul, work together in achieving our more visionary flights of consciousness.

The orchestra can be seen as one colossal BRAIN - a working total of 100 billion neurons per player, multiplied by 110 to account for the number of musicians performing: a simply astronomical number of creatively energized neurons. The overseeing, directing, informing power of MIND is epitomized in the person of the conductor whose insightful comprehension and interpretation of the musical score, brings to life its acoustical form and content. And with the performance underway I would say we are hearing (and witnessing) a manifestation of SOUL - channeled through SPIRIT - as it informed the life of the composer.

In a sense, the life of each one of us is played out in a similar symphonic way. Soul and spirit write the score that determines our responses to life - a score we read and perform as an orchestra of one, as well as acting as the maestro giving a daily performance. We conduct serene passages bathed in light; others dark and disquieting. In an ideal world and an ideal life, the overall presentation of this internal score would be akin to the musical journey created by Beethoven in the Ninth Symphony culminating in the Ode to Joy. Home at last.

Daily, on our television screens, we see men and women stalked by tragedy - victims of starvation, genocide, acute breakdowns in health, unforeseen accidents, and the destruction of their lives in natural disasters... Yet somehow there are those who manage to transcend the dark hopelessness of their personal symphony of pain and hopelessness... yet find peace of mind, and direction of spirit, in the conducting of it.

> *No, what it (my mind) is really most like is a spider's web, insecurely hung on leaves and twigs, quivering in every wind, and sprinkled with dewdrops and dead flies. And at its geometric centre, pondering for ever the Problem of Existence, sits motionless and spider-like the uncanny Soul.*
> Logan Pearsall Smith: *Trivia, "The Spider"*

18

THE AGE-OLD CONCEPT OF SOUL: ITS VITAL ROLE IN CONSCIOUSNESS

> Our birth is but a sleep and a forgetting:
> The soul that rises with us, our life's star,
> Hath had elsewhere its setting
> And cometh from afar.
> William Wordsworth (1770-1850):
> *Intimations of Immortality*

When the Headman of the Sakuddei led his people out from the hardwood forests of Java it was not a happy occasion. For countless years the tribe had lived deep within these luxurious woodlands, perfectly adapted, knowing no other habitat. Now the Indonesian government were 'harvesting' the tall trees of teak - clearcutting over vast areas - and the Sakuddei were in the way. Consequently, they were being moved - 'relocated' to an area where they would live beyond the forest in a commune-like situation with other displaced persons who were not of the Sakuddei. It was the end of their life as an autonomous, isolated society - the only way of life they had known for generations.

As the long column emerged from the trees - lean and upright men and women carrying their few possessions and walking with great dignity - they were confronted by the cameras of a television news team led by a reporter accompanied by an interpreter.

"We have brought only those things we need," said the Headman in response to questions. "Those things which mediate between the world and the soul. And in addition we will require only a river for the baptism of new-born children. Water is purification: it is essential if we are to dance with the souls..."

"How do you feel about this move?", asked the journalist. "Do you intend to join any resistance to the government?

"No: that is out of the question. I know our group-life, our ways... cannot survive intact in the future we now face. Yet we must behave well. We must always behave well."

"Even when you are dispossessed of your home and land?"

"Yes: more than land, more than possessions, it is important that we behave well so that our souls will like to stay with us. If we do not they will leave. And then we have lost everything."

The Headman of the Sadukkei was expressing his belief in the age-old and deeply held conviction that a uniquely personal and spiritual principle lies at the center of human being: a psychical power outside the sphere of physical science, generally referred to as the soul, by which the cellular material of the body is imbued with the vital attributes of spirit, conscience, and the force we call love - and without which we would live a purposeless and heartless existence. Furthermore, the belief has persisted that on the death of the body the soul survives to go its own way. For as Wordsworth's lines intimate, he envisages the soul as a preternatural force beyond the scope of words like 'temporal' or 'material' to define it. Neither is the thought that not to 'behave well' will cause the soul to abandon one.... peculiar to the Sakuddei. The belief that the soul will become estranged and depart if one succumbs to thoughts and actions that work against such principles as wholeness, goodness, benevolence and truth, has long been a central tenet of many of the world's religions. It is the penalty paid, for example, in not heeding the moral dangers presented by Saint Thomas Aquinas' Seven Deadly Sins. And one recalls the advice given in Saint Mark's Gospel, *'What shall it profit a man, if he shall gain the whole world, and lose his soul?'*.

'Twenty-six souls lost at sea' was the newspaper headline when a freighter sank recently in a North Atlantic storm. Not 'twenty-six *bodies...*' which would surely, in this rational age, be a more acceptable statement of fact. The traditional use of the word 'souls' when reporting life and death situations implies that in the loss of a human being more is involved than simply physical death - that something surpassing the mortal frame, a soul, has also gone from the world. S.O.S. - popularly thought to be short for *'Save Our Souls'*- was the international distress signal, until telegraphing in Morse Code became technically obsolete some years ago. But it is interesting to note that when people are in great danger, or experiencing levels of intense suffering and privation, they are referred to as 'souls in distress': an appellation which might well be considered out-of-date, and untenable in the rational atmosphere of today's secular, science-oriented societies

- where it is certainly not taken for granted that a human being does, in fact, comprise a unity of body and soul. And I wonder sometimes - even in Christian societies - how often a cleric of any denomination will talk about ministering 'to a soul in distress'? Yet in the remote villages of the upper Amazon one of the mystical roles of the shaman is still to mediate with a 'troubled soul', both in life and at the time of death. Their spiritual powers are seen as vitally necessary in not letting the soul get discouraged and independently seek its spiritual home.

'*My mind,*' wrote Thomas Alva Edison, '*is incapable of conceiving such a thing as a soul. I may be in error, and man may have a soul; but I simply do not believe it.*' Edison (1847-1931), the American inventor of the incandescent electric lamp, the microphone and phonograph... patented over 1000 inventions. As a scientist, intent on uncovering the secrets of electrical energy, reason and objectivity served him - and us - well. In a newspaper interview he made the famous remark that, '*Genius is one percent inspiration and ninety-nine percent perspiration.*' The significance of this statement depends on how Edison would define 'genius' and 'inspiration'. Yet I think that despite his insistence on the importance of 'perspiration', there can be little doubt that his inventive genius resulted from the constancy and inspired nature of his intuitive, imaginative insights. But obviously he was not prepared to see such creative gifts as anything more than the application of strictly rational and objective processes of intellect. And his inability to move into metaphysical speculation and conceive of such an abstraction as the soul results - in my view at least - from the complete suppression of any conscious recognition that such a subjective mental activity as intuition could be at work, so effective was the stranglehold exercised by a supremely rational way of looking at things. This was not a dilemma shared by his near-contemporary Einstein who, like Aristotle, believed that any serious pursuit of the physical sciences brings one inevitably to contemplate metaphysical issues. The ability to be a speculative philosopher in terms of pondering the underlying principles that might explain *why* things are the way they are and, at the same time, be an objective scientist searching out *how* things are the way they are... are not mutually exclusive: in fact, as we have seen, both left and right (rational and intuitive) hemispheres of the brain work in unison to secure a fully balanced consciousness.

There are two modes of acquiring knowledge - wrote Roger Bacon, English philosopher and scientist, in his Opus Majus of A.D. 1266 - *namely by reasoning and experience. Reasoning draws a conclusion and makes us grant the conclusion, but does not make the conclusion certain, nor does it remove doubt so that the mind may rest on the intuition of the truth...*

Similarly for Einstein, the most significant way of working as a scientist was not one in which reason and factual knowledge dominated, but one where personal vision and insight played an integral part in the process of discovery. Writing in *On Science,* he said, *Imagination is more important than knowledge.* He rarely talked about the concept of soul when responding to questions concerning religion or spirituality in general, but in *Ideas and Opinions* he did write 'It is only to the individual that a soul is given...' And responding to a question about God he said, 'My comprehension of God comes from the deeply felt conviction of a superior intelligence that reveals itself in the knowable world...' And, 'Science without religion is lame, religion without science is blind.'

I wonder how Edison would have responded to these statements by Einstein? And, speculating further, what Aristotle might have said if some contemporary had proposed a corollary to his thoughts on the validity of metaphysics as a discipline, by declaring: 'If, after physics... comes metaphysics; then after body... comes soul'?

One should browse constantly through the writings of famous men. There is no better way to ascertain the general drift of human thought when inquiring into the workings of nature, or when preoccupied with spiritual issues and the idea of an individual soul. And while acknowledging the contribution of Ancient Egyptian thought in these areas, it was during the civilization of Classical Greece that the disciplines of science, philosophy, and metaphysics really got under way. Leucippus developed his atomist theory, and Aristotle his laws of natural history. The Greek mind seemed to naturally recognize the creative outcome of the partnership between rational deduction and intuitive flashes of insight. Words such as 'soul', 'spirit', 'heart', 'virtue'... are used to express abstract concepts and appear as natural to the writer as do the descriptive, objective terminologies of science. The *Meditations* of Marcus Aurelius - Roman Emperor from A.D.161-180 - and written during the seven years spent campaigning on the Danube and Rhine striving to preserve Rome's crumbling frontiers against attacks by the Parthians in the East and Germanic tribes to the North, comprises a sophisticated and enlightened treatise. It combines his observations on human behavior; on the unpredictable circumstances of fate against which the philosophy of the Stoics provides the only effective response; together with his thoughts on how to live humanely, exercising wisdom, compassion, selflessness, in order not to shame the soul. It is a wonderfully challenging read - probably more relevant today than it was even two millennia ago

In Book VI of the *Nicomachean Ethics,* Aristotle writes that, '*... the states of virtue by which the soul possesses truth by way of affirmation or denial are five in number, i.e., art, scientific knowledge, practical wisdom, philosophic wisdom, intuitive reason...*' In this statement Aristotle makes it clear that he regards the entity he calls the soul to be the repository of all the truths a human being can know; and that for truths to be made known the prime avenues of exploration to be followed are found in the disciplines of the arts and sciences - where both objective knowledge and intuitive reason must be employed to work together. The unmistakable inference is that the soul is an immanent and persuasive psychic force - one which illuminates the truly significant aspects of human experience, possessing the power of revelation and moving one into a metaphysical and preeminent dimensions of being. If this is so, then no wonder the Sadukkei were so intent on holding onto their souls.

One of the most famous musings concerning the presence of the soul is to found in the Roman Emperor Hadrian's *Ad animan.* A learned ruler and writer of genuine gifts (antedating Marcus Aurelius' rule by 23 years) he wrote: *O fleeting soul of mine, my body's friend and guest, whither goest thou, pale, fearful, and pensive one? Why laugh not as of old?* And one should bear in mind that neither Hadrian nor Marcus Aurelius were Christians.

In the late 15th century, Leonardo da Vinci, an extraordinarily inventive genius as both scientist and artist, wrote: *That figure is most praiseworthy which, by its action, best expresses the passions of the soul.*

I realize it would be possible to fill the succeeding pages with thoughts and sentiments from every epoch - a great many supporting the 'reality' of the soul; others dubious about the validity of the supposition. Certainly up to the 18th century - before the emergence of the so-called Age of Reason in the Western World - it would seem that the populace at large accepted the belief that the soul was the spiritual center, a touch of the Divine in every human being. And I wonder, for example, if anyone has ever counted the number of occasions when Shakespeare uses the word 'soul' - not to merely semantically identify a deep-seated moral center - be it strong or weak - in his characters, but to infer that in the psychological dramas played out between the forces of good and evil, it is the soul which is involved in a struggle leading either to spiritual death, or to spiritual confirmation and invigoration.

I was about sixteen when I first encountered the witty and pithy aphorisms of Blaise Pascal (1623-62), French physician, physicist and philosopher. Shakespeare died in 1616, and one finds that many of Pascal's terse sayings echo the Bard's insights concerning human nature. Pascal poses questions and juxtaposes antithetical images which highlight the paradoxical double life we lead: that while the senses tell us of the material reality of the world, the imagination intuitively suggests that there are truths 'out there' not available to the senses. Such a two-track system of awareness helps to ensure that we remain somewhat of a mystery even to ourselves.

Here is Pascal at his best: *The Mind bends down to wash its hands in the wash basin/ The Soul stoops to tie up its shoelaces.* As a lad, this caught my fancy: the words raised a mental picture of a marionette-like figure bent low, alternately wringing its hands and flopping unsteadily over its feet in a Chaplinesque sort of way, as if body and limbs were weightless, and gravity was not helping in the washing of hands or the tying of laces. This legacy of Pascal's has remained with me ever since. On shaving and washing my hands I frequently wish my mind 'good morning'; put on my shoes - with or without laces - and murmur, 'Hi, soul - everything O.K. today?' And feel strangely better for it.

Within the last two years or so three large jet aircraft have gone down in the Atlantic shortly after taking-off from New York. In every case all on board have died. Yet despite the fact that - as we have previously noted - the Morse Code and its tapped-out signal S.O.S. is no longer in use, the press reports of these tragedies still employed the old, popular version of Morse: *'More than two-hundred and fifty souls lost...' was* the headline appearing in a number of newspaper and television accounts. Perhaps old practices die hard. Or perhaps, even in this age when the physiological aspect of Darwinian evolutionary theory is so well supported by scientific developments, the essential principle of life is still felt to be the soul, the *psyche* as named in the ancient Greek world. *Music is a higher revelation than philosophy,* wrote Beethoven in a letter to Bettina von Arnim in 1810, fifty-four years before Darwin's theory saw the light of day - a conviction that was to be taken further by Freud some fifty years after Darwin's death when he wrote that *Music is the high road to the soul.* Both these statements assume the existence of an inner awareness, an inner intelligence, able to be activated by music and move one through intense feeling, and the insights of a disembodied mind, beyond the confines of time and space. Freud was certainly familiar with the relatively new science of physical evolution, yet it did not cause him to modify his view that to be human is to live an integrated existence - one in which a physical life is governed in large part by a mental life which, in turn, is home to the spiritual force of the mysterious, persuasive power of an immaterial soul. And he was able to ally the presentiment that the soul constitutes a truth in its own right with his objective researches as a doctor and scientist.

Ludwig Wittgenstein, the Austrian-born philosopher who died in 1951, remarked that, *'When the problems of science seem to be answered...the questions of life remain totally unanswered'.* I suspect that if a similar statement had come to the attention of Socrates or Plato they would have substituted... questions of the *psyche* for Wittgenstein's *'questions of life'.* And I would assume that Wittgenstein, recognizing the validity of such editing, would give the nod to Socrates and say, *'life, psyche, soul...whatever...!*

In the early years of the 20th century the principles of contemporary psychiatry were pioneered by two doctors: Sigmund Freud who died in 1939, and Carl Gustav Jung who died in 1961. These are the two men whose researches into the workings of the human psyche have, in my opinion, led us the most effectively to understand ourselves. (Jung, in particular, was followed by adherents such as Adler and Neumann in Europe, and Erich Fromm and Rollo May in the United States.) Central to both the Freudian and Jungian process of mental healing was the practice of depth-analysis requiring long and frequent sessions on the 'psychiatrist's couch'. The procedure called for relaxed and confidential exchanges between doctor (analyst) and patient, during which the analyst sought to uncover, and bring the patient to recognize, deep-seated and disturbing undercurrents of his or her mental life - fears, anxieties, phobias, obsessions, frustrations, passions, and fleeting dark thoughts.... constantly besieging day-to-day consciousness as it encounters the world. (Wittgenstein - a contemporary of them both - would surely have gained some answers to his questions had he been familiar with their theories concerning the psychological dynamics created by an objective and worldly consciousness harried by 'underground', unconscious mental insurgents.)

Jung laid great emphasis on a spiritual component operating from these relatively unconscious undercurrents - seeing the soul as an immaterial and vital life force of the psyche driving the creative, moral, and purposive aspects of human existence. And he believed that to bring patients to a *complete* state of wholeness it was necessary to go beyond unveiling indwelling phobias and other anxieties... and work to take patients into the deeper regions of self to feel the influence of 'the other' - the force traditionally thought of as spirit: - the often called *'still, small voice'* (inspired 'breath' as the Greeks would have it...), hinting at 'truths' not limited to this world. And if this was brought about, Jung considered a patient to have become completely individuated: that is to say, 'whole' and therefore 'healthy', in the fullest sense of the word. Jung's many successes as an analyst brought him worldwide fame; a reputation that caused him to be known as 'a healer of souls'.

In contrast, many of Freud's major hypotheses were made after studying patterns of behavior - the most enduring and influential theory being that infantile sexuality and the Oedipus complex, depending on the degree to which it was accepted or repressed by the individual, provided the libidinous energy-drives of adult life. This is a rather broad generalization of Freudian theory, but it will serve to indicate why it has been more commonly acceptable in psychiatric circles, inasmuch as Freud's observations of *behavioral* development give a scientific credibility to his method and conclusions.

However, in his book entitled *Freud and Mans' Soul* (published by Albert A. Knopf in 1983) Bruno Bettelheim who also studied at the University of Vienna, reveals that Freud's writings and beliefs were not without their spiritual depth - a factor he considers has constantly been lost in translation from the German. In Chapter 1, Bettelheim writes:

> In <u>The Interpretation of Dreams</u> 1900), which opened to our understanding not just the meaning of dreams but also the nature and power of the unconscious, Freud told about his arduous struggle to achieve ever greater self-awareness. In other books he told why he felt it necessary for the rest of us to do the same. In a way, all his writings are gentle, persuasive, often brilliantly worded intimations that we, his readers, would benefit from a similar spiritual journey of self-discovery. Freud showed us how the soul could become aware of itself. To become acquainted with the lowest depths of the soul - to explore whatever personal hell we may suffer from - is not an easy undertaking. Freud's findings and, even more, the way he presents them to us give us the confidence that this demanding and potentially dangerous voyage of self-discovery will result in our becoming more fully human, so that we may no longer be enslaved without knowing it to the dark forces that reside within us. By exploring and understanding the origins and potency of these forces, we not only become much better able to cope with them but also gain a much deeper and more compassionate understanding of our fellow man. In his work and in his writings, Freud often spoke of the soul - of its nature and structure, its developments, its attributes, how it reveals itself in all we do and dream. Unfortunately, nobody who reads him in English could guess this, because all his many references to the soul, and to matters pertaining to the soul, have been excised in translation.
>
> This fact, combined with erroneous or inadequate translation of many of the most important original concepts of psychoanalysis, makes Freud's direct and always deeply personal appeals to our common humanity appear to readers of English as abstract, depersonalized, highly theoretical, erudite, and mechanized - in short, 'scientific' - statements about the strange and very complex workings of our mind. Instead of instilling a deep feeling for what is most human in all of us, the translations attempt to lure the reader into developing a 'scientific attitude' toward man and his actions, a 'scientific' understanding of the unconscious and how it conditions much of our behavior.

I have quoted extensively in Chapter 7, *Time and the Whirlpool*, from Dr. Robert O. Becker's book *The Body Electric: Electromagnetism and the Foundation of Life*. Here are two other statements from this very significant book which lend support to the view that sometimes to be on the cutting edge of science is to be skirting the boundaries of metaphysical country - the soul's domain.

> On page 264: *Following the curious dogma that what we don't understand can't exist, mainstream science has dismissed psychical*

phenomena as delusions or hoaxes simply because they're rarer than sleep, dreams, memory, growth, pain, or consciousness, which are all inexplicable in traditional terms but are too common to be denied...

On page 181: *However, chemical reactions and the passage of compounds from cell to cell can't account for structure, such as the alignment of muscle fiber bundles, the proper shape of the whole muscle, and its precise attachment to bones. Molecular dynamics, the simple gradients of diffusion, can't explain anatomy. The control system we're seeking unites all levels of organization, from the idiosyncratic yet regular outline of the whole organism to the precisely engineered traceries of its microstructure. The DNA-RNA apparatus isn't the whole secret of life but a sort of computor program by which the real secret, the control system, expresses its pattern in terms of living cells.*

This pattern is part of what many people mean by the soul, which so many philosophies have tried to explicate. However, most of the proposed answers haven't been connected with the physical world of biology in a way that offered a toehold for experiment...

But let us return to Jung for a moment. He was born in Switzerland in 1875 and died in 1961 some fifteen years after the end of the Second World War. Consequently, his life straddled a most turbulent epoch: years during which he witnessed the range of mental and behavioral responses to life that European 'civilized' societies displayed - from great creative achievements in music, art, literature, medicine... to the breakup of Kingdoms and two calamitous World Wars. It was a period he saw as accelerating the loss of religious faith - of trust in any such spiritual concept as the soul. (Souls killing each other in the millions?)

The preceding 19th century had seen major political, cultural, and religious changes - not to mention scientific advances - taking place in the European world: turbulent events that some historians see as the inevitable result of the passionate blood-letting drama that characterized the French Revolution in the last years of the 18th century.

Old political and national divisions were beginning to break down as small Principalities and Kingdoms struggled to maintain an independent existence while the major Powers - Britain, France, Prussia, the Austro-Hungarian Monarchy, and to a lesser extent Russia - worked (and fought) to extend their boundaries and spheres of influence. Also, as the industrial revolution began to spread across the face of Europe, the exodus from the countryside to the cities took people away from the land - away from their roots - concentrating them in factories to work, and to live in cheek-by-jowl cheap housing. And then Darwinian theory rattled the Church, challenging its doctrine, casting doubt on its theological veracity. Scientific discovery in the fields of physics, chemistry, biology, medicine and its new branch psychology, brought old beliefs into question, creating a new philosophical climate... a general sense of uncertainty regarding the 'how's' and the 'why's' of life. Mobility increased by leaps and bounds as steam engines and railways, followed by the invention of the internal combustion engine, broke down hitherto localized, static, and hidebound ways of living. Change and progress brought moral and political instability. France and Prussia, longtime opponents in the territorial stakes, finally commenced hostilities in the Franco-Prussian War of 1870-71, resulting in the Siege of Paris by the Prussians - a military blockade in which the citizens of Paris suffered several terrible months of starvation.

A breathing space of some forty years before the start of World War I in 1914 - a conflict which engulfed Europe in four mad and disastrous years of trench warfare resulting in millions of casualties on both sides. By the time I was growing up in the thirties the magnitude of the slaughter in the First World War finally hit home. It could now be seen as a cataclysm so obviously inhuman, so totally unnecessary: a human disaster resulting from the arrogance and stupidity of a ruling aristocratic elite, militaristic politicians, and the hysterical nationalistic fervor so prevalent at the time. And belief in 'soul' and human spirituality was even then becoming something of a lost cause.

And then, in 1939 followed World War II. I will not go into the multiplication of horrors Germany and Japan imposed on the world. They are still close enough in time for the Nazi genocidal atrocities and Japanese catalog of barbaric cruelties to be known and remembered. This was a war in which hundreds and thousands of civilians - men women and children - became the victims.... falling before the German and Japanese advancing armies, perishing in German air attacks on London, Southampton, Coventry, Birmingham... and in Allied raids by day and night on Berlin, Essen, Dortmund and all the major cities of the Third Reich, climaxing in the destruction of Dresden... and the fiery holocaust of Hamburg. And with Germany defeated in 1945, the Japanese finally surrendered after two atom bombs destroyed the cities of Nagasaki and Hiroshima.

The nuclear age. The epoch that Jung believed would present mankind with its ultimate test. For he understood that in the fierce light of such destructive power, the only way by which the planet and mankind could, in the long run, be saved from nuclear destruction would be through the universal recognition and experience of our common humanity. He had found in his practice as a psychiatrist that when patients came to be aware of some divine principle present in their being they understood the need to love... others as well as themselves. And he warned constantly that without the spirit of compassion and empathy - largesse of soul - we will continue to kill both each other and the planet.

Jung devoted his life - all his considerable powers of reason and intuition - to probing the evolutionary and neurological foundations of how and why we came to be driven by such complex, polarized, positive and negative, mental forces - both conscious and unconscious. The extent of his research resulted in a prodigious and exhaustive literary output - a vast body of work having profound implications for the new science (or art) of psychiatry, all of which brought him renown as both scientist and philosopher, but particularly as a doctor and healer. And if there is one essential conclusion to be garnered from his thoughts about humanness, it is that the most vital competing mind-forces with which each one of us must deal are those which occur between the demands of an overblown or *gross* ego and those of the soul. This psychological struggle is 'built' into our consciousness, and the outcome determines whether or not we attain the fulfillment that attends a spiritually aware selfhood.

On several occasions Jung spent time living with the surviving members of ancient communities in many parts of the world, being admitted into their councils and talking with those elders who still acted as the channels for keeping alive traditional wisdoms and beliefs - dialogues which confirmed the veracity of his own insights regarding the mental problems afflicting so many people in the modern world: problems which he considered in so many cases to stem from the silence of the inner self - a condition that has often been referred to as 'the dark night of the soul'. And I think he would have agreed that despite the sophisticated learning and theories that support contemporary psychiatric practices, this loss could not be better illustrated than in the words of a Native American Indian, a Cherokee elder teaching his grandchildren about life who said:

> 'A fight is going on inside me... it is a terrible fight and it is between two wolves. One wolf represents fear, anger, envy, sorrow, regret, greed, arrogance, self-pity, guilt, resentment, inferiority, lies, false pride, superiority, and ego.
> The other stands for: joy, peace, love, hope, sharing, serenity, humility, kindness, benevolence, friendship, empathy, generosity, truth, compassion, and faith.
> This same fight is going on inside you, and inside every other person too.
> They thought about it and then one child asked his grandfather,' Which wolf will win'?
> The old Cherokee simply replied, 'The one you feed'.

Also, had Jung been alive in 1971 to witness the Headman of the Sakuddei walking out from the Java forest leading the few hundred surviving members of his tribe, and had heard his remarks concerning the conditions under which the soul might be disposed to vacate the body... I can imagine Jung nodding his head approvingly, for the Sakuddei leader's words supported the views Jung expressed in Modern Man in Search of a Soul (published in America in 1933) where he wrote of the spiritual plight of man lost in the 'progress' of modern times - lost to the profound glimpses of truth and wisdom that reflect the influence of the soul. He writes:

> Being that has soul is living being. Soul is the _living_ in man, that which lives of itself and causes life... Were it not for the motion and colour-play of the soul, man would suffocate and rot away in his greatest passion, idleness.

Both Freud and Jung came to the conclusion that in the days of primitive man - between two and four million years ago before the brain had attained its fully evolved form - the mechanical processes of instinct governed most of his behavioral responses to life; particularly when he faced uncertain or dangerous situations when he was prompted simply to either fight or flee. And they considered that the ease with which we still, more or less spontaneously, adopt an aggressive stance when events seem to be potentially threatening, can be seen as the subconsciously persisting influence of primordial man's traumatic life: a hair-triggered mental throwback to the time before the advent of the powers of reason, the faculties of intuition and conscience, together with moral and spiritual sensitivity... superseded the control of instinct. It is interesting to note that contemporary neuroscience has determined that the cerebellum - the brain's repository of residual memory - has tripled in size over the last million years: a finding that lends credibility to the Freudian and Jungian proposition that such ancient and potent responses to potentially hostile situations can persist in the contemporary consciousness.

When Jung talks of spiritual sensitivity playing a vital role along the road to psychical wholeness, he is not addressing only the reasonably sophisticated patients who constituted the majority of his clientele. He is not suggesting that spirituality goes hand in hand with any particular form of culture, religious persuasion, or sophisticated levels of learning. During the time spent with tribesmen in Africa, and with Hindus, Muslims, and Buddhists in India, for example, he found that the belief in an immaterial spiritual force at work in human life, answering to the dictates of the soul, was commonly accepted. Some communities with whom he lived would be considered completely unschooled. Throughout his life Jung was always urging us to remember that the concept of the soul is universally present in human consciousness; that no one religious faith or denomination - neither Christian, Muslim, Jewish, Hindu, Buddhist... - can claim that they possess exclusive and certain knowledge pertaining to spiritual Truth.

Over two thousand years ago the Roman orator Cicero (whom I quoted previously in Chapter 14), talked about the soul as '... *bringing knowledge of things human and divine and of the causes by which those things are controlled.*' And in expressing my own thoughts concerning the *'divine'* knowledge of which Cicero speaks, and which Jung particularly saw as essential to mental health, I would say that soul is the metaphysical instigator of such spiritual insights, evoking:

> ... *a strong personal sensibility to conscience, moral issues, the power of love and compassion, the presence of a guiding force leading to values that serve a higher purpose than biological imperatives (even survival); that surpass the reality offered by the senses; and that transcend self-interest, self-gratification.*

Since Carl Jung's death in 1961 great advances have been made in the discovery and production of chemical and synthetic drugs intended to treat all kinds of psychological problems. It has resulted in a drug-therapy approach to mental illness which may well alleviate and, in the long term, diminish many of the symptoms that attend neurotic and psychotic complaints. Drugs are used to redress chemical and electrical imbalances in the brain, restoring neural pathways that were physiologically damaged or otherwise malfunctioning. Yet at the same time it has been observed that analytical therapy - now often called 'counselling' the patient - when pursued with Jung-like insight and patience, can also work on the physical brain by restoring old neural pathways or opening up new ones. The fact that we can now use chemical substances to effect changes in the chemical and electrical activity of the brain represents a remarkable medical advance. But even more remarkable is the discovery that bringing patients closer to their metaphysical center by way of analytical therapy.... can also produce similar alterations in brain physiology. The fact that such *verbal* counselling can create *physiological* changes in brain functioning, is surely a significant phenomenon for medicine in particular and science in general.

Consequently, a combination of drug therapy and psychiatric analysis is now thought to be the most effective way of treating mental illness. Yet it must be said that the clinical psychologists and psychiatrists entering medical practice today go for the drugs; talk is not high on the list of their priorities - neither are they schooled well enough, nor have the time and patience enough, to be effective analysts. Neither are the insurance companies prepared to pay for 'soul-chemistry'.

The Soul
We know we're not allowed to use your name. We know you're inexpressible, anemic, frail, and suspect for mysterious offences as a child. We know that you're not allowed to live now in music or in trees at sunset. We know - or at least we've been told - that you do not exist at all, anywhere.

*And yet we still keep hearing your weary voice in an echo, a complaint, in the letters we receive from Antigone in the Greek desert.**

---

* Adam Zagajewski, from *Without End: New and Selected Poems*, trans. by Clare Cavanagh (New York: Farrar, Straus, Giroux, 2002.)

Edith Hamilton, the renowned Greek scholar writes in *The Greek Way*: *A single sentence of Socrates, spoken when he was condemned to death, shows how the Greek could use his mind upon religion, and by means of human wisdom joined to spiritual insight could sweep aside all the superficialities and see through to the thing that is ultimate in religion: 'Think this certain, that to a good man no evil can happen, either in life or in death.' These words are the final expression of faith in the soul.*

While the poet Emily Dickinson, wrote: *'Hope.. the thing with feathers that perches in the soul...'*

THE ANALOGY OF THE ANISEED BALL: When I was a lad you could buy ten aniseed balls for a penny: ten long-lasting, gum-sucking treats purchased at the corner sweet shop on the way home from school. On the final lap of the walk the four of us would have competitions to see which of us could make the first ball last the longest. The best technique was to hold the dark red sphere - roughly half-an-inch in diameter - pinned at the side of the lower jaw between cheek and gum; repress the urge to lick with the tongue and leave the ptyalin in the saliva to work unaided in slowly dissolving the ball. Talking between ourselves speeded things up so we walked the final half-mile in silence - the well known Gang of Four, faces totally devoid of expression, numbering forty-four years between them, and a seeming menace to the world at large. At the bottom of Crimicar Lane - where we dispersed to go to our separate houses - the aniseed balls were solemnly extracted and compared to see who was left with the largest specimen.

One standard test - eminently scientific to our way of thinking - determined the winner. This was the test of color, for the smaller the ball became, the whiter it got; therefore the one whose specimen remained the most obviously reddish would have exercised the greatest sucking restraint and won the contest for the day - a victory recorded on the scorecard for the week. Judging the degree of coloration demanded an eye skilled in discriminating between shades of pinkish-red and pinkish-white - a faculty each of us claimed to have developed to the highest degree. And then, one afternoon, on some spontaneous collective whim, we decided to reverse the process and see which one of us could get to Crimicar Lane with the *smallest* core of aniseed remaining.

Chewing was out. Pressure-sucking only allowed. Splintered fragments of the original were disqualified. The remnant must be perfectly circular - a micro-aniseed-ball, bead-like and absolutely white. It was much less easy to judge the winner of these competitions, the offerings being miniscule and their respective sizes difficult to assess when held up between finger and thumb for scrutiny. Even when they were gingerly placed on the palm of the hand the eye was not really up to it. We needed calipers if diameters were to be accurately measured. So when it came to determining the winner of these new contests, judging usually disintegrated into scuffles and slanging matches, and the weekly scorecard was forgotten.

Then came the day when, subjecting my aniseed ball to the erosive action of a relentlessly rhythmical suction, I reduced it to the point when not a vestige of its globular morphology remained. Never before had I worn the whole away... to be left with only a tiny wood-like splinter of concentrated aniseed - the hitherto undiscovered pith of the whole aniseed phenomenon. Needless to say, this small woody residue was not accepted as a valid entry by my companions. They argued that it was no longer an aniseed *ball*; the fact that it wreaked of aniseed, was obviously the concentration of *aniseedness* to which the ball owed its essential nature, counted for nothing. It was no good protesting that my skill in being the first to discover this hitherto invisible and unsuspected heart of the aniseed phenomenon represented a real breakthrough. So far as my friends were concerned, there was no 'aniseed reality' left once all the rounded substance of the ball had disappeared.

In the end we gave up on both games - it being generally decided that in trying to control the way in which we consumed these daily treats, we lost out on the simple enjoyment afforded by just relishing them without being subject to any controlling inhibition. Personal, unselfconscious pleasure triumphed over theory and discipline.

Jack Longden was the only one who seemed to prefer the long drawn-out approach, turning to ask as we approached Crimicar Lane, "How many balls've you got left?" The response was usually, "Two, haven't you?" "I mean *aniseed* balls," he would say humorlessly, knowing full well that the rest of us had been delving repeatedly into our paper bags. "Well, I'm still on my first; got nine left," he would announce smugly with a great air of superiority.

I have only once been asked to deliver a church sermon, and that was for Evensong one Sunday at Connecticut College for Women in 1964. The title I chose for the address was *The New Existentialism*, and for my text I quoted a little verse of e.e. Cummings: *We have eyes to see/ And lips to kiss with/ Who cares if a one-eyed son-of-a bitch/ Invents a machine to measure Spring with?* And while I thought that this bit of doggerel would provide a good introduction to my talk, sorry to say it did not seem to have the desired effect.

Since then I have preached only silent sermons, usually while cutting the lawn. But if I was ever to be asked again to speak in church, I would choose to discuss the concept of the human soul. And my text would be: *Life is an aniseed ball: Suck it and see.* I would then proceed to relate how four small boys devised competitions which, unbeknown to them at the time, came to invest the daily sucking of aniseed balls with a special metaphysical significance. If I told my story convincingly, there would be those members of the congregation who would see - without need of explanation - how the tale of the Gang of Four becomes a parable relating to the mystery of the hidden core we think of as the soul - the extra-biological factor representing the *essence* of humanness. However, not wishing to leave any of the congregation in the dark regarding the symbolic import of sucking an aniseed ball until arriving at its pithy essence, I would conclude the address as follows:

> *To partake of the aniseed ball is analogous to partaking of life. One should choose to savor the confection by taking as much time as is required to appreciate each unfolding color and flavor, rather than devouring it like a dog with a bone. Thus one comes to experience the aniseed phenomenon to the full - its changes of color, subtle variations of taste, and the skill required to keep the constantly diminishing materiality of the ball balanced in mid-tongue in order to partake evenly of its rotundity. And then, at the end of it all comes the bonus - the revelation which is only to be discovered after such a patient and discriminating involvement. For when the globule of boiled sugar and butter has been reduced to nothingness, the foreign element at the core is revealed: a sharp, woody splinter charged with 'aniseedness' - an element or nucleus quite different in constitution from the parent body; yet, nevertheless, responsible for its erstwhile essential characteristics.*
>
> *My 'sermon' - which might seem irreverent to some of you - therefore likens the savoring of our brief life to that of judiciously*

relishing the aniseed ball... patiently... waiting for the revelations... So we partake of life calmly and perceptively, doing our best to stoically work through all its vicissitudes until the body goes the way of all flesh to dissolution, and the essential, enduring pith of humanness at its core is revealed - that which has been identified for thousands of years as the soul, the home of the human spirit. And the fact that the immaterial soul (which can really only be thought of as a pure, energy force) has no discernible material nature comparable to the woody pith of the aniseed ball, does not, I believe, invalidate the analogy. For the aniseed splinter is a foreign element when compared to the sugar-butter amalgam it inhabits - as is the soul a foreign (if immaterial) element in the flesh and bone body it temporarily occupies. So do we make our choices: either sampling the aniseed ball of life wisely by calmly accepting the involuntary flow and pace of events as we encounter them, appraising the value of each in turn - or, conversely, giving it an injudicious heck of a sucking, in which case sheer appetite obliterates the subtle shades of meaning and the ultimate discovery of any hidden truth.

SOMEWHERE IN LOVING

On a long flight across the United States a few months ago, the man seated on my right spent a great deal of time working on his laptop computer. When he finally tired of this he was ready for a little conversation, introduced himself, and we exchanged the usual pleasantries. I was reading Elizabeth Tate's book, *Somewhere in Loving*, and my travelling companion was curious about the title. I told him that the book was about her brother, a Royal Air Force pilot who was lost on a night operation early in the Second World War; that it was a moving testimony to his strength of character and vitality of mind, and revealed the strength of the love between them. I went on to say that she wrote of the persistence of this bond after his death - of the sense of his presence she experienced from time to time, his frequent intrusion in her thoughts, and the certain feeling - which no amount of rational intervention could dispel - that his soul had gone on, and that with her own death they would be reunited.

A bemused half-smile, part incomprehension, part disbelief, greeted my remarks. And I realized that what seemed to me a perfectly understandable statement of belief on the author's part - one based on the conviction that love may only be experienced at such a profound level when the soul is aroused - lay beyond my companion's frame of reference. I was not really surprised at this. A constant preoccupation with business, commerce, and money - facilitated by the omnipresent computer - is seen by many nowadays as the most significant way to invest life with meaning. Of course I can only assume that such was the case with my fellow traveller. Yet the impression he gave was that in discussing Elizabeth Tate's account of an affinity between souls, and a love that 'passeth all understanding', he regarded her as being at least fifty years out of date. And I did not see fit to ask him whether or not he had ever been devastated by any major personal loss in his life.

"You find a book like this stretching credulity a bit?", I asked him (not aggressively).

To his credit he did not seem to be uncomfortable with the question; thought for a moment before responding. "Well, you know, if that's how you feel... that's how you feel, and I'm not going to question anyone's sincerity. But I don't know... there's something too mystical, too naïve about that kind of stuff. Because, you know, it seems we're moving beyond these old romantic and religious attitudes - props, I think, in a way; part of the old superstitions about the 'hereafter' - that kind of thing, you know..."

I nodded, suddenly feeling mildly depressed, a fish out of water: he reminded me of a 'with it' priest I knew who would have nodded tolerantly at my acquaintance's comments and talked about 'the manifestations of humanitarian feelings' - anything save the enigma of love and the human soul.

So I turned the conversation to the topic of travel, and for the remaining hour we talked about the interesting places in the world we had visited.

'Somewhere in loving': what an evocative and lovely phrase. 'Somewhere...' whether it lie in the past or be part of the present, the love of which Elizabeth Tate writes would seem to be indestructible and persists to light the future with hope. What is this state of consciousness which stands time on its head and can make our brief existence so ideally beautiful, so significant and purposive despite the transient and uncertain nature of our biological destiny... the certain fate to which we march? What are we to make of an all-pervading emotion which can possess such a numinous quality that to experience it at its most intense level - the high plateau which Rudolph Otto described in *The Idea of the Holy* as the 'mysterium tremendum' - has one hovering somewhere on a hyperplane of existence: on 'cloud nine' as the song writers call it.

> *It is love, not reason, that is stronger than death.*
> Thomas Mann, *The Magic Mountain*

I have spoken many times in these pages of the transforming power of love in the life of the world: how it inspires courage and acts of selfless heroism; brings light and meaning into the bleakest heart, and can cause the will to live to fade away when the object of love is lost. It would seem that the need to love and be loved is psychically invested in most human beings at every level of society - 'rich man, poor man, beggar man, thief'...' The emotion of love is to the psyche what oxygen is to the body. And the world suffers grievously at the hands of those so spiritually crippled that their hearts can never go out to others or to the world at large.

In *De l'Amour (1822)*, the French novelist and essayist Stendhal described four different kinds of love: *Passion-love, Sympathy-love, Sensual love, and Vanity love*. In the first instance he identifies the great and compelling strength of feeling we call passion as the vital drive to love. In the second he recognizes sympathetic feelings of goodwill towards, and affinity with, others. Thirdly, he names the sensual experiences of life: the emotional and bodily arousals that occur when we perceive what we consider to be tangibly, physically, beautiful or erotic in the world... and desire to possess he, she, or it. And fourthly comes 'vanity love': that which is enjoyed because it brings an inflated pride in oneself in attracting the love of another - an attitude that can easily degenerates into a level of narcissism where one is oneself the sole object of existence, with little or no consideration for others; awash in all the excesses of *gross*-ego.

Yet, as Elizabeth Tate reminds us, there is a level 'somewhere in loving' where devotion to those themes, things and persons nearest and dearest to us, attains a degree of transcendence which is not really accounted for in Stendhal's tally: exalted plateaus of feeling which the poet Wordsworth characterizes as sublime... taking one beyond the habitual limits of emotional response to reach a heightened awareness of what is true and precious in life - purposive and trailing gleams of fulfillment. *Sublime love*, Stendhal might have called it.

Throughout the centuries the tellers of tales, writers of ballads, poets, philosophers, composers... and all we would-be Romeos and Juliets, have been intrigued by the theme of perfect or transcendent love: the mystical bonding between human beings, between us and other creatures (Saint Francis and the Birds), or even between a person and a perceived object (D.H. Lawrence described Vincent van Gogh's sunflower paintings as resulting from the 'perfected relationship between a man and a sunflower'). Such is the loving which brings serenity and certitude to the psyche - an inexplicable equanimity arising from what Antoine de Saint-Exupéry called 'the heart's intelligence' - a lyrical way of referring to the spiritual center of our being, to the soul, traditionally seen as the source of such sublimity of feeling.

> *The mind has a thousand eyes,*
> *And the heart but one;*
> *Yet the light of a whole life dies,*
> *When love is done.*
> Francis William Bourdillon: *Light*

One could amend the title of Elizabeth Tate's book to read, *Somewhere... the Soul in Loving*, and I am sure she would have no objection.

I have often thought how useful it would be to carry in one's pocket a few slim cards on which are written snippets of poems, fragments of philosophy, and a few chosen sayings of great men and women. Then, being well equipped to illustrate one's position on the issues which arise when in conversation with temporary acquaintances, pull out the appropriate card and pass it over to be read. If, for example, I had been able to present my computer-working companion on the aircraft with Elizabeth Barrett Browning's famous lines from one of her sonnets, there would have been little need for me to say anything.

> *How do I love thee? Let me count the ways.*
> *I love thee to the depth and breadth and height My soul can reach,*
> *when feeling out of sight For the ends of Being and ideal Grace.*
> *I love thee to the level of everyday's*
> *Most quiet need, by sun and candle-light.*
> *I love thee freely, as men strive for Right;*
> *I love thee purely, as they turn from Praise.*
> *I love thee with the passion put to use*

> *In my old griefs, and with my childhood's faith.*
> *I love the with a love I seemed to lose*
> *With my lost saints, - I love thee with the breath,*
> *Smiles, tears, of all my life! - and, if God choose,*
> *I shall but love thee better after death.*
>   Elizabeth Barrett Browning
>   *Sonnets from the Portuguese, 43*

I cannot imagine a more evocative use of language - of words eliciting such a powerful mental image of what it is like to experience a profound, sublime, state of loving. They bring to mind the romantic term 'soul mates', conventionally used to denote such a totally committed involvement. It has become an expression used somewhat superficially in this day and age, and likely to raise a few amused, if not cynical, smiles from the 'cool-culture' generation. In fact, I wonder how many of the population at large would grasp that Elizabeth Barrett Browning is telling of a love embracing as much a spiritual reality as the physical one presented by the sheer presence of the other. Certainly the gossip magazines and the tabloid press featuring the ever-changing 'love life' of celebrities, give the impression of only equating love with sex. And the lyrics of many pop songs intimate that 'love' is either simply an uncomplicated form of rutting, or a mindless and monosyllabic exchange of platitudes.

I feel sure that the majority of my readers know that however prevalent are such relatively slight and uninformed attitudes to love and loving in contemporary culture, they do not represent the kind of deep emotional and spiritual bonds I have attempted to describe. In the middle years of Plato's life - (the famed Greek philosopher lived from c427 to c347 BC) - he wrote *Symposium*. Socrates is used in this dialogue as the spokesman for Plato's views, and it is here that we find the earliest surviving writings which discuss how it is possible, through the power of love, to bring about the ideal or perfect union between lover and the object of his or her affections. It requires a veritable alchemy of love - a magical mix of the senses, intellect and emotions brought to fruition by the infusion of a spiritual element. In *Symposium*, at a banquet given in honor of Eros, the guests exchange ideas about love while Socrates, Plato's teacher and confidant, acts as the devil's advocate in pointing out flaws in this or that theory. But what emerges for the reader is that love is seen as an omnipresent human need which, when *fully* realized, brings to each person a new sense of wholeness - body, mind, soul, and spirit come together as one. Consequently, in all actions born of all true love, the soul finds its place in earthly life.

One very positive result of on-the-spot television transmission of world events is that one sees first hand how universal is the deep undercurrent of loving in human life: the pain and shock on the strained face of a firefighter bringing the broken body of a child from a bombed building in Oklahoma; the empty hopelessness in the eyes of an old man holding the dead body of a grandchild after an earthquake in Mexico; the overwhelming sadness that cocoons the bereaved at the graveside just about everywhere... And then, in stark contrast, one is moved by the sight of happiness: the gaze of devotion passing between a man and wife contentedly together for over fifty years; the rapture of the young in heart and in love; the shield of fiercely protective care given by a wife to a crippled husband; the delicacy of the kiss and gentleness of embrace offered by loved ones reunited after absence, or after having survived a near disaster.

What is implied by the expression, 'To die of a broken heart'? It is, I would suggest, a metaphor for the death which results when the will to live is lost: when the psyche's vital drives of love, hope, and a sense of purpose have departed - and here I join the Sakuddei in believing that such desolation is experienced when the soul has taken its leave.

From time to time we hear of people who have lived devotedly together for many years and of how, when one dies, the survivor, still seemingly in reasonable health, succumbs in a very short time. To continue living when the soul is lost is to have all light turned into darkness; each day become 1,440 minutes of terrible loneliness, bringing a debilitating, aching sorrow of mind and body. And yet, paradoxically, if such pain is endured, and the sufferer finally walks out into the light again, it seems in many cases that he or she returns to a way of life in which faith in some ultimate purpose of continued existence is actually strengthened. *'Better to have loved and lost/ Than never to have loved at all'*, wrote Byron in *Child Harolde*.

The mysterious and compelling persuasions of the psyche which introduce us to the profundity of loving are, in my view, the most telling experiences in our mental world. They bring us to ponder this 'other' within - the presence of an authority bringing us to face truths not in the power of the senses to single-handedly deliver. *'Where love rules, there is no will to power, and where power predominates, there love is lacking',* wrote Carl Jung in *Psychological Reflections*. And writing in the same vein almost thirty-five years later, the famed American psychologist Erich Fromm said, *'Love is the only satisfactory answer to the problem of human existence'*. Would Emily Dickinson object, I wonder, if one substituted 'love' for 'hope' and rewrote the first two lines of her lovely quatrain to read, *'Love' is the thing with feathers/ That perches in the soul...*

Language is a mysterious and wonderful human faculty. The utterances we call words - speech sounds by which we became able to give verbal expression to observations, thoughts and feelings - are the principal means by which we fashion the mind's encyclopedic-like build-up of knowledge. Without vocabulary we would have no means of understanding our own needs and motivations, or of consciously formulating what we think we know of the phenomenological world. Yet, as constructs of our own mental powers - for which we cannot claim infallibility - they can only be seen as limited instruments for achieving any absolute comprehension of our own lives, or of life beyond ourselves. This limitation is particularly evident when we come to describe animal behavior, of necessity using words which symbolize the abstract nature of *our own* thoughts and feelings. Words such as 'soul', 'devotion', 'love...' may be totally irrelevant if applied to the life of animals and if used lay one open to the charge of being grossly anthropomorphic. When, for example, we come across cases where creatures appear to bond together - 'even unto death', as the poet might say - can we say that they are *'devoted'* to each other? Is it a reasonable conclusion to reach, for example, in the case of the great twelve-foot-wingspan aviators of Antarctica's Southern Ocean - the wandering albatrosses - who, it is recorded, generally mate for life, forging a union so strong that the death of one bird will often result in the decline and demise of its partner? Arctic snow geese appear to be similarly *'devoted'*, in that male and female create bonds equally binding and long standing as those formed by the large albatrosses. And I would think that anyone who saw the documentary made by the Gorilla Foundation of Woodside, California, featuring the female gorilla, Koko, had to be deeply moved and found it all too easy to see human-like qualities in her behavior. This large and powerful animal 'adopted' a small kitten, treating it with a remarkable gentleness and concern. To my eyes there was no suggestion that she was simply 'playing' with this tiny creature. Too much exuberant play and she would have killed it in an instant. She had no progeny of her own; yet her treatment of the kitten - an animal not even of her own species - suggests that Koko was moved to express deep feelings of affection that went beyond normal in-species caring. For she made no attempt to suckle the kitten, suggesting that she was not simply indulging a maternal instinct to nurture an infant, no matter of what kind.

Given such examples of animal mentality and behavior, is it misguided and sentimental to describe their behavior as 'loving'? I wonder how many veterinarians would say with any certainty that the response of a dog to its owner is driven entirely by the mechanics of instinct - that it does not result from an altogether more complex consciousness capable of communicating with our own? And go on to suggest that it is mere anthropomorphic wishful thinking to say that a dog 'misses' its owner in his or her absence; or responds sensitively to human moods and feelings; or experiences 'grief' at an owner's death...? Of course, to talk about ourselves as loving a dog is fine - we are human after all - but we should not assume that the dog is incapable of loving us back. (In the next few pages I will recount some instances concerning dogs that give the lie to the instinct-only argument.)

Miguel Unamuno - the distinguished Rector of the University of Salamanca (who stood up strongly against General Franco, the fascist dictator who governed Spain after his victory in the Spanish Civil War) - pointed out how presumptuous we are in thinking of ourselves as the supreme species. ' For all we know,' he said, 'the dog lying there on the carpet may be performing simultaneous quadratic equations in its head right now....'

There are times when we empathize strongly with certain higher animals, seeming to identify with thoughts and emotions that obviously surpass the mechanical, behavioral patterns governed solely by instinct. The problem is that in seeking to describe such relatively lofty qualities of their mental life we have no recourse but to use words symbolically rather than literally: as indeed is the case when we attempt to describe the nature of our own inner sensibilities. Anthropomorphic though this may be, we have no other way to comprehend and portray examples of the cognitive and affective psychologic life of those creatures whose complex attitude and behavior intrigues us. Consequently, I would suggest that those zoologists who dismiss 'anthropomorphic interpretations' out-of-hand come to appreciate their possible validity. 'Of course', they say, 'dogs (animals) don't have souls!'

Some of the most poignant expressions of human anguish and love have been written on the loss of a dog, as in this epitaph from antiquity for a Greek dog of the house.

> *Stranger by the roadside, do not smile*
> *When you see this grave, though it is only a dog's:*
> *My master wept when I died, and by his own hand*
> *Laid me in earth and wrote these lines on my tomb.*
> Anonymous: from *The Greek Anthology*
> edited by Dudley Fitts

About twenty-five hundred years later - Robinson Jeffers one of America's master poets - wrote the following moving poem commemorating his dog. It could well be named *'Somewhere in Loving'* for it is all here - the joy and the spiritual nature of such deep affection; the pain and sadness which comes when it is lost. And so, across a span of two and a half thousand years we find human beings, soulfully expressing the depths of their feelings at the death of a dog.

> *I've changed my ways a little; I cannot now*
> *Run with you in the evenings along the shore.*
> *Except in a kind of dream; and you, if you dream a moment,*
> *You see me there.*
>
> *So leave awhile the paw marks on the front door*
> *Where I used to scratch to go out or in,*
> *And you'd soon open; leave on the kitchen floor*
> *The marks of my drinking pan.*
>
> *I cannot lie by your fire as I used to do*
> *On the warm stone*
> *Nor at the foot of your bed; no, all the nights through*
> *I lie alone.*
>
> *But your kind thought has laid me less than six feet*
> *Outside your window where firelight so often plays,*
> *And where you sit to read - and I fear often grieving for me --*
> *Every night your lamplight lies on my place.*

THE WIRE-DANGLED HUMAN RACE

You, man and woman, live so long, it is hard
To think of you ever dying.
A little dog would get tired, living so long.
I hope that when you are lying

Under the ground like me your lives will appear
As good and joyful as mine.
No, dears, that's too much hope: you are not so well cared for
As I have been.

And never have known the passionate undivided
Fidelities that I knew.
Your minds are perhaps too active, too many-sided...
But to me you were true.

You were never masters, but friends. I was your true friend.
I loved you well, and was loved. Deep love endures
To the end and far past the end. If this is my end,
I am not lonely. I am not afraid. I am still yours.
> Robinson Jeffers: *The House Dog's Grave.* From
> the Vintage Book's Edition of Selected Poems

If one ever needs reminding of what it is to be truly human - of the grace which flows from the soul in loving - think of Robinson Jeffers and his dog. For the loss of those we love 'cuts to the quick', as that somewhat archaic phrase puts it, and to which the poet so frequently attests. The word 'quick' (Old English and Scandinavian) has a compound meaning, signifying on the one hand the physiological and mental vigor of life; on the other the deepest of feelings and sensibilities - the territory of the soul. Therefore to say that someone is cut to the quick by the death of a loved one, implies that not only is physical vitality significantly diminished, but that there is also a grieving at the spiritual center which cannot be treated by doctors and medicines - only by faith: a faith which enables Jeffers to write, 'Deep love endures/To the end and far past the end... I am still yours'.

So we can be comforted in bereavement if we have faith in the soul as the ground of our spiritual indestructibility... for where two souls have been as one - somewhere in loving - they will seek each other out 'far past the end'. For myself, I have no difficulty in accepting the possibility that a dog so significantly attuned to a human being - as that depicted in

Jeffers' *The House Dog's Grave* - can have come to 'love' its owner. And this because the kindness and loving a dog receives may arouse some latent, soul-like essence to reciprocate in kind. For ourselves, we do not have to be religious (in a church-going way) to recognize the profound sense of wholeness, pervading all aspects of our being, when we love or are being loved. Or, conversely, to realise how psychically fragmented we become when plunged into the emotional abyss sometimes called 'the dark night of the soul...' when love is lost.

> *Loves mysteries in soules doe grow*
> *But yet the body is his booke.*
> John Donne (1573-1631) *The Extasie*

I wonder how Robinson Jeffers would have responded to the following account of a remarkable dog called Sally. We are all aware of the high level of intelligence possessed by some dogs, but Sally, a golden retriever, displayed extraordinary powers of awareness with regard to events involving her master taking place hundreds of miles away. He and she were inseparable. I never saw him on the airfield without her. And she seemed always to be attuned to his every thought and feeling, anticipating everything he wanted her to do without need of any word of command. All of us serving on 207 Squadron, Royal Air Force, during the Second World War - air and ground crews alike - were always greeted with a thrashing tail by Sally who knew nothing of the distinctions conferred by rank. Yet it was only during the post-war years, sharing life with dogs of my own, that I came to fully comprehend - and wonder at - the nature of the remarkable relationship Squadron Leader Beaumont shared with his dog. For the love she bestowed on her 'owner' transcended the bounds of the usual master and 'pet' relationship.

As the constant companion of the Squadron Commander she enjoyed a privileged status on the aerodrome. Every night when the squadron was scheduled to be operational over Germany she accompanied him to the Flight Hangar where the bomber crews drew their flying gear and parachutes. Sally then stayed in the Flight Office in the charge of the Duty Flight Sergeant. The target on this particular night in November, 1942, was, once again, Berlin. The dog always became very restless when the throbbing roar of the first Lancaster racing down the flarepath reverberated through the hanger. After that she would settle down beneath the office desk for the long wait. But on the night to which I refer, Sally surprised Flight Sergeant White by suddenly, at 11.05 p.m. - just about four hours after takeoff - running from the Flight Office into the hangar and out onto the nearby airfield perimeter track where she stood, tail between her legs, howling inconsolably into the night. White described the sound as the most desolate, soul-wracking cry he had ever heard. She could not be induced to return to the hanger; just lay unmoving, head down on her paws until the early hours after midnight when the first of the returning aircraft landed. Then she got to her feet, ears up, tail ready to wag... only to suddenly stand dejectedly, unmoving. Even after the last machine had landed and the aerodrome became relatively silent, Sally stood there, waiting; could not be induced to return to the hanger.

As the hours dragged on it became obvious that two of our Lancasters were missing: Squadron Leader Beaumont and his crew; and Flying Officer Greene and crew. Those who made it home went through the usual debriefing routines - questions concerning weather over the route, navigational problems, observations made over the target, and so on. Two crews reported seeing the Squadron Leader's aircraft receive a direct hit by anti-aircraft fire and go down in flames. No parachutes were seen to open. The rear-gunner of one crew reported that the bomber was stricken at 23:00 hours: 11 p.m. English time; the flight-engineer of the other gave the time as 23:03 hours.

Finally, at 3 a.m., Flight Sergeant White was able to approach Sally and carry her back into the Flight Office. Now, ears flat to her head, body limp, she offered no resistance but would not walk of her own volition. Once inside she crouched beneath the desk and refused to move, listless, breathing shallowly. 'Chalky' White, as we called him, stayed there with her for the rest of the night. He was a career peacetime member of the Service - an airframe-fitter and excellent all-round aircraft engineer not given to sentiment or casual conversation - and it was only at first light that he managed to coax her to walk with him to the Sergeant's Mess.

I spoke to him the following afternoon after he had delivered Sally to the Station Commander who adopted her, and I will never forget the way he said, 'Her heart's just broken, you know.'

Writing about Sally so many years after the event, I recall a statement made by Sir James Jeans, physicist and Astronomer Royal, on the occasion of his lecture to a group of Royal Air Force officers at the end of the war: (an event about which I spoke in the first chapter of this book). He said, *"The world is not just stranger than we think; it may be stranger than we can think. We must allow not only for countless' scientific' fresh truths but equally countless new evaluations of their comparative importance."*

The fact that animals - particularly primates, dogs and elephants - would appear to grieve for each other is well documented; and dogs mainly, perhaps uniquely, suffer the loss of the humans to whom they have attached themselves. While we ourselves can experience acute distress when deprived of a favorite animal's company. The story of 'Greyfriars' Bobby - the name given to a Skye terrier who daily accompanied his master to work in Edinburgh - is a famous account of remarkable canine devotion. From the day his master, 'Auld' Jock, died, and was buried in the cemetery of Greyfriar's parish church, Bobby regarded the grave as his home, and for the first few days following the funeral would not move more than a few yards away from the graveside. Leanor Atkinson in her book, *Greyfriars Bobby,* tells of the dog's continued devotion to 'Auld' Jock - of the days when *'he faced starvation rather than desert that grave...the days when he lay cramped under the table-tomb... His never-broken silence in the kirkyard was only to be explained by the unforgotten orders of his dead master.'* Yet being much loved by all in the parish he was gradually encouraged to set forth alone every day on the familiar rounds formerly made by he and his master - a neighborhood circuit on which 'Auld' Jock's skills as an odd-job man had always been in great demand. Bobby made this daily peregrination for the rest of his long life, being fed and cared for by one or other of the households he visited. There was always fresh water in his bowl by the grave when he returned at night to sleep alongside his late owner, whatever the weather was like. And the custodian of the cemetery would never lock up and leave before visiting the burial site to bid Bobby 'good night'. The Lord Provost of Edinburgh had a collar made for him, with the inscription:

GREYFRIARS BOBBY
FROM THE LORD PROVOST
1867 Licensed

He personally buckled it around Bobby's neck. The little dog guarded Auld Jock's grave for eight years, was visited by the famous and the unknown alike, and revered by hundreds of ordinary folk. He died, and is buried, next to the remains of his master.

So what price Sally as a strange canine phenomenon - with her apparent ability to be aware, more or less simultaneously, of the disaster that had overtaken her master some 800 miles away? And what price the haunting devotion of Greyfriars Bobby? Those who talk so condescendingly about 'dumb animals' should think again in the light of such extraordinary faculties - unless, of course, they wish to continue revealing their own dumbness. For the manifestation of such compelling forces in the world - mental powers creating the affective bonds we attempt to explain by coining such words as 'love' and 'devotion', and which, furthermore, are not bound by the biological frontiers between species - support Sir James Jean's views that the world is 'stranger than we think...' even 'stranger than we *can* think.' And if we consider that our own ability to love is due to a mysterious emanation from a specifically human soul, then should we not attribute to the non-human higher animals capable of demonstrating a high order of devotion... the presence of their own kind of animal soul? In fact, the latest reports concerning the mental life of animals published by animal welfare and husbandry scientists, reveal that animals possess a far more complex consciousness than was thought.

Jonathan Leake, Science Editor for *The Sunday Times* of London, commences an article in the issue for February 27, 2005 with the headline, '*The secret life of moody cows*'; and goes on to write as follows:

> *Once they were a byword for mindless docility. But cows have a secret mental life in which they bear grudges, nurture friendships and become excited over intellectual challenges, scientists have found.*
>
> *Cows are also capable of feeling strong emotions such as pain, fear and even anxiety - they worry about the future. But if farmers provide the right conditions, they can also feel great happiness.*
>
> *The findings have emerged from studies of farm animals that have found similar traits in pigs, goats, chickens and other livestock. They suggest that such animals may be so emotionally similar to humans that welfare laws need to be rethought.*
>
> *Christine Nicol, professor of animal welfare at Bristol University, said even chickens may have to be treated as individuals with needs*

> and problems. "Remarkable cognitive abilities and cultural innovations have been revealed," she said...
>
> John Webster, professor of animal husbandry at Bristol... and his colleagues have documented how cows within a herd form smaller friendship groups of between two and four animals with whom they spend most of their time, often grooming and licking each other. They will also dislike other cows and can bear grudges for months or years...
>
> Keith Kendrick, professor of neurobiology at the Babraham Institute in Cambridge, England, has found that even sheep are far more complex than realised and can remember 50 ovine faces - even in profile. They can recognise another sheep after a year apart. Kendrick has also described how sheep can form strong affections for particular humans, becoming depressed by long separations and greeting them enthusiastically even after three years...

Consequently, these reports indicate that one is not necessarily being anthropomorphic when expressing views concerning an animal's motivations or emotions, particularly when it can be scientifically shown that the creature's themselves are - in their own right - experiencing cognitive and emotional forms of awareness not totally dissimilar to our own. Neither should we forget that for thousands of years throughout our history, belief in the idea of the 'animal spirit' or the 'animal soul' has been widely held. We talked earlier about the cave paintings of the Palaeolithic period in Europe that were made some 15,000 years (or more) ago to induce a state of 'sympathetic magic' linking hunter and animal: a ritual - that in the opinion of some twentieth-century anthropologists - was believed by the hunter-artist to give him access to, and power over, the spirit-essence or soul of his prey - a mystical form of possession that would ensure success in the hunt. In addition... in the view of experts like the Abbé Breuil and Professor S. Giedion, the empathic sensitivity displayed in some of the paintings suggests that the hunters were also attempting, through the ritual, to assure the animal spirit that it would not be hunted without mercy. In other words - as Giedion points out in his, *The Eternal Present: The Beginnings of Art* - the animal was seen as giving up its life in order that the hunter might live: it was not being killed simply for the sake of killing.

When I pointed this out in the course of a lecture on Palaeolithic art, a student remarked: 'Tell that to the animals jammed into trucks heading for the abattoirs... or to the trophy hunter who kills just to get the antlers to put on the wall.'

But back to the dog: here are the first two lines of St. John Lucas' poem inscribed 'To the Eternal Dog' and entitled *The Curate Thinks:*

> *The curate thinks you have no soul;*
> *I know that he has none.*

In the absence of any material evidence to account for this noumenal entity we call the soul, I would suggest that the most credible testimony to its authority is that provided by the example of one person's life that seems almost too good to be true: one not dominated by material concerns, or even much concerned by the normal demands of an acceptable and necessary level of ego'. And, in my view, the graveside oration delivered at the funeral of Anna Liszt is an impressive testimony to the life of a truly good soul. Anna Liszt, mother of the great pianist and composer Frans Liszt, was buried in Paris' Montparnasse cemetery on February 8, 1866. The funeral oration was delivered by Liszt's son-in-law, Emile Ollivier, whose own wife, Liszt's daughter Blandine, had died in 1862. I have taken the following lines spoken by Ollivier, from *Frans Liszt-The Final Years,* by Alan Walker.

> *... She deserves that we do not take our leave of her without expressing a feeling of deep sorrow. She had all the gifts which inspire affection: a lively and serious intelligence, a lovable character, always even tempered and always kindly; a goodness whose depths one could never plumb; and overlaying it all, a serenity which even her simplicity did not prevent from being impressive, and which she owed to the nobility of her thoughts, to the elevation of her feelings, and to that admirable purity which, during the whole of a long life, was never even sullied for a moment by a fleeting temptation.*

I have talked previously about my stepfather - a chemical and metallurgical engineer by profession - who believed that a unique synergenic activity between chemical and electrical systems is responsible for the animation of the body's cellular structure. He called it the élan *vital* (from the French; literally, vital force). In this belief he belonged to the Bergsonian school of Vitalist philosophy which postulated that such a vital impulse sustains organic life, and that death results when it ceases to be. I can still picture the white and blue-tiled kitchen where we had our usual Sunday verbal duel as to whether or not I should go to Sunday School. Perched on a high stool I would advance my argument against attendance, to which he always listened patiently before responding to tell me of the benefits I would receive. And I still remember the particular Sunday when he said, 'Spirit is energy, my boy: cosmic chemical and electrical energy, the very élan of your little existence. Very necessary that you receive some instruction about what it's up to.' I protested that spirit and energy were never discussed in Sunday School, and that therefore there was no reason for me to go and waste my time. 'Touché," he said. "Alright: let's talk about it, and that'll take the place of Sunday School for today."

Towards the end of his life he came to live with me in northeast Yorkshire. Unfortunately, it was not to be for long. He died about two years later. But going through his notebooks I found he had kept a record of our Sunday skirmishes. And there was his definition of the soul as presented at our kitchen table one Sunday, twenty-four years earlier. I'm sure that at the tender age of eleven I could not have understood very much of it. Yet there must have been some discussion because he noted at the top of the page that, 'The discussion went on for twenty-five minutes'.

> THE SOUL - as I told the lad - cannot be scientifically proven for it is not a physical phenomenon and so has no known physical location in the body. Yet it is a word coined to signify the seat of a spiritual force in human life to which peoples around the world testify - and although I'm a scientist I'm not going to scoff at it simply because the whole idea defies reason and can't be proven.
>
> So I told the lad to think of the soul as being like a wireless set receiving messages from God, and that to do this without missing some important information it was necessary to tune it in from time to time. And Sunday School was the place to do it.
>
> I might have been talking over his head when I said that the soul needed the brain with its millions of cells - neurons - to decode its communication.
>
> Yet he seemed to take this thought in his stride - perhaps because for a couple of weeks he's been looking at that book, The Life of the Cell, describing the workings of a cell's negative and positive electrical charges - particularly the nerve cells or neurons of the brain.
>
> "You mean God-charged neurons," he said.
>
> "Not a bad idea," said I, impressed by this response.
>
> "Well, they don't say anything like that about the soul and God in Sunday School; so it really is a waste of my time going, isn't it?"

> "Yes: it suppose it is..." What else could I say? "So why don't you do some reading for today then?
> "Right," said the lad in a triumphant kind of way.
>
> After lunch he came up to me with The Life of the Cell. "Did you know, Dad, that a Frenchman called Descartes said he thought the seat of the soul was in the pineal body of the brain? Right here..." He held the book out for me to see. "Yes, I remember that," I responded, "but nobody really knows much about what goes on in the pineal gland."
> "Mm," was his response. Then he drifted off; went outside to kick a ball against the side of the garage.
>
> After this usual Sunday skirmish I began to wonder if age can have anything to do with developing an awareness of soul? Are we all born equal in terms of a soul's strength to effectively break through into consciousness? After all, people come in all sorts of physical shapes and sizes; are endowed with greater or lesser mental capabilities. Are there 'weak' souls and 'strong' souls?
> I should have suggested to him that some souls are so weak at the outset that they might depart if not nourished by life; that some may start as apprentice-souls, setting out to learn the ropes of physical incarceration; on the other hand, some may be master-souls approaching the end of a long journey. And that he may be in the latter category, and I in the former. But it's a moot point; it's not likely I'll be here to find out where he belongs in the soul stakes.

When reading his notes I remember thinking what a pity it is that we youngsters who are indeed 'apprentice' souls' and would ask questions of the 'master-souls' among us, are not old enough to know just what are the crucial questions to ask before the wiser ones die. Such was the case with my stepfather and myself. But one thing he impressed upon me: a principle which I believe to be important if life is to be lived to some purpose - and that is never to retreat from the metaphysical question: give no quarter to those who see life solely in mechanical terms, apparently cut off from the complex offerings and mystery of their own inner life... and so unknowing of soul. The last conversation we had concerning science and the arts the day before I left Yorkshire was, he announced, intended to 'stiffen my spine as an artist.' 'Tell them,' he said (referring to the extreme rationalists), 'that reason is only one among the many conscious processes with which we are endowed. And don't worry about being labelled a mystic - remember you're in good company, lad. If a physicist like James Jeans can say, *The universe begins to look more like a great thought than a great machine...* and a philosopher-mathematician like Leibniz announce that, *The soul is the mirror of the indestructible universe...* then that's good enough for me. Remember that science is not infallible... scientific objectivity suffers from the sensory and intellectual limitations built into our own consciousness. We are part of it all. To be truly objective and dispassionate we would have to stand outside the cosmos.'

On 12 February 1564 Michelangelo was working all day long on a 'Pietà' - Mary grieving over the body of Jesus taken from the cross. It was to be the last work of his long life. Six days later, just short of his 89th birthday, Michelangelo died. It is assumed that he commenced the work some ten years earlier, and that the legs of the Christ and the free-standing right arm are relics of the first version. All the accounts of those who witnessed his last day of work on the sculpture testify to the fury of his 'attack' on the marble - the chips were flying all over the place. With every blow of hammer and chisel the physical bodies of Jesus and Mary were being reduced in both volume and detail. And one is entitled to wonder just what would have been left of Jesus and his mother had Michelangelo lived for just a few more days.

Throughout Michelangelo's life one particular spiritual conviction influenced the course of his work: namely, that the goal of life's journey is to eventually realize that behind the physical immediacy and plausibility of things and events occupying one in the world, lies a spiritual and greater truth. His arrangement of the central panels of the Sistine Chapel ceiling frescos, painted between 1508 and 1512, testify to this religious philosophy. The panels narrate the three origins of the world and of mankind. First, the Creation, involving the separation of light from darkness, of water from land, and the birthing of the stars. Second, the creation of man and woman and the subsequent Fall. Third, the origin of sin; the Flood, and Noah's drunkenness.... Yet these frescoes are not positioned in chronological order. For as one enters the chapel by the west door and looks upward, it is the drunkenness of Noah which meets one's eyes; not the *first* act of God in separating the light from the darkness. Only as one proceeds eastward towards the altar do the panels depicting mankind's beginnings and shortcomings give way to the spirit-genesis of the world at the hand of God. One must arrive at the altar, where the mysteries of the Mass are celebrated, before looking up to see God separating the light from the dark. The message is clear. One moves physically and temporally through the stages of life - experiencing oneself in the world from infancy to maturity - only in the end to come face to face with a spiritual truth.

And so here is Michelangelo, at the end of his own life, reducing the physique of Mary and Jesus to minimal proportions: chipping away at the marble to create the slightest of sculptural forms possible - figures that would still allow Jesus and his mother to be identifiable, yet reveal them as but ghosts, their souls laid bare, as it were. For how else to portray the holiness of mother and son save by '...*disarranging the body from the soul*'? (His own words from his allegorical madrigal 'Costei pur si delibra'.)

Michelangelo was recognized as a man of great and heroic spirit in his own lifetime - a creative force seen to be directed and inspired from the center of his own being.

Not too many years ago I remember showing a particularly good photographic slide of the Rondanini Pietà to a large university audience, and talking about Michelangelo's heart-felt beliefs concerning the dual nature of human existence. They were a passive, somewhat inscrutable lot, and I couldn't really gauge their response either to the sculpture or to the story of Michelangelo's philosophy. Finally, I posed the question, 'Do any of you think of yourself as having a soul, or not?', and asked the women and then the men. More than half the women responded affirmatively, whereas comparatively few male hands were to be seen. Yet, paradoxically, when I asked how many of the group found the Pieta moving, or powerful, just about everybody - men and women both - said they did. Which, of course, may indicate that many of the men found it embarrassing to their sense of masculinity to publicly reveal that they could be moved in this way by such a thing as a work of art - however sublime.

It is difficult and usually not very productive to induce discussion in a large group such as this - there were more than 300 people in the hall - for young people are easily inhibited when surrounded by a sea of faces and tend not to speak up. But a few lingered afterwards and responded personally to my question. It transpired that, generally speaking, the women present believed that to love deeply and to find the world taking on a new vividness and beauty - while time either stood still or ran amok - was to encounter a hidden and radiant self they naturally thought of as the soul.

WHAT THE HELL ARE THE NEURONS UP TO?

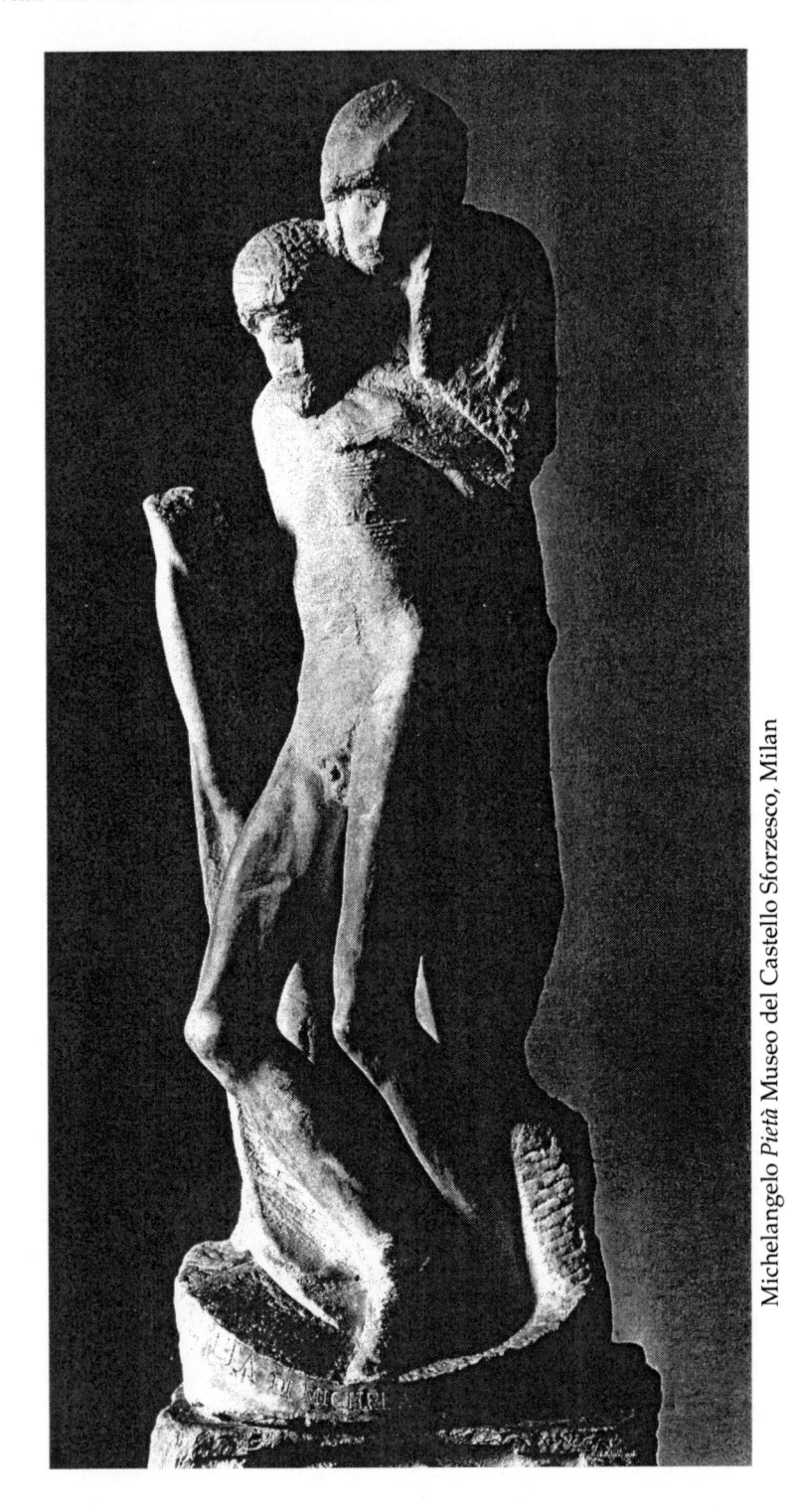

Michelangelo *Pietà* Museo del Castello Sforzesco, Milan

"It's like going walkabout," said a fair-haired girl who hailed from Australia, "and when you do that the Abos say you're in soul-time." One of the five men present nodded his head in approval at this remark. The others smiled indulgently. The gist of the male collective view was that in the days when religion was mans' only refuge against natural forces, to believe in a soul was a big help in getting through a harsh, and often brief, life. But now, with science advancing longevity and offering all kinds of new and exciting discoveries and experiences, life could be lived 'sufficient unto the day'.

"You mean that today's way of life offers the outgoing personality enough interest and excitement, in and of itself, to provide all the meaning and satisfaction necessary to live without metaphysical questions?", I said.

"Exactly," responded the male spokesman.

"And what about those less fortunate - those whose backgrounds have not helped them along, either socially, economically, or intellectually, to take part in this new opportunistic world of yours?"

"Too bad. Maybe they should attend church and hope the soul will turn up."

"And what about older people who may have psychologically passed beyond pursuing such an extroverted life?"

"Well, if that's the case they should go to church too."

"And yourself in forty year's time? Church for you?"

"No way. I'll make sure that what the world has to offer is enough."

He paused, probably expecting me to advance the usual argument that when one becomes more advanced in years the psyche demands more than external experiences to find meaning in life, but I said nothing.

"Anyway," he went on, "I still think Michelangelo's pretty cool."

Writing now, well into the first decade of the 21st century, when even more people's interests are seemingly directed to things outside themselves; success measured largely in material and financial terms - while a constant round of spoon-fed entertainment causes the brain and its neurons to gradually abandon their self-contained world of insightful reflection and 'word-play' - I wonder how many young people in university would be intrigued by Michelangelo's last work; and how they would respond to his transcendent philosophy of human life? And how many would not be affected - finding themselves siding with Nietzsche who wrote in 1882 that...

> *The greatest modern event - that God is dead, that the belief in the Christian God has become unworthy of belief - has now begun to cast its first shadows over Europe.*
>
> Extracted from *The Gay Science*

If a majority were to agree that 'God is dead' - at least in terms of the Christian notion of God as it has evolved in the West - then Nietzsche's statement was indeed prophetic. In which case there would seem to be very little point in discussing the concept of soul as a spiritual force... unless it is made clear that our reflections are not based on conventional western dogma alone, but embrace thoughts and attitudes representing the views of many world religions: beliefs from well beyond Western borders with which Nietzsche might have felt more at ease. Moreover, Nietzsche did not have the advantage we have today of being exposed - courtesy of modern science - to the extraordinary structural and dynamic complexities underlying all natural phenomena. Just think, for example, of the amazing photographs of the cosmos taken by the Hubble telescope. They reveal absolutely incredible dimensions of both time and space, and indicate extraordinary outbursts of energy, which beggar belief: mind-expanding phenomena, facts to which Nietzsche was not privy, and which must surely renew in ourselves a sense of awe concerning this universe in which our planet plays such an infinitesimal part.

Consequently, I see science as an ally pushing us to keep an open mind regarding these questions, precisely because it exposes such an incredible cosmic drama at both macro and micro levels. However, the ambiguities and paradoxes which science encounters lead one to realize that in many instances no one truth supplies the answer to a particular problem. In physics, for example, it is perplexing to be told that the energy of light (and perhaps of energy systems as yet unknown) can be said to manifest as both wave and particle. And, psychobiologically, a similar intellectual incongruity is presented by describing a human being as the union of a physical body and an incorporeal soul.

It will serve my argument here if I expand somewhat on the impact of quantum mechanics we discussed briefly in the opening pages of Chapter 12. This new theory of physics developed in 1926-27 by Werner Heisenberg and Erwin Schrodinger defined the nature and activity of small entities like electrons: a development that led Heisenberg to formulate the principle of indeterminacy - often called the *uncertainty principle* - where it is said that in the case of a small thing like an electron, it is impossible to ascertain, with any certain degree of accuracy, the simultaneous momentum and position of that electron. The quantum physicist advances the suggestion that our own involvement as observer, as experimenter, affects the electron's behavior: in other words that human consciousness cannot stand outside the event as a detached and totally objective knowledge-gathering system looking 'in' on nature... because the chemical and electrical activity of the brain's neurons is itself *part* of nature's own energy-field-system - we are participants in the dynamic activity of electron behaviour.

Consequently, in the post-quantum mechanics era the claims of science to be objective are seen to be qualified by the fact that our own mental processes share in nature's universal energy-fields. We cannot, therefore, in God-like fashion, stand beyond them - which we must needs do if we are to know absolute truths concerning the phenomenon of matter. And even less definitively can we hope to account for such immaterial concepts as soul and spirit.

Human consciousness is bedeviled by uncertainty. Whether it be in the empirical and objective strivings of science searching out the *how* of things, or in the more abstract ponderings of philosophy as to the *why*... Both disciplines depend upon keen powers of observation, shrewd reasoning, and imaginative insight, but the reflections of philosophy - as to what reason or purpose underlies everything in a state of being - are inevitably the more subjective, being generally prompted by that puzzling human need to ask 'why'.

Those who find themselves curious about the many directions taken by their mental life, and by the fact that often it wanders off by itself... might find it helpful to ponder Aldous Huxley's pithy observation, made in his *Themes and Variations* that, *'Man is an intelligence in servitude to his organs.'* The key question here is obviously one asking what Huxley means by *intelligence* in this context. Knowing the strains of mysticism that pervaded his later writings in the 1940's and '50's, I would say that he is referring to the four human resources which have figured prominently throughout these pages - the mental images radiating from soul, spirit, mind, and brain - and which *together* can cause us to ponder whether or not we are more than the sum of our physical parts.

My old friend the Bishop, who initiated my theological education on the flight to Dublin so many years ago, was not familiar with Aldous Huxley's writings - but I felt at the time that he and Huxley would have had a lot to say to each other had they met. For in one of his letters to me, written after his retirement, my flying companion added a postscript: *'Keep the mind open; the spirit in earshot; and the soul pure by recognizing the spiritual dimension to which it belongs, and do not take its name in vain by using its authority spuriously, self-servingly, to justify worldly interests.'*

But even a cursory study of mankind's history reveals that such an ideal balance between spiritual and worldly values have not often been achieved. The Benedictine Order, following the Rule of their founder Saint Benedict, came very close to accomplishing it between the 8th and 12th centuries in Europe. As we have discussed previously, the Benedictines ministered to both the physical and spiritual needs of those in their respective neighborhoods - a holistic view of human existence that gave them great influence and authority in the early Middle Ages. However, as some historians have noted, this balanced understanding of human existence was gradually lost during the years of the later Middle Ages. The Church generally became more 'earthbound' as wealth accumulated from its extensive land-holdings and social and political status took precedence over the mystical life. The world became an end unto itself. Sir Arnold Toynbee, the eminent English historian (who died in 1975), suggested that the spiritual side of life's equation was - as early as 1000 A.D - becoming eclipsed by the importance attached to the pleasures and successes of the 'here and now'. And as evidence for this opinion he points out that after this date a small, but significant, change began to take place in the sculptured figure of Christ - the 'Christ in Glory' motif - traditionally occupying the tympanum, the space above the main doors and beneath the stone ribs defining the entrance-arch of Romanesque and early Gothic churches. The difference lay in the way the sculptor treated the eyes of Christ. Prior to 1000 A.D. - an approximate date, of course - the eyes are looking out into space, beyond the world. But then Toynbee noticed that in succeeding years

Christ was glancing downwards, looking intently to the left or the right, engaged solely in observing the world: involved with the world.

As the years heralding the European Renaissance advanced, concern with the mystery inherent in spiritual experience was less relevant as scientific inquiry into the nature of the physical world became the order of the day. Yet even during the secular years of the Renaissance and High Renaissance, and on into the 18th century's Age of Enlightenment, the compelling presence of an inner life taking consciousness beyond the world remained an influential force in the lives of men like Leonardo da Vinci, Michelangelo, Shakespeare, Donne, Pope, Pascal, Descartes, Voltaire... And when the 18th century had run its scientifically-oriented course, the visionary, spiritual and poetic insights heralding the so-called Romantic Rebellion of the early 19th century moved in to restore the balance between the mental powers of creative intuition and rational deduction. From France Germany, and Russia came the great composers, poets, novelists, painters... While in England, a pragmatic, Empire-dominated, political and material outlook existed comfortably not only alongside the romantic attitudes of poets and painters paying homage to the intensity of feeling and the sublimity of landscape, but also with the work of novelists who sought to reveal moral truths in their storytelling. Also, by the time Queen Victoria died in 1901, just about every level of society was interested in, if not enthused with, a new faith called Spiritualism: a belief that the dead survive in spirit form and may be contacted in the afterlife by using the appropriate rituals. Mediums gave séances to communicate with departed souls - the 'spirits of the dead' as they were called - in the hope that they would 'come through', thus enabling the bereaved to make contact with their deceased loved ones. Yet, while such non-rational and religiously persuasive ideas were widely accepted, the objective pursuits of science were yielding significant results in aviation, general engineering, biology, medicine, physics, astronomy... It would seem that later Victorian and Edwardian society in Britain accepted the irrational nature of Spiritualism's beliefs and practices, no less positively than they accepted the rational and practical findings of science.

Such tolerance is a form of wisdom in that it involved 'keeping an open mind'. It is a mental position taken by the wisest societies in exercising sound and sagacious judgment.

I have enlisted the help of Cicero (Marcus Tullius) the great Roman statesman, advocate, philosopher and orator, several times throughout these pages. For if ever a man had the gift of succinctly expressing the profoundest of thoughts it was he. Here is his definition of wisdom from *De Officiis, Book ii*: *Wisdom is the knowledge of things human and divine and of the causes by which those things are controlled.* It is a statement that both defines and supports the ideal *wholeness* of the kind of consciousness we have been discussing.

In surveying the philosophical views that have historically prevailed in many world cultures, warnings are frequently delivered concerning the psychological poverty of a life lived totally materialistically: one unable to feel and recognize the glimmerings of a 'spirit-truth' - a faculty some neuroscientists suggest are 'built-in' to the genetic makeup of some individuals. Thomas Merton, the Trappist monk who died suddenly and tragically in 1968, and whose life and writings had - and continue to have - a powerful influence on the lives of many of us, was a man as wise to the ways of the world as he was to those of the soul. And he became increasingly concerned about our future as he perceived the constantly increasing drift towards a materialistic secularism that was taking place in the world. Monica Furlong, in her finely woven story of his life, writes: 'Nothing, Merton felt, could restore modern man, lost in technology, in depersonalizing societies, in fierce activism, except a new contemplative vision.' And then she quotes the following passage from Merton's own book, *The Living Bread*, published in 1952.

> *We find ourselves living in a society of men who have discovered their own nonentity where they least expected to - in the midst of power and technological achievement**.

But over a hundred years before Thomas Merton wrote these words, the opening lines of Wordsworth's poem, *The World Is Too Much With Us*, indicate that already the Industrial Revolution in England was inducing a similar retreat from the inner world of contemplative thought concerning the 'why's' and 'wherefore's' of life. So far as the workers were concerned, the long hours of labor that kept their noses to the grindstone, left little

---

* Monica Furlong, '*Merton: A Biography*' (New York: Harper & Row, 1980), p. 265.

time for the 'soul' part of the 'body and soul' equation to make its presence felt. While the so-called 'captains of industry' saw life in terms of balance sheets and profits, and moving up the social ladder - for the workers it was a matter of physical survival. Since Wordsworth wrote the following words in the early years of the 19th century, the Industrial Revolution has become a global phenomenon - introducing the material benefits that can alleviate hardship and suffering; yet, at the same time, creating consumer societies where, as Wordsworth suggests, a life geared only to *'getting and spending...'* is a life wasted.

> *The world is too much with us; late and soon,*
> *Getting and spending, we lay waste our powers:*
> *Little we see in nature that is ours,*
> *We have given our hearts away, a sordid boon!*

Writing closer to the present day in 1921, the Irish poet William Butler Yeats describes a world that appears even more spiritually bleak, for in addition to the time-serving dreariness of commercial and industrial life, the moral bankruptcy of Europe following the end of the First World War has cast further doubt on the believability of spiritual values and ideals. The images of *The Second Coming* are more specific, dire and Armageddon-like, than those that shape Wordsworth's and Merton's concerns. The well-known line,' *the centre cannot hold...'* provides the key to Yeats' vision. For he fears that the *'centre'* (seen as a synonym for soul), will not be able to fulfill its purpose as a moral and spiritual counterweight in the aftermath of the war - a conflict resulting in the slaughter of millions, followed by years of desperately 'living it up' for those with the means to do so, and of acute suffering for the average working class members of society.

Yeats is writing during the 'sad' years after World War I. He died in 1939, and although the war clouds had been gathering in Europe once again, he did not live to see the outbreak of World War II. Even had he done so, the first lines of *The Second Coming* could not have been more prophetic and despairing:

> *Turning and turning in the widening gyre*
> *The falcon cannot hear the falconner;*
> *Things fall apart; the centre cannot hold;*
> *Mere anarchy is loosed upon the world,*

> *The blood-dimmed tide is loosed, and everywhere*
> *The ceremony of innocence is drowned;*
> *The best lack all conviction, while the worst*
> *Are full of passionate intensity.*

In a speech on Armistice Day, 1948, General Omar Bradley, the second most senior American commander in the invasion to liberate Europe, echoed the sentiments of both the priest Merton, and the poet Yeats, when he said, *'The world has achieved brilliance without conscience. Ours is a world of nuclear giants and ethical infants.'*

Over a thousand years ago the Benedictines referred to the *mysterium conunctionum* of body and soul - a phrase used frequently by Carl Jung in his many writings. But in our modern world it is all too easy to forget the soul's role in such a partnership. Dealing with religious, political, and social problems on a global scale; with technology advancing so rapidly that at times it seems to eclipse 'human' decision making... it becomes increasingly difficult to maintain some sense of this inexplicable union of the spiritual and the physical. Many among my contemporaries smile condescendingly at the premise. They generally respond by pointing out that we have now moved into a 'brave new world' where mankind at last can find its true destiny in following a realistic, secular and objectively-sensate way of life. An existence, they say, infinitely more genuine and honest than one complicated by the presence of a so-called soul - an illusion which only serves to put a spoke in the advancing wheel of progress and the 'good life'.

Yet, curiously enough, some of the big corporations are now actively seeking to hire 'right-brain people', thereby acknowledging the significance of Jung's description of introverted and extroverted personality types: (the words introverted and extroverted being coined by Jung in his analysis of patient characteristics.) That the business and commercial world should see the need to balance the rational deliberations of extroverted employees by taking on personnel more intuitively, creatively gifted... is a wise move in itself. And while it may not represent the onset of a Benedictine-like wisdom in recognizing the metaphysical reality of the soul - nevertheless, it certainly suggests that executives and managers recognize that we possess an imaginative vitality of mind capable of going beyond the objective deliberations achieved by rational processes alone.

More indicative, however, of the resurgence of a deeply embedded and intuitive spirituality - of the soul's presence and stirrings in contemporary society - is revealed by the millions who stood overnight in the streets of Rome, waiting their turn in the slow-moving procession to file past the body of Pope John Paul the Second and pay homage to the unwavering spiritual example his life has provided in a sad, sad, world... For as he travelled from country to country his presence and his message convinced millions that a great soul had come among them, speaking out fearlessly against the sufferings wrought by violent political and religious dissensions; the merciless killings of terrorist movements; and the futility, insanity, and inhumanity of war...

I will let Carl Jung and Iris Orego have the last word: Jung for his great ability to - as the German philosopher Max Scheler put it - *'see into the whole of the external world and the soul...';* and Iris Orego because her writings express so evocatively the hope and the dream emanating from what W.B.Yeats describes as our spiritual *'centre'*.

> *In spite of the materialistic tendency to conceive of the soul mainly as a mere result of physical and chemical processes, there is not a single proof of this hypothesis. On the contrary, innumerable facts show that the soul translates the physical processes into images which frequently bear hardly any recognizable relationship to the objective process. The materialistic hypothesis is much too bold and oversteps the limits of experience with 'metaphysical' presumption. There is no reason whatever to picture the soul as something secondary or as an epiphenomenon; there are indeed sufficient reasons for regarding it, at least hypothetically, as a factor sui generis, and for continuing so to regard it until it be adequately proved that psychic processes can also be made in the test tube.*
>
> C.G.Jung: Psychological Reflections

Iris Orego in her Introductory Note to *The Vagabond Path*, writes:

> *Where is 'Far'? And where the road"*
> *For some people, it is the path that leads back to childhood: the road taken by Traherne, Blake, Wordsworth, Rilke, Alain-Fournier. 'There was a time when meadow, grove, and stream... It is a country,' wrote Forrest Reid, 'whose image was stamped upon our soul before we opened our eyes on earth, and all our life is little more than a trying to get back there, our art (little more) than a mapping of its forests and its streams.'*

> For others, it is the road that seeks the unfamiliar, the exotic, 'by a silent shore, by a far distant sea': the road to Babylon, to the Gardens of the Hesperides, to the lost Atlantis, to Kubla Khan...
>
> These were the hopes, too, of De Quincey, Sir Thomas Browne, John Aubrey, Gerard de Nerval, Coleridge - all seeking the world of dreams... For such men, this subliminal world is often more vivid, more real than their daily life, 'the slumber of the body seems to be but the waking of the soul.'

She then goes on to quote from Coleridge's *Anima Poetae*:

> *If a man could pass through Paradise in a dream, and have a flower presented to him as a pledge that his soul had really been there, and if he found that flower in his hand when he awoke - Aye and what then?* *

Of all the highly imaginative images occupying consciousness, the presence of such an archetypal one as that of the soul is, in Shakespeare's word, 'wondrous strange'. Yet the fact that there is a word signifying 'soul' in many of the world's historic languages testifies to the enduring - therefore one could say meaningful - presence of the concept in human thought. In Greek it is the *psyche,* Latin the *anima,* Hebrew *Nepesh,* Sanskrit *Atman,* Goth *Saiwala,* and in German *Seele*...

---

* Iris Orego, *The Vagabond Path* (Charles Scribner & Sons: New York, 1972) p.3.

19

A PYRAMID OF SOULS?
A SENSE of the HOLY: MUSIC of the SPHERES

> There was a little Man, and he had a little Soul;
> And he said, 'Little Soul, let us try, try, try!'
> Thomas Moore, *Little Man and Little Soul*

Don Quixote de la Mancha is surely the most engaging dreamer in the history of literature. A benign and ageing country gentleman of spare and bony physique, his obsessive preoccupation with stories about the early medieval ideals and ways of knighthood leads him to develop a phantasy life - one in which he sees himself called upon to set out and 'right' all the 'wrongs' he comes across on his peregrinations. Thus possessed of 'knightly' visions he embarks on a series of imaginary adventures - illusions of his own making: windmills he imagines to be threatening giants, for example, and so charges them, lance to the fore in tilting position as in contests between combative 12th century knights. His failure to unhorse these 'opponents', or indeed to prevail in similar chivalric delusions, leads him to doubt his belief that the knightly attempt to oust evil will win out in the end; and he is further disillusioned by the shallowness and heartlessness of those he encounters on his way, and by their inability to comprehend the power for good that inspires his missions. After many humiliations - (which the reader might well regard symbolically as glorious failures) - he returns finally to La Mancha, where he renounced all books, and all ideals of knight-errantry, shortly before he died.

His very ingenuousness carries with it a nobility of spirit: a grandeur of character that exemplifies the struggle (forlorn and doomed as it is) to reconcile the high ideals that have taken over his mind with life as he found it to be lived in the 'real' world. The Age of Chivalry was long gone when Miguel de Cervantes wrote his novel *Don Qixote* in 1605 - the single work on which his reputation as one of the world's greatest writers can be said to rest. For the contrasts the novel makes between virtue and unscrupulousness, knowledge and ignorance, graciousness and vulgarity...

reveal Don Quixote to be a man for whom the essentially spiritual nature of knighthood becomes the truly human way to live - one inspired by a sense of the holy. The parable-like nature of this long and complex novel has excited literary criticism and psychological analysis from the day it first appeared. And I would say that, essentially, it is the pathos of Don Quixote's knightly progress that has tugged at the heartstrings of generations of readers. For here is a man who, towards the end of his life, becomes acutely sensitive to the human principles of virtue and morality - impassioned and inspired to attain the kind of transcendent wisdom that Plato believed could only be achieved when a man was so detached from time and the world, as to be virtually *'out of his senses...'*

Poor old Don Quixote - taken over by a childlike simplicity, believing in the supreme goodness of all things: a saintlike and uncorrupted outlook on life that has often been seen as the mark of the child-soul in all its innocence: (*'Heaven lies about us in our infancy!* wrote Wordsworth in the poem *Intimations of Immortality*). Some would describe him as an *idiot savant* - a person mentally afflicted but nevertheless brilliantly inspired in one particular area of truth. Or, as on old saying (the origin of which I cannot place) puts it: *Blessed are the cracked... for they shall let in the light.*

The story of Don Quixote has a particular relevance to the suggestion made in this Chapter that we are not all equal in the soul department. For from what we know of the constant struggle throughout our history between the creative and destructive forces in human nature; and taking into account contemporary experiences of the horrors the conscienceless members of humankind are capable of visiting on the world - it would seem obvious that if the 'soul factor' plays any part in human destiny, its influence is extremely variable. Its authority as a force for good may run the gamut from being extremely strong to extremely weak - or, indeed, being virtually nonexistent. And it is this premise that no two souls are quite the same - that they differ in the constancy and pervasiveness of their presence and in the luminosity of their power to flood consciousness - that gives rise to the concept of a 'Pyramid of Souls'.

One might assume that Don Quixote would agree with such a presupposition, for in talking to his squire-in-training, Sancho Panza - a paunchy little peasant retainer, uneducated and with no knowledge of the world beyond the confines of La Mancha, yet devoted and faithful to his master

- Quixote the 'knight' remarks *'Everyone is as God made him, and often-times a great deal worse.'*

Just what is Thomas Moore (1779-1852) - who came to be regarded as the national poet of Ireland - getting at when he talks about '... *a Little Man*' who had a *'Little Soul'*? Does he mean that both are small in size? The factor of size could indeed apply to his *'Little Man' where* physical factors such as height and girth are measurable. But he certainly cannot be inferring that such an ephemeral entity as the soul can be assessed according to substantiality and dimension. No: I would say that the word *'Little'* is used symbolically to imply that there is nothing particular noteworthy about the quality of this man's character or the resources of his soul. He is an average 'little' man possessed of an average 'little' soul making an average 'little' job of life - average in that he would be considered to have no philosophical point of view concerning the nature of the world and his own presence in it, and is intellectually and emotionally neutral when it comes to determining right and wrong or being influenced by ideals. Such pedestrian character traits are sometimes attributed by poets, writers and philosophers to an ineffectualness of soul.

Yet there is another way of looking at the two lines of the poem I quote. A far more positive and encouraging interpretation. For the very fact that Moore's *'Little Man'* is aware of some essential deficiency in his life, and should exhort this inner power called soul - (a resource which he identifies and obviously knows to be an essential part of his own being) - to *'try, try, try...'* and help out, is the mark of a spiritually sensitive person. Consequently, although physically a small man, he possessed a certain spiritual stature. Therefore I would change the first line of Moore's verse to read: *There was a Grand little Man, and he had a Grand little Soul...*

But having thus promoted the poet's Little Man to give him a higher rating on this hypothetical 'soul standard', the reader must be wondering how any such evaluation can be taken seriously. After all, the soul is not like an internal combustion engine, the power of which depends on the known 'horsepower' the machine is capable of generating. It is not a material, biological entity capable of having its condition and performance measured and checked out and therefore is not subject to the testing methods of science. Nevertheless, throughout our history, shamans, wise men, and those given to the pursuit of philosophy and religion, have never considered all souls to be equal in inspirational power; one priest of my acquaintance - who was a very funny man - said that they come in a range of sizes from deluxe to economy to bargain basement! However, the yardstick employed to ascertain their worth has been the notion that a person's *actions* mirror the quality of his or her soul. *'By their works shalt thou known them...'*, declares the old Biblical criteria. *'Work is the sustenance of noble minds'*, wrote Seneca, the famed Roman statesman and philosopher.

But in such a rational, hedonistic, and a-mystical society as our own - where I think it unlikely that many people would ever ponder this assumed link between soul and behaviour - characters such as Cervantes' *Don Quixote* and Thomas Moore's *Little Man* would generally be dismissed as out-of-this-world crackpots who can tell us nothing about how to be materially successful and enjoy life. (*Blessed are the cracked, for they shall let in the light...* I find myself silently whispering this pithy old saying from time to time when conversation cannot move beyond the catalog of daily comings and goings.)

Blessedness implies a state of grace - the serenity and apparent spiritual surety that can be said to emanate from grandeur of soul - and which can usually be discerned in the countenance of those possessing it. The converse may also hold true: for sometimes an individual turns up and one promptly has the impression that some essential level of inner integrity is missing. Could it be that there goes someone whose soul is altogether out of it, so weak as to be unknown to consciousness: incommunicado to the extent that its 'owner' is left   morally and spiritually neutered - a relative stranger to love, compassion, charity, wisdom...? Or might the soul have fled the coop altogether - departed out of sheer spiritual incompatibility with a totally earthbound ego'? This is the kind of loss that the Headman of the Sakuddei feared could happen, one that would leave a person utterly benighted - meaning literally to be encompassed by darkness or night... as the Middle English term *binighten* signifies.

But having to contemplate a ranking of souls presents one with tough metaphysical problems. For to paraphrase Gertrude Stein's famous remark about roses -'a rose is a rose is a rose...' one could just as well say 'a soul is a soul is a soul...' in which case it seems reasonable enough to think that everyone of us should start life with an equal 'quantum' of it. The reality, however, would seem to be that such is not the case.

Florence Nightingale - who was born in 1820 and died in 1910 - was the kind of person who, then and now, would be described as a 'good soul': one of that notable company of human beings who are always ready to help others, and work to improve the lot of mankind in general. It was during the Crimean War of 1853-56 - when Britain, France and Turkey finally defeated Russia in the struggle for domination of South East Europe - that she and her nurses brought the best medical care available to the makeshift and primitive battlefield 'hospitals', caring for the wounded and the dying at the expense of their own safety and health. It was the first time in the history of British arms that such medical facilities to help ease the physical and psychological sufferings of war's casualties were brought to the fringes of the fighting.

After the war she continued the struggle against the medical establishment to improve the basic nature of nursing conditions in British hospitals - leading to her being regarded as the founder of modern nursing. She brought middle-class women into the profession, was a compassionate yet tough disciplinarian, and is regarded as one of the great personages of the 19th century. (However, this is not to say she was a saint, being less than charitable to Mary Seacole, a black nurse who also made a significant contribution to the growth of frontline nursing in the mid-nineteenth century and whose portrait hangs in the National Portrait Gallery in London.)

Yet also living during the same period in England was a man who, in 1888, became known as 'Jack the Ripper' after murdering six women in London. He was never caught and his identity was never established - although there has been much speculation over the years and many theories put forward as to who he actually was. The killings were agonizingly brutal. His victims were disemboweled expertly and left to die on the street - causing the police to think the killer may have been a doctor by profession. In any event, the murderer stands as a grim example of a chronically psychopathic personality - malevolently butchering unfortunate 'ladies of the night' seemingly without, or at least unmoved by, any moral or spiritual scruple. One is left wondering if there could ever have been a soul residing in that hollow psyche and, if so, why should it be so weak or what had caused it to flee?

The contrast in human and in spiritual terms between Florence Nightingale and Jack-the-Ripper is absolute and stark; yet they were contemporaries and quite likely members of the same middle-class culture. Yet it is downright obvious that 'The Lady of the Lamp' - as Florence Nightingale came to be called due to the frequency of her nightly rounds in the field hospital - represents the best in us; the vicious nightly killings by 'The Ripper' in the back alleys of London, the worst. One said to be a *good* soul; the other a *lost* soul. Such extreme examples of human conduct mark the moral heights and depths on a scale on which we can all be measured.

Florence Nightingale would occupy a high rung on such a ladder. Jack the Ripper might not even have a foot on it. 'Nightingales' and 'Rippers' - a strangely apt play of names... Yet, it is surprising to discover how a name so often fits a character. We have talked a great deal about character (particularly in Chapter 12) and discussed how its makeup determines whether we nurture life or destroy it mercilessly. For character comes about through a grand *melange* of psychologic forces determining the way a person thinks and feels - an admixture that Shakespeare described as *'wondrous strange.'* And, as I am suggesting here, the *'wondrous'* aspect is that the most influential of all these forces - capable of overriding the strongest of motives and emotions - is the voice of conscience: that invocation of ultimate human values which can be thought of as resulting from the presence of an active soul.

Thomas Aquinas, the 13th century theologian and philosopher and founder of Catholic natural law, held that the soul mysteriously invaded the fetus to sow the psychical seeds of humanness at its 'quickening' time... during the first trimester of pregnancy. Medically, this time of significant change for the embryo is still generally assumed to be the stage at which the rudimentary clutch of embryonic cells becomes a human being: human in the sense that a central nervous system is responsible for the foetus' movements, and can be said to denote the first signs of an individual consciousness - the onset of a rudimentary *self*-awareness. As a mark of what constitutes humanness, this neuromuscular development falls somewhat short of the spiritual intervention envisaged by Aquinas.

'Is man an ape or an angel? Now I am on the side of the angels', remarked Benjamin Disraeli, British author and prime minister at a meeting in Oxford in 1864: (a statement I have mentioned earlier and which amused some of his friends - and opponents - who would not have placed him on the side to which he assigned himself!). Now bearing in mind the tongue-in-cheek sharpness of Disraeli's wit one should not perhaps take

this particular remark too literally, but the underlying sentiments would not strike Victorian society as being ingenuous. Supranatural eventualities were not ruled out in the grand scheme of things, despite (or because of) Darwin's shadow. But here we are, some hundred and fifty years after Disraeli, and it seems to me we are no nearer to understanding the relation between body and soul which would put us on the side of the angels: (angel defined here as the lowest spiritual being in a celestial hierarchy.) In fact, I doubt that the very notion of angelic beings would be taken very seriously nowadays, particularly by the younger members of society. And that one would be met by blank looks if asking for an opinion as to whether some souls could be considered more potent than others.

One wonders how Thomas Aquinas would have responded if asked whether all souls, bequeathed to the embryo as he suggests, would be possessed of equal spiritual potential.

But nowadays, as the Age of Science moves rapidly ahead, the Thomist embryogenetic doctrine is at an increasing disadvantage in having to confront advances made in the fields of genetics and embryology. For as science becomes ever more revelatory in understanding our physiological selves... it would surely arouse a skeptically tolerant smile among neuroscientists to advance the metaphysical argument that the soul - as a potential fountainhead of spiritual wisdom - is a totally immaterial force unrelated to our biological circumstances, and becomes an indwelling psychical influence at an early stage of embryonic life. Even so, the central question in any discussion such as this, would be whether or not the advent of this soul is to be considered a strictly supranatural event in the Thomist sense... or accepted as a form of spiritual awareness incurred somehow through the complexity of our biological systems.

ENTER THE GENE

Even as I write, advances in the fields of genetics and neurology are shedding some light on this question concerning the supranatural or natural origins of the soul. In addition they are provoking more thought on the statement made by the late Wilder Penfield, the preeminent Canadian neurologist when asked as to how he would distinguish between Brain and Mind. In his response - quoted in Chapter 17 - he said that Mind can only be described as a *'non-temporal, non-spatial entity...'* and I suggested that by this definition he was referring to a psychologic energy-force having the ability to... 'direct the brain down pathways of knowledge other than those chartered by the senses'. In other words, imbuing it with the creative power responsible for our imaginative breakthroughs in art, science, philosophy... and, more mysteriously, to act as the soul's agent by introducing moral and spiritual persuasions to consciousness. Yet not everyone will be comfortable with this supposition. For it lies beyond the boundaries of science.

However, some geneticists are now alluding to the presence of a gene that affects brain chemistry in a way enabling it to bring 'spiritual experience' to consciousness. Apparently it is not a gene that is commonly held by us all - leaving the question unanswered as to why it should be so unevenly distributed. Yet the very possibility that there are those amongst us who do *not* possess it is, in itself, a reason for considering the validity of the spiritual 'grading' implicit in the symbolic image... a Pyramid of Souls. Such a gene could also be regarded as a 'missing link' in brain function if it is not present to neurally bring abstract, metaphysical insights to consciousness - leading one to ask whether the presence of such a gene can be taken as a likely indication of a spiritual presence such as the soul; thus giving credence to the concept of consciousness as a *holistic* system in which spiritual and material aspects of the life-experience come together.

If such is the case, then - as the mystery-writer would put it - 'the plot thickens...' After all, a gene is a mere segment of a DNA strand, utilizing primarily DNA and protein to produce a particular state of biological organization within the chromosome that carries it. And it is an untold number of differing combinations of these 'bio-states' - each regarded as an informational entity in its own right - that determine the nature of an organism. In our own case, the differing physical and psychological characteristics that distinguish each one of us.

When talking earlier about both the creative and saintly lives led by many great men and women throughout history, I borrowed from Sir James Jeans' remarks on the cosmos - as discussed in Chapter 1 - to speculate that there may be many unknown forms of *hyper*physical, irradiating energy 'out there' - (the 'music of the spheres...'?) - to which we are subject: vibrating strings of energy that could possibly engage the mental traffic of the cortex's neural pathways to activate a range of insights that would seem to be beyond the power of the senses alone to engage - and so might account for the most profound and revelatory imagery released by soul and mind to consciousness. And this without need of any biological intermediary such as a gene. However, should the geneticists be right in their theory, then one could speculate that the 'spirit-gene' - if it may be described as such - facilitates, amplifies, the reception of such insights: and that without it, the soul would be incommunicado, and the Mind thoroughly earthbound.

The current level of genetic research is now becoming particularly significant in the long standing debate between science and philosophy, inasmuch as it brings to the case for the 'body -mind- soul' hypothesis a certain degree of scientific verifiability. And as knowledge of our genetic complexity progresses, so might we come to understand more of the physical and metaphysical conundrum that drives our lives.

But suppose the neuroscientists and geneticists are moving towards the day when they will be able to show that the spiritual quality inherent in the concept of Soul, together with the imaginative feats exerted by Mind… are nothing more than the Brain's *own* mental creation - the product of its *own* neural chemical and electrical interchanges? If this were to be established then the superior status generally accorded Mind - and especially the credence it possesses as the soul's authorized agent, will come under heavy fire. A totally materialistic view of life would most likely prevail if biology is ever able to demonstrate that the brain - 'feeding on itself', so to speak - is the *only* source of all the mind's most inspired creative and spiritual experiences. And any belief in a transcendent, cosmic realm of 'hyperphysical irradiating energy' would have to be put on hold; as would the long-held theory that to be inspired is to be subject to unconscious and intuitive sources beyond the reach of the Brain's sensory and rational operations. The dualist position would collapse. And Shakespeare's lines from the *Sonnets* CVII - *The prophetic soul/ Of the wide world dreaming on/ things to come…* - would have little poetic and metaphysical significance.

'Aha… the spirit-gene at work, think you…?' might be a question a modern time-travelling geneticist would put to a no doubt nonplussed Bard.

However, in presenting this new gene theory, science is inadvertently supporting the proposition put forward in this chapter that not all souls are of equal potency - indeed, implying that some are not even capable of making their presence felt at all. For to state that the 'spirit-gene' is not present in everyone's DNA is one way to account for the fact that there are those amongst us who are apparently unaware of what Addison called, *'The Divinity that stirs within us.'* And even for those who have the gene, it should be born in mind that it is in the nature of genes to be either dominant at the top end of performance, or recessive at the bottom - which may explain why so many differing levels of sensitivity are to found amongst even those genetically predisposed to realize the spiritual power that *'...'stirs within us'*.

A geneticist would go further in explaining differences in 'spirit-gene' performance by pointing out that particular genes are subject to interactions with hosts of other genes in the overall biological system, and that in this mix the ability of one gene strain to 'stand up for itself', so to speak, is easily compromised; especially when one considers the incredible number of gene-combinations involved. So it is reasonable to suggest that in those whose lives are strongly influenced by greatness of soul, a dominant 'spirit-gene' of unusually strong biological constitution could be playing a vital role.... and that the combination of other genes was contributing to its influence.

And Jack the Ripper? Well, if the gene is it, he obviously didn't have it. A physician of Ancient Greece, schooled in Classical wisdom, would say that the murderer's total psyche was in a mess: his soul weak and ineffective - lost, a spent force - incapable of holding its own with a deluded and libidinous *gross*-ego' which knew nothing of conscience or spiritual authority beyond its own self-imposed judgments as to right and wrong. And contemporary psychiatrists dealing with such a pathological character would likely see the problem as purely biological. No soul-mind forces in the picture to be reached through analysis. Drugs the only hope.

All of which leads one to wonder how Thomas Aquinas might have responded had he been privy to the genetic knowledge we possess today. Whether he might not have said, *'See... is it not wonderful that we are thus made, and so can come to experience the mysterious workings of the inward soul?'*

BARAKA

Some years ago now, in the 1960's, the English novelist and poet Robert Graves commenced his address to the Oxford University Poetry Society with the following words:

'This is a critical, not a poetic age,' I am told. 'Inspiration is out. Contemporary poems must reflect the prevailing analytic spirit.' But I am old-fashioned enough to demand 'baraka', an inspirational gift not yet extinct, which defies critical analysis.

Graves' 'baraka' is Arabic in origin and has connotations of 'blessedness', of the 'gift of divine favor'... And so in his use of the word he is quite clearly intimating that inspiration represents an exalted gift of insight - a creative flash of awareness indicative of an inner-directed breakthrough from the unconscious of the soul's 'instant wisdom': Graves would have us know that poetic inspiration *'defies critical analysis'*, thereby surpassing the objective understanding gained by purely cerebral, rational processes. Certainly, in his case, he was sufficiently 'blessed' with poetic inspiration to take a leading place in any hierarchy of poets. And I know that Robert Graves - with whom I corresponded from time to time in this vein - believed that 'baraka' is a term that may be applied to many kinds of inspirational breakthroughs other than those involving an encounter with the poetic Muse; that Einstein's ability in resolving intractable problems of physics, for example, might equally well be gained from a similarly inspired source. Writing for a journal on Modern Art in 1921, he discussed how artistic and scientific sensibilities may all be fed from differing channels of a single mainstream of creative insight - namely, the 'gift of divine favor'.

Graves was always disinclined to talk or write about the soul. And it would seem that the concept of 'baraka' - with its connotations of 'blessedness' - was sufficient in itself to suggest the degree of transcendent awareness he believed was responsible for the profound nature of the poet's 'inspirational gift'. But it is a gift which, some forty years ago, he suggested is 'out'.

Yet one is then prompted to ask: 'How long has this 'gift' been out?" Or, 'What degree of absenteeism is meant by 'out?' And, 'Has it ever been universally *in?'* Such questions draw attention to the fact that the ranks of the *great* creative innovators in science, art, philosophy, music, poetry, religion... have always been pretty thin on the ground - more so at some times, and in some places, than others. One does not have to be a serious student of history to gather that the truly revelatory 'inspiration' of which Graves speaks has never been munificently and widely distributed. And although Graves does not refer to the soul as such, to suggest that poetic inspiration comes as the gift of 'baraka', can be seen as pretty close to saying that such moments of vision emanate from some source akin to the soul.

Which brings us back to the central question of this chapter: why is the 'baraka experience' not more evident, given the roughly seven billion (soon to be nine million) so-called souls on this planet? Certainly it is relevant to point out that in the physical sciences at least there have been a great number of inspired breakthroughs over the course of the last two hundred years - indicating that in these fields the baraka phenomenon is certainly *'not yet extinct.'* And it is worth remembering that the concept of inspiration as a 'divine gift', linked to the activity of the soul, is not new. If Aristotle had known of the word 'baraka' and its significance, I think he would have seen it as a term particularly relevant to support his own thoughts concerning the vitality of the soul, and how it asserts itself by inspiring understanding in five major areas of human endeavor. In the last chapter I quoted from his *Nicomachean Ethics* where he talked about the approach to truth: '...*the states of virtue by which the soul possesses truth by way of affirmation or denial are five in number, i.e., art, scientific knowledge, practical wisdom, philosophic wisdom, intuitive reason...*'

Yet the questions remain: why should there be such vast differences between souls in their respective '...*states of virtue'*?; and can we see any evidence that the number of souls seeking to possess truth by '... *way of affirmation'* is steadily increasing? Certainly it could be argued that the 'baraka of *'scientific knowledge'* has served us better in recent decades than the 'baraka of *'wisdom.* We have affirmed our inspirational gift in the scientific pursuit of empirical truths. Physics and biology in particular are taking us ever further into the mysteries of the universe, and bringing us to understand the complex structure of our own bodies - this latter achievement especially helping to ameliorate some of the physical sufferings that attend human life. Yet are we making such notable progress as a species in developing the refinement of thought and feeling that is the ultimate mark of an advanced soul... leading to a decrease in violence, killing (for religious reasons...!), or a rise in levels of compassion, unselfishness and moral accountability. Are there signs of any real wisdom, sanity... the hint of a growing humanitarian sensibility in the international political air, nations collaborating to put down oppression and 'tribal' violence wherever it occurs? Do we see differing regional doctrines and sectarian ways of life being subordinated to a common cause for the general good - one that transcends religious differences; political and egocentric ambitions driven by a desire for power? Or, to use more traditional (if archaic) terminology, are we arriving at a 'world-consciousness' that recognizes human responsibility in working for the forces of good rather than those of evil?

There are certainly numbers of 'good souls' both working in, and financially supporting, relief organizations throughout the world, as well as those urging nations and their governments to act in protecting the environment from the burgeoning demands of commercial and industrial expansion. We should take hope from them. But whether, in looking back over our history through the centuries, it can be said that a greater charitableness or altruistic spirit prevails in today's world - is obviously difficult to determine. Yet it must be taken into account that nowadays television news and instant satellite communication blankets the planet, bringing tragedies and disasters to the attention of a vast audience - consequently providing the opportunity for more people than ever before to be compassionately moved, discover some degree of fellow-feeling. Neither should one forget the unselfconscious moral goodness displayed daily by so many 'ordinary' and 'decent' men and women - by those whose '....*little, nameless, unremembered acts/ Of kindness and of love*' as Wordsworth wrote, have gone (and continue to go) largely unrecorded. But, as John Milton - writing in *Areopagitica* in 1644 - reminds us: 'Good and evil, we know, in the field of this world grow up together almost inseparably...'

And here we are, at the start of the twenty-first century, and Robert Graves - were he alive - might wonder whether the spirit of 'baraka' has not finally succumbed to extinction.

A SENSE OF THE HOLY...

is a very special kind of awareness - and one I would add to Aristotle's list of five 'soul-possessed' truths. Many times throughout these pages I have relied on the profoundly insightful writings of the late Loren Eiseley, the distinguished American anthropologist and paleontologist. He has illuminated and given substance to my own but half-formed thoughts on the many wonders and mysteries attending all natural phenomena - and I owe him a great debt of gratitude. In his essay *Science and the Sense of the Holy*, he writes of our evolution and particularly of the growth of the brain and of how, *'without the sense of the holy...'* - (the vital qualification he makes) - we can destroy ourselves and create havoc in the world. For given that the evolution of the brain has provided mankind with an intelligence that replaces instinct, and a reasoning intellect coupled to strong egocentric desires, Eiseley intimates that unless this new intelligent consciousness - free to act perversely, willfully, and solely out of *self*-interest - is not influenced by *'even the faint stirrings of some kind of religious compassion'*, then we lose our humanness: go through life capable of committing all kinds of atrocities, violating the human person and other advanced forms of life.

Well, here he is:

> *Somewhere in the far past of man something strange happened in his evolutionary development. His skull has enhanced its youthful globularity; he has lost most of his body hair and what remains grows strangely. He demands, because of his immature emergence into the world, a lengthy and protected childhood. Without prolonged familial attendance he would not survive, yet in him reposes the capacity for great art, inventiveness, and his first mental tool, speech, which creates his humanity. He is without doubt the oddest and most unusual evolutionary product that this planet has yet seen...*
>
> *It cannot, in the beginning, be recognized clearly because it is not a matter of molar teeth and seeds, or killer instincts and ill-interpreted pebbles. Rather it was something happening in the brain, some blinding irradiating thing. Until the quantity of that gray matter reached the threshold of human proportions no one could be sure whether the creature saw with a human eye or looked upon life with even the faint stirrings of some kind of religious compassion...*
>
> *When man cast off his fur and placed his trust in that remarkable brain linked by neural pathways to his tongue... all else would live by his toleration - even the earth from which he sprang. Perhaps this is the hardest, most expensive lesson the layers of the fungus brain have yet to learn: that man is not as other creatures and that without the sense of the holy, without compassion, his brain can become a gray stalking horror - the deviser of Belsen.* *

When I first read these words of Eiseley's I recalled a phrase my stepfather had written in his somewhat fragmentary record of our weekly discussions (which had gone on long after my attendance at Sunday School

---

* Loren Eiseley, from *'The Star Thrower'* (New York: Harcourt Brace (Harvest Edition), 1978.)

was an issue). For when discussing his theory that the soul - in acting as a moral and spiritual compass - could not be regarded as simply a psychological phenomenon resulting from the countless interconnections between billions of neurons... he talked about *'the ethereality of the soul's signals'*: thus implying that such a state of super eminent awareness is brought about by the intrusion of an externally-sourced spiritual intelligence which he would surely have equated with Eiseley's *'sense of the holy'*. And I suspect that he would have seized onto Eiseley's *'some blinding, irradiating thing'* as the force we call the soul that vitalized the brain, enabling it to receive such a transcendent form of intelligence.

Neither would he have been unaware of the significant implications of the line *'... that remarkable brain linked by neural pathways to his tongue'* - agreeing with Eiseley's suggestion that the onset of the consciousness-expanding *irradiating* force invested us with the powers of imagination... allowing primitive man to conceive of human souls, and of spirit-forces animating nature: mental experiences that exacted verbal expression, forcing the grunts to become words, and the words a vocabulary... by which he named and identified the mysterious forces he imagined lay within himself and beyond the tangible manifestations of the natural world.

Thus were sown the seeds of a *naturally* spiritual frame of mind: Eiseley's *'... sense of the holy'* - feelings of awe and veneration for the wonders, mysteries, and hazards of the elemental world in which early humankind found itself. (Incidentally, he might just as well have said 'sense of the sacred', for the Latin root 'sacer' also means holy.)

Yet there is a note of urgency in his essay. For while *'the fungus brain'* (which I would interpret as Eiseley's way of referring to the 'little gray cells' or neurons of the cortex), provided mankind with the intellect to reason cogently and work beneficently for the general good... it must also claim responsibility for the mentality of the brute and the megalomaniac. During the span of our recorded history, the *'horror'* of which Eiseley speaks has manifested itself in the devising of many 'Belsens', has presided over many genocidal rampages and is still at it, warring, bombing, and massacring its way around the world. (Can anyone who has seen Goya's *Désastres de la Guerre* drawings and etchings - images depicting the brutal and evil acts of violence suffered by the civilian population during the Napoleonic invasion of Spain in the early nineteenth century - ever forget them?)

Eiseley's words echo those of Carl Jung in emphasizing the nature of mankind's most pressing problem: the fact that the *'stirrings of some kind of religious compassion'* - and particularly *'the sense of the holy'* - can be outdone by an insistent mental proclivity for violence. We constantly see this disastrous psychological shortcoming resulting in a resort to force and destructive action against those who are labelled - for whatever reason - the enemy: those who, if not for you, must be *against* you. Behavior occasioned sometimes by blind hate, or more often justified as a Cause - religious, political, nationalistic, ethnic... or simply, and more misleadingly, is propagated in the name of self-defence. Violence unmitigated by any sympathy for another's' distress, any humane sensibility. The *'gray stalking horror'* is all too often loosed on the world when men are gripped by rigid ideologies which become excuses for an insatiable hunger for power, domination, and self-gratification - all vices which have been variously described as serving 'the evil shadows of aspiration'; 'the avarice of power'.

Belsen - or Bergen-Belsen as this most notorious of German concentration camps was called - is a name (in company with Auschwitz and Treblinka...) that immediately tells how far down the road to hell human beings are prepared, and able, to go. Opened in 1940 as a prison for Belgium and French prisoners of war, it became a concentration camp for Jewish and political detainees in 1943. Originally built to house ten thousand, there were closer to sixty thousand prisoners incarcerated when it was liberated in 1945. More than 35,000 inmates died from starvation, overwork, brutality, disease, and sadistic medical experiments. Even after liberation by members of the British Royal Artillery 63rd Anti-Tank Regiment on April 15, 1945, about 500 people continued to die daily of starvation and typhus, ultimately reaching a total of almost 14,000. That was Belsen. That is an example of the horror to which Eiseley refers.

The camps S.S. Commandant, Josef Kramer, who came to be known as the 'Beast of Belsen', was tried and found guilty of crimes against humanity by a British military court and was later hanged. Forty-five members of his staff were also tried; fourteen were acquitted. Belsen and camps like it provided an environment where the darkest sides of human nature could take over - where 'beasts' of commandants like Kramer, and 'doctors' such as Mengele had free reign to indulge their sadistic impulses. Such places create a moral vacuum which brings out the best and the worst in mankind - an arena in which evil flourishes. And as only fourteen of the jailors at Belsen were found not guilty of war crimes, the majority of them, willingly or simply obediently, participated in the trade of suffering and death. Yet one wonders how many of those convicted were pathological sadists, their brains diseased in that the neural 'wiring' did not activate either the empathic or altruistic areas of the frontal cortex, or the more highly placed region of conscience. And without such neural connections, perhaps even the most active souls remain silent.

As to the others who were found guilty, it may be that fear of the consequences should they not carry out orders was responsible for their acquiescence; or a misguided sense of duty, overcame conscience - understandable perhaps... when one considers the rigidity of the military discipline and the intensity of the 'master race' brainwashing to which they had been exposed? Yet after their release some survivors told of a few brave individuals working for the authorities at Belsen and other such camps - guards who tried to show a little kindness to the prisoners when the opportunity arose... despite being aware of the danger to their own lives were they to be found out. And it is difficult not to wonder if one, oneself, were to be similarly conscripted as a jailor, whether or not would extend a helping hand to those in need.

However, while among those incarcerated in these terrible places there were a number who acted as lackeys for the Germans, there were many prisoners who managed to maintain a high level of fortitude and stoicism, and would always move to help their suffering colleagues. In interviews after the war some survivors described how they experienced the rise of an inner presence: a 'self' that took consciousness out of Belsen, beyond time.... kept hope alive, enabling them to transcend physical and mental suffering.

BELSEN and the PYRAMID

When the population of Belsen was close to 60,000 it is likely that every type of human being was represented. The extremes of good and evil walked side-by-side there: a community hedged in by wire and towered machine-gun nests, isolate and insular. And I find myself thinking about the camp not only in terms of its existential reality - the physical and mental agonies suffered by the victims, and the nature of the cruelties inflicted on them by the persecutors - but also in a more abstract sense... a ghostlike environment where the human spirit is tested to the utmost: a Great Pyramid of Souls in which the position occupied by each individual soul testifies to its spiritual integrity and strength - qualities that have little to do with whatever temporal authority, personal power, and material 'success' were enjoyed by its 'owner' when in the land of the living.

This may be an overly simplistic way of visually symbolizing our spiritual future - if any such profound and mysterious metamorphosis occurs - but it does provide us with a 'diagrammatic' image that can help one to ponder the current state of one's little psyche in terms of where the soul comes to be 'seated'. In Chapter 4 - the Chart of Consciousness - I mentioned that the great German playwright and poet Goethe, said to a friend that he found diagrams to be more helpful than words when it came to comprehending abstract principles. Therefore in any sketch of the Belsen Pyramid, the 'seats' indicating the souls of the most compassionless and degenerate members of the S.S. guards (and those fellow-prisoners deputed to act as such), would be located in the lowest regions of the Pyramid's wide base. And then would come an ascending order of souls moving up and through the central zone - those having the greatest propensity for good occupying the higher levels. While the rapidly narrowing space at the apex of the triangle - only faintly discernible at those ethereal heights - would be the souls of those who lived totally selfless and benevolent lives.

Such symbolic associations suggested by Pyramidal structures go back a long way. Seen as huge three-dimensional architectural triangles they are found in many parts of the Ancient World, from South America, to Egypt, to parts of Asia. They have been described as 'sacred mountains' - structures by which man could give tangible form to his latent and imaginative religiosity: enabling him to symbolically forge a link between the horizontal base-line of earth, and the vertically ascending spiritual realm of the vast cosmos where he believed his destiny to lie. In the case of the Egyptians, for example, between themselves and the realm of the sun.

This archetypal urge to identify with, and reach up towards an extra-terrestrial realm, when translated into Christian culture may well have been one factor contributing to the achievements of mediaeval stonemasons in the High Gothic period - driving the ambition to go as high as stone would allow and resulting in the creation of those great churches and cathedrals: vertically-thrusting (even if not triangular) lanterns of stone and glass created to the glory of God. And this ancient motivation to practically and symbolically escape the gravitation of earth may still be subliminally present in contemporary architectural practice, showing up in the number of high triangulate structures that dominate the skyline of a modern city. (The idea of reaching for the stars - *Per Ardua ad Astra* - is the motto of England's Royal Air Force.)

With the advent of the geometric abstract movement in painting early in the 20th century, the triangle is a dominant motif. It certainly had great significance for Wassily Kandinsky, one of the movement's founders, who might well be called 'The Master of the Triangle', so constantly do triangles of varying angular characteristics, and both two and three dimensionality, appear in his work. In 1912 he published an article entitled, *Concerning the Spiritual in Art,* where he talks about the *'forward and upward'* spiritual progress that inspires the imagination of the artist - moving him or her *'above and beyond'* - in the following words:

> *The spiritual life to which art belongs, and of which it is one of the mightiest of agents, is a complex but definite movement above and beyond... . Although it may take different forms, it holds basically to the same internal meaning and purpose,*
> *The causes of the necessity to move forward and upward ... are obscure. The path often seems blocked or destroyed. But someone always comes to the rescue - someone like ourselves in everything, but with a secretly implanted power of "vision."*
> *He sees and points out. This high gift (often a heavy burden) at times he would gladly relinquish. But he cannot. Scorned and disliked, he drags the heavy weight of resisting humanity forward and upward.*

And so one moves in the imagination from the Belsen Pyramid of 60,000 souls to the World Pyramid of over 6 billion (and counting) - wondering whether, in these first years of the Twenty-first century, we would be able to assess our progress as a species in terms of the growth of wisdom and compassion... find evidence of any mass *'forward and upward'* movement of souls in our make-believe planetary pyramid?

In reading the books of Loren Eiseley it becomes obvious that his *'...sense of the holy'* is derived in large part from the sense of awe that results when he confronts the sheer complexity of the structure and workability of life on our planet. And we have seen that as a scientist he is not alone in this. In an interview published in the June 11th 2006 edition of The Sunday Times of London, Francis Collins, the director of the U.S. National Human Genome Research Institute - whose team of scientists finally uncovered the *complete* genetic structure, the 'genome' so-called of a human being, describes the result as follows: *'... an instruction book of 3.1 billion letters of the DNA code arrayed across twenty-four chromosomes...'* And he goes on to say, *'...you can't survey that, going through page after page, without a sense of awe...'*

I would say that these words place him firmly in Eiseley's camp; particularly when they are followed by: *'This most beautiful system could only proceed from the dominion of an intelligent and powerful being.'* Throughout the interview he stresses that science and faith are not incompatible; that scientific discoveries of great moment bring their own form of transcendent awareness with them. This has been the premise underlying much of what has been written throughout these pages. Yet there are undoubtedly those able scientists who do not experience Eiseley's *'sense of the holy'*, or feel the *'awe'* of which Francis Collins speaks. Albert Einstein affirms the gist of what both Eiseley and Collins are getting at when he said in a speech presented on the occasion of Max Planck's sixtieth birthday:

> The state of mind which enables a man to do work of this kind... is akin to that of the religious worshipper or the lover; the daily effort comes from no deliberate intention or program, but straight from the heart.

Even in these days of technological innovation - with efficient systems of artificial intelligence and data-processing on hand to provide factual information on just about everything - when attempting to describing abstract *human* qualities, the symbolic implications inherent in language are still unique in this regard. There are key words in every tongue that have commonly been used throughout the years to convey aspects of mankind's inner mental life that cannot be objectively described. 'Double-edged' words, you might say, that can work both literally and symbolically, amongst which 'heart' is a good example. For it is used not only to refer to the heart as a physical organ, but also as the seat of powerful moods, attitudes, make-up of character... and so on. When it is said of a person that 'their *heart's* in the right place', it could be referring to the fact that the heart is where it should be - within the rib cage. But far more likely it is meant to imply that an individual has integrity, is trustworthy, intends well; that there is a 'trueness' at the center of their being - a capacity for insight and revelation. All personal qualities that are seen to be the mark of a worthy individual - of a 'good soul', as is frequently said.

Consequently, when Einstein uses the words *'straight from the heart'*, the implication is that a scientist's work only becomes truly creative when inspired from such an inspired center. This is not a far cry from saying that, *'the daily effort comes... straight from the soul'* And certainly one gathers from his writings and statements that he experiences a similar sense of wonder to that described by Francis Collins. Also, it would seem that Einstein shared Loren Eiseley's belief in the operation of intuitive insight ('... *some blinding irradiating thing...'*) when, in the same speech given on the occasion of Max Planck's birthday celebration, he said: *'The supreme task of the physicist is to arrive at those universal elementary laws from which the cosmos can be built up by pure deduction. There is no logical path to these laws; only intuition resting on the sympathetic understanding of experience can reach them'*.

When the Nazis came to power in Germany in 1933 Einstein, who was outspokenly opposed to Hitler, had his property confiscated and German citizenship revoked in 1934. He accordingly went to the Institute for Advanced Studies in Princeton to which he had been invited in the prior year. One cannot but wonder what might have happened to him had he remained in Germany. As a Jew, he could have been incarcerated in Dachau or Belsen... And, if so, how would he have fared? His physical health would have suffered grievously, but spiritually he would likely have occupied the apex of the pyramid, and dragged as many detainees as he could to be up there with him.

One way to deal with this whole metaphysical dilemma would be to simply dismiss the concept of soul (and spirit, its active intermediary to consciousness) outright. And accept the fact that one hundred billion neurons are, of their *own* volition, capable of inducing the kind of free play of imagination that results in all kinds of so-called mystical ideas and experiences. But then one is left to wonder if such an explanation can account for the fact that so many widely separated and differing societies throughout history - from the most primitive to the most advanced - should have arrived at the common belief that some kind of spiritual essence permeates human life. For the Ancient Egyptians it was the *ka*; in the classical world of Greece the soul is discussed by Pythagoras, Epictetus, Socrates, and Plato; in that of Rome by Lucretius, Plutarch (a Greek), Ovid, Horace, Seneca, Cicero, Virgil... not to mention the emperors Hadrian and Marcus Aurelius. The central tenet of Christianity is the immortality of the soul. Islamic theology has long wrestled with the issue of how much freedom is possessed by an individual soul totally in thrall to God. While the Upanishads of Hindu history and culture developed the doctrine of a universal soul or being to which individual souls will be reunited after 'maya' - the illusion of time and space - is conquered. And it is worth noting that even the ancient Oracle of Delphi - (the site of the sanctuary on the lower slopes of Mount Parnassus was established at the beginning of the fourth century B.C.) - emphasizes the central significance of the soul. In *Tusculanarum Disputationum*, the Roman statesman and philosopher Cicero gives us the *full* text of the famous exhortation, *'Know Thyself'* (Nosce te), pointing out that this is incomplete; that the Oracle, spoken by a priestess named Pythia, then goes on to say, *'Know thy Soul'* (Nosce animum tuum).

Yet the great Greek dramatists of the 5th century BC - Sophocles, Aeschylus, Euripides - whose profound understanding of human nature shaped the philosophical and psychological ethos that permeated later Western culture - were seemingly disinterested in the collective level of virtuousness, its rise or decline, *in the population as a whole* (as envisaged in the concept of the 'World Pyramid'). They took the view that the life of the individual constituted the ultimate reality - each person forging his or her own fate on a most personal journey, down a most private road; in the course of which the three major forces of *destiny, tragedy, and nemesis* were encountered in order to forge and test strength of character. They put the soul on 'the back burner', so to speak, as a spiritual entity that was more an observer than a participant and to which, in the last resort, one could appeal to do its best and *'try, try, try!* ' It represents a view of the individual psyche and its destiny very much in keeping with contemporary psychiatric practice. Here is the opening paragraph of the second chapter in a book entitled *The Transparent Self* by a well-known American psychologist:

> *The soul of which poets speak, and which philosophers and theologians concern themselves with, is now operationally defined by psychologists and called the Self. That the soul, or self, is real in the sense of existing, few can doubt. At least few would doubt its reality when we define the self as the subjective side of man - that which is private and personal, which he experiences immediately and spontaneously...**

This is a contemporary statement with which Plato would certainly not have quarrelled, and which also testifies to the influence of Sigmund Freud and Carl Jung who saw this concept of the 'soul-self' as a crucial issue in psychiatric healing. Jung in particular heeded the Greek warning of the damage inflicted on character by *hybris* - 'hubris' as we call it: meaning excessive self-absorption, exorbitant pride, conceit, arrogance... that can all lead to the ruthless pursuit of any personal whim, desire, or ambition - and result in the supremely selfish and morally castrated individual we described in Chapter 16 as driven by *gross*-ego. In this sense the two terms - hubris and *gross*-ego can become interchangeable. Jung was constantly pointing out that when such extreme self-indulgence comes to totally dominate consciousness, one is psychically crippled. For then those soundings of the intuition that dispense notions of right and wrong, love, compassion, beauty, truth... cannot break through to enlarge the range of consciousness beyond the purely sensate and hedonistically self-serving. Which makes it extremely difficult for the nascent soul to gain a foothold in life.

Gross-ego is the soul's true enemy. And Jung knew the dangers Western society faced in this regard as he saw consumerism and purely existential values becoming the order of the day. He states in *Psychological Reflections*, a comprehensive selection of his voluminous writings edited by Joland Jacobi, that... *Western man is held in thrall by the 'ten thousand things'; he sees only particulars, is ego-bound and thing-bound and unaware of the deep root of all being.*

---

* Sidney M Jourard, *The Transparent Self* (New York: D.Van Nostrand Company, 1964), p, 9.

So, this being the case (as I think it has proved to be), what price Eiseley's *'sense of the holy'* in these early years of the 21st century? Has anybody - particularly the religious leaders - got it right: Christians, Muslims, Jews...? Some would say they are all grubbing around at the base of the Pyramid, missing the point that to experience holiness is to 'sense' the presence of the 'divine': a transcendent event that must surely be the same for all of us. Can the concept of the divine be locally defined and redefined in order to suit the secular interests of one particular religious group? I would say that men and women like those who work in organizations such as Medecins Sans Frontières, occupy higher positions in the Pyramid than many a Bishop, Ayatolla, or Rabbi...

The influential religious philosopher Emanuel Swedenborg, who died in 1772, preached that the quality of a soul is determined by *'...the character of its love'*. If so, one can only conclude that given the warring tribal, religious and political ideologies that give rise to the global terrorism responsible for so much human suffering, any general upward movement in the world Pyramid of souls would seem to be stalled... (if, in fact, there ever has been a time when humanity at large has made a significant climb to escape earthbound limbo in one form or another.) And especially after Belsen, Auschwitz... and two World Wars, when it might be supposed that some fallout from the terrible loss of life and the brutality inflicted on soldier and civilian alike, would cause succeeding generations to raise their moral sights and identify more with the plight of others - follow up in large numbers on the popular injunction appearing on the rear bumpers of automobiles during the years of the Cold War and Vietnam: 'Make love, not War'.

Christ talked about 'brotherly love', and while there may have been more of it about at some periods in our history, and less at others, it would appear that in our contemporary world there is still not enough of it to break the age-old pattern of nations and factions resorting to violence, and taking the lives of the innocent and helpless. Given the geographical, cultural, and psychological mix that constitutes the world population, the idea of universal love would seem to be an impossible dream. And yet there are instances when surpassing demonstrations of compassion and fellow-feeling - loving in its most absolute form - make startling forays into the dehumanized, barren zones of conflict. When this happens one cannot but be intensely moved by the humanity, the spiritual force, they present. I read of a Palestinian mother whose young son was shot by a Jewish soldier in Gaza; yet she donated one of the boy's organs to a Jewish family to save the life of their own youngster. There is a *'sense of the holy...'* in such inspiring magnanimity. And one is awed and humbled by the example of a great soul, high in the Pyramid, acting according to humane principles that completely transcend the local strictures of sectarian and political conditioning. Yet there was criticism of this mother's supremely altruistic gesture - disapproval implied in the remarks made by those who considered how unfortunate was the lot of a Jewish child who had to live his life with a 'Palestinian kidney' in him. Is this not indicative of the mental stupidity, the utter vacuousness of intellect and poverty of soul that distinguishes the zealot?

'Love is the only cure for the world...' said the Russian novelist Aleksandr Solzhenitsyn in his remarks on being awarded the Nobel Prize for Literature in 1970. The poets and seers of many lands have always written and talked about the humanizing power of loving. And since research into the mental forces driving the whole psyche was begun in earnest towards the end of the 19th century, it has been recognized that it is through the experience of love, the ability to love, that we become whole: physically, intellectually, emotionally, and spiritually whole. Erich Fromm, the distinguished American (German-born psychiatrist who died in 1980), writes of the important role played by love and conscience in the course of becoming a complete human being as follows:

> *Man must relate himself to his fellow men lovingly. If he has no love, he is an empty shell, even if his were all power, wealth, and intelligence. Man must know the difference between good and evil, he must learn to live with the voice of his conscience and to be able to follow it.*
> Erich Fromm: *The Art of Loving*, p. 4.

Together with other noted psychiatrists of his generation he had no compunction about relating mental health to the state of the soul - and that if asked to discuss the symbolic implications of a world Pyramid of Souls, I have little doubt that he would have responded with his conviction that the measure of a person's ability to love, and to act on conscience, determines the effectiveness of their soul and its place in the Pyramidal scheme of things. I wonder how many psychiatrists practising today would agree with the advice given by the ancient Delphic Oracle and the similar - if more sophisticated - counsel offered by doctors like Freud, Jung, Neumann, Rollo May, Fromm, Jourard.... writing but a few decades ago.

MUSIC of the SPHERES: THE WINDS of MARS

Pythagoras' discovery that the pitch of notes was dependent on the speed of vibrations, and that the planets travel at varying rates of motion, led him to conclude that they must create sounds as they move - sounds that vary in tone according to the velocity of their progression. Furthermore, as he believed that all natural phenomena work together in some degree of harmony, then so the differing sounds of planetary movement must produce harmonic 'sound effects'. Thus developed the old theory of the 'harmony of the spheres'. Johannes Kepler, the distinguished German astronomer who was very much influenced by the teachings of Copernicus and became involved in friendly scientific exchanges with Galileo, wrote a summary of Copernican astronomy in 1618, rounding it off in 1619 with *'De cometis'* and *'Harmonice mundi'* - two related treatises which may be translated as, *'Of heavenly bodies'* and *'Music of the universe'*.

It seems inevitable that Einstein would be amongst the scientific luminaries who have talked about the cosmos in similar terms and, indeed, so he was. Here is his response to a questioner - (as recalled by Peter Bucky in *'The Private Albert Einstein')* - following his contribution to a symposium on Science, Philosophy, and Religion: *'There are the fanatical atheists whose intolerance is the same as that of the religious fanatics, and it springs from the same source.... They are creatures who can't hear the music of the spheres.'* And again, in answer to another question after the same symposium, Einstein is quoted as saying: *'Mozart's music is so pure and beautiful that I see it as a reflection of the inner beauty of the universe'* - which is really synonymous with describing it as 'the music of the spheres'. The Irish poet Thomas Moore (whose *'Little soul'* verse was quoted at the beginning of this chapter) had his own witty way of describing the effect on one of music that attains the sublime heights of which Einstein speaks...

> "This <u>must</u> be the music," said he, "of the <u>spears</u>, For I am curst if each note of it doesn't run through one!"

H.D. Thoreau, writing in his *Journal* on 3 September 1851, was somewhat more specific, in that he identifies the source of the sound he regards as the 'music' of cosmic energy; and he makes no bones about the 'sense of the holy' that it induces:

> *As I went under the new telegraph-wire, I heard it vibrating like a harp high overhead. It was the sound of a far-off glorious life, a supernal life, which came down to us, and vibrated the lattice-work of this life of ours.*

It would be possible to fill several pages of this chapter by using quotations from across the board of human history testifying to a widespread belief that music touches a spiritual nerve. Over two thousand years earlier, Plato had written in *The Republic* that, *Music and rhythm find their way into the secret places of the soul:* a belief that the many so-called 'music-lovers' today will certainly not quarrel with. And in wondering how Mozart's music (and that of Bach, Beethoven, and the rest) might come to reflect '...'*the inner beauty of the universe*', I remember Arnold Franchetti, a well known teacher of theory and musical composition in the Eastern United States (of whom I have written in Chapter 4), regularly declaring to students that '... *it's all out there: all the possible combinations of tone, pitch, rhythm... just sail with heart, mind and ear close to the wind at all times and inspiration will strike!'*

So one may reflect on whether Mozart was unconsciously tuning into the Pythagorean sounds and silences of the cosmos, and then transforming them into the instrumental forms that captivated Einstein - who was himself an accomplished violinist. After all, one remembers how Mozart described the nature of his inspiration: 'When I am, as it were, *completely myself, entirely alone, and of good cheer* - say, *travelling in a carriage, or walking after a good meal, or during the night when I cannot sleep; it is on such occasions that my ideas flow best and most abundantly.* <u>Whence</u> and <u>how</u> *they come I know not; nor can I force them... Nor do I hear in my imagination the parts* <u>successively</u>, *but I hear them, as it were, all at once...'*

In an earlier chapter I commented on Michelangelo's belief that music is not so much a matter of the ear... as it is of the soul. And I wonder how he would have responded to the electronic range of contemporary instrumental sound, delivered at high volume and following a regular and repetitive beat, to which nowadays an audience jives more or less automatically. It is an experience that can certainly induce a group camaraderie - a collective emotion of 'togetherness' that has the effect of temporarily releasing one from a strong sense of personal identity. My own view is that he would likely have considered a contemporary rock concert to constitute a major assault on the ear - one that produces such an overwhelming auditory dominance of consciousness that the more psychical lines of communication to the inner self or soul would become effectively 'jammed'.

Michelangelo, of course, could have had no concept of electronically produced music. In his time the sounds of lute, harp, recorder, flute... were intimate and pensive, whether expressing the feelings of love, the dramas of life, or the mysteries of religion. He would consider their tones to constitute music because they possessed the melody and harmony of ordered sounds in succession, in unity and continuity; giving specific acoustical form to those experiences of life that engaged people powerfully and spiritually at the deeper levels of their being. And he would surely have considered that the English poet John Milton - who was born in 1608 (forty four years after Michelangelo's death) - had penned the lines in *Paradise Lost* that truly defined music:

> *Music, the greatest good that mortals know,*
> *And all of heaven we have below.*

The phrase, Music of the Spheres, is sometimes written as the Music of Heavenly Bodies - a variation that echoes the wording of Milton's couplet. Yet whatever fine philosophical or semantic distinctions may be made between them, the fact is that both Spheres and Heavenly Bodies are high-altitude - extremely high-flying - celestial entities. So any music conjecturally associated with their velocities may - not necessarily flippantly - be thought of as 'high-altitude music'; resonating only in the higher reaches of our imaginary, symbolic Pyramid. This is a metaphorical way of talking about the power of great music - supporting Einstein's thought that Mozart's compositions reflected the inner beauty of the universe, and Plato's statement that music finds its way into the soul's secret places.

But having said this, how is one to account for the fact that conscienceless, iniquitous characters like Hitler, and many senior Nazi ministers and generals who frequently attended symphonic and operatic concerts, were not more 'humanized' by the experience? Beethoven, Brahms, Wagner... were frequently heard in Berlin - perhaps even in Belsen - without, seemingly, any Nazi soul being very much affected. As might be expected, selections from Wagner were very popular in National Socialism circles. The young Wagner was, after all, a revolutionary, speaking out for the man-in-the-street in the developing struggle against the traditionally powerful aristocratic and monied political Establishment. And his music - inspired by the Norse myths found in the Scandinavian group of Germanic languages, and by the military exploits of the mediaeval Teutonic Knights - was romantically stirring. All music has the power to engage the emotions in one way or another and Wagnerian sound is as good as it gets in this regard, running the gamut of feeling from the most tender to the most martial and warrior like. One can see how it would appeal to the Nazi concept of a Nordic master race in which the family values shaped by German motherhood, are married to the masculine ideal and proselytizing fervor of latter-day Teutonic supremacy. But, Wagner aside, all symphony concerts were well attended by the Nazi elite. Perhaps such concert-going invested the Party with a certain cultural status - which it certainly needed to counter its brutal methods of dealing with the 'Jewish problem', and in silencing dissent generally. And music is one of the best ways to induce the psychological bonding that comes with the sharing of heightened feelings - the sense of 'togetherness', of 'belonging'... the warm and comfortable glow of inclusion that comes from being accepted as a member of, say, the Establishment.

Yet one has to wonder how the true 'music lovers' among them would cope as they felt themselves uplifted, feelings and spiritual sensitivity challenged by an adagio swell of the strings or the plaintive cry of oboe or flute... bringing them as individuals to privately question the inhumanity of the regime they served, and the moral culpability of their own part in it.

Perhaps they tried to assuage conscience by trying - whenever the opportunity presented itself - to alleviate the inhuman treatment meted out to Jews and political prisoners. However, it would not seem that the majority of Nazi music-goers experienced this level of transcendent response. And so, echoing Michelangelo, it could be said that music for them was, more mundanely, a matter only of the ear: simply an interesting auditory experience of rhythms and instrumental sounds... sensorily stimulating and inducing those emotional reactions that are a natural response to music - mental images and feelings very much *of the moment* and which might be described as being but 'skin-deep'. It was not given to the average Nazi to move from the world of the senses and encounter the Pyramidal 'high-altitude', soul-reaching potential of symphony or sonata.

And yet... the 'monster' Caliban, one of the most paradoxical and pathos-ridden of Shakespeare's creations - the very antithesis of everything supposed to constitute humanness and particularly the Nazi view of the Teutonic ideal - reveals how deeply he can be moved by music. He appears in *The Tempest* as the deformed son of the *'foul witch Sycorax'*, and is kept as a sub-human slave by the wizard-master Prospero on his castaway island. Yet it becomes apparent not long after his first appearance in the play that although he is a primitive creature, he nevertheless feels strongly about his condition and is deeply resentful of Prospero's hold over him. After all, it was Prospero who taught him language and endowed him with the power of speech... thus providing him with the most significant of human faculties - that of verbally shaping thought and feeling: the means of becoming self-aware. Consequently, this unnatural creature develops a sense of what it is like to be of humankind. Yet the mental conflict aroused in attempting to reconcile the mechanical nature of an existence governed by primal instinct, with one involving a thinking, feeling 'self'... leaves him confused and frustrated. He constantly cries out to be placed on an equal footing with the newly arrived, shipwrecked people around him. And then Shakespeare reveals the level of humanness the 'monster' has attained since Prospero's intervention. He has Caliban declaim eloquently on the power of music to move him - and in words that imply he was affected more profoundly than was possible were it simply a matter of the 'ear'. So now we have a being - once a sub-human oddity - discovering the delights of 'sounds and sweet airs'... hearing the 'voices' they evoke... and 'dreaming' the dreams that ensue. Also, when he speaks of 'riches' dropping from the clouds, it seems unlikely - given the psychical nature of the pleasures he derives from his musical, dream-experience - that he is referring to material wealth. Here is Caliban on the music that spoke so movingly to him:

> *Be not afeard, the isle is full of noises,*
> *Sounds and sweet airs that give delight and hurt not.*
> *Sometimes a thousand twangling instruments*
> *Will hum about mine ears, and sometimes voices*
> *That if I then had wak'd after long sleep*
> *Will make me sleep again; and then in dreaming*
> *The clouds me thought would open and show riches*
> *Ready to drop upon me, that when I wak'd*
> *I cried to dream again.*

As Caliban in his pre-Prospero state knew nothing of civilization and the sounds created by musical instruments, might one consider that in talking about sounds as *'sweet airs that give delight...'* his transformation had caused him to become tuned-in to the sounds of the cosmos - to the Music of the Spheres? Might it be that Prospero's magic had awakened dormant areas of that primitive brain to make it susceptible to Loren Eiseley's *irradiating thing'...* thus permitting musical contact to be made with the universe, and the first stirrings of a human soul?

The Mars Pathfinder and Mars Global Surveyor landed on Mars on July 4, 1997. It was a small aluminum vehicle equipped with cameras and many sensors designed to collect visual and meteorological data about the red planet. There was no microphone on Pathfinder, but among the instruments on board were 12 very accurate meteorological devices, and during its roughly three month mission the probe transmitted back to earth 4,000,000 temperature, wind, and barometric pressure measurements. One file of data, representing a continuous 24 hour period during which readings were taken from each of the 12 sensors every four seconds, consisted of 22,194 lines of data, 13 columns wide. The project group responsible for creating the tape, *The Winds of Mars*, describe their venture as follows:

> *We assimilated that Mars wind data, and combined it with what we know about the physical properties of wind on earth. Factored in were different varieties of terrain, temperature, barometric pressure and gravitational pull (for instance grains of sand blowing across terrain at 10 miles an hour on Earth do not behave the same as those moving 10 miles an hour on Mars). Fortunately, the laws of physics are the same on Mars as on Earth. Put this all into a computer, mix it with the music of Bach, and the result is Winds of Mars. Our ultimate goal: one day,*

when a microphone arrives on Mars and when the project scientists turn up the volume, this is what they'll hear.

The music of Johann Sebastian Bach has long been regarded as a perfect combination of mathematical precision in rhythmic and tonal modulation serving a wonderfully melodic line of instrumental sound: music that can stay in head and heart for life, holding 'ear' and 'soul' captive, so to speak. Bach's writing for the piano in particular is considered by many to epitomize the transcendent power of music. The *Winds of Mars* is, in effect, a duet; involving the music of Bach - played on a grand piano by a noted concert pianist - performed in company with the wind data from Mars, the musical form of which is revealed through the masterly analysis performed by computers. The musical compatibility between the sounds from a planet some 48,682,000 millions of miles from Earth and a piano score written here over 300 years ago is extraordinary.

As Pythagoras - and the much later Johannes Kepler - ascertained, sound is the result of vibrating motions, and sound waves can be reflected and refracted as can light waves. Vibrations that can simply be described as noise are irregular and disordered, their frequencies (high or low) having neither modulation nor sequential affinity. Whereas vibrations that produce the sounds we deem to be musical are ordered, creating in their progression some kind of unified relativity in terms of tonal pitch, rhythm, harmony, and melody: the musical 'sound effects' it would seem that Pythagoras had in mind when he pondered the nature of extraterrestrial sound* encircling the high summit of the Pyramid.

Heard melodies are sweet, but those unheard
Are sweeter; therefore, ye soft pipes, play on;
Not to their sensual ear, but, more endear'd,
Pipe to the spirit ditties of no tone.
    John Keats (1795-1821) Ode on a Grecian Urn

---

* The information concerning the Winds of Mars is reproduced by permission of Music Crest Productions and Colvin Miller at 141 South 7th Street, Minneapolis, Minnesota 55402.

20

SOUL and the WANDERING SPIRIT

*Hands of invisible spirits touch the strings
Of that mysterious instrument, the soul,
And play the prelude of our fate. We hear
The voice prophetic, and are not alone.*
Longfellow, *The Spanish Student.* Act I, sc. 3

In the quatrain quoted above, the American poet Henry Wadsworth Longfellow (1807-82) uses a musical image to differentiate between spirit and soul - likening the soul to an instrument that requires the energy of spirit to *'touch'* its strings if it is to resonate. The analogy illustrates the dual nature of our spiritual life. The *'mysterious'* entity we call the soul can be seen as the 'command center' - the spiritual *force majeure* capable of illuminating consciousness in its search for truth at both physical and metaphysical levels. Whereas the role of spirit, as a vitalistic energy force, is to *engage* the soul and channel its revelatory intelligence to consciousness. The idea that soul and spirit play this kind of symbiotic role in our overall understanding of existence has been presented by both eastern and western philosophers.

It is a concept that plays a central role in the discourses of thinkers like Plato, Saints Thomas Aquinas and Augustine, Telesio, Spinoza, Fichte, Hegel, Schopenhauer... And one French philosopher and mathematician - Henri Bergson (1859-1941) - proposed that a creative and dynamic force representing the energy of spirit which he called the *Élan Vital*, was responsible for bringing about the evolution of life-forms and their functions. One might extend this Bergsonian theory by suggesting that the *human* spirit represents a particular endowment of this universal vitalistic principle - a personal and individual quantum of spirit-energy that not only animates and supports the cellular life of the body, but is also accountable for the presence of that immaterial essence we think of as soul.

A central tenet in Plato's philosophical system is that our most revelatory thoughts are initiated by the soul... in that it plays a mediating role between sensory experience and the questing curiosity as to 'how' and 'why' life. In ancient Egypt, a thousand years or more before Plato, the spirit was seen as a disembodied self - a double that would wander around for a time after death until finally uniting with the departed soul - leading to the ritualistic (if paradoxical) practice of leaving food and drink outside the tomb to nurture the transient spirit; in addition to providing a statue called a *ka* into which it might enter:

> Somewhere - in desolate wind-swept space -
> In Twilight land - in No-man's-land
> Two hurrying Shapes met face to face,
> And bade each other stand.
>
> 'And who are you?' cried one a-gape,
> Shuddering in the gloaming light.
> 'I know not,' said the second Shape,
> 'I only died last night!'
> Thomas Bailey Aldrich, *Identity*.

Bruce Chatwin, the writer who traveled some of the world's most remote regions, tells the story in his *Notes* of the bearers on one African journey who refused to go on after moving for days at a fairly hectic pace - and this despite being offered extra money. 'We have to wait for our spirits to keep up with our souls,' was the reason they gave for not proceeding. Such a belief in the separation of soul and spirit is found in many parts of the world - it being thought that the spirit wanders around a bit, whereas the soul is always in residence.

One often hears people - especially those who are involved in intellectually or creatively demanding jobs - bemoaning that they are devoid of ideas, are 'stuck', unable to envision the next stage in the development of their work (or even their life). Writers refer to this mental wilderness as 'writer's block; artists and poets call it the loss of the 'Muse'; scientists lose the 'creative edge'. It would seem that inspiration has left them: that they are not *in spiritus*. The messenger has no message. The soul is not available. Chatwin's African bearers, if asked, might have explained their attitude by saying that although their spirits have been wandering off, unable to stay the pace, they always return once the stress suffered by the body has alleviated and the psychic circumstances are favorable. In our own contemporary culture we express somewhat similar sentiments when we describe someone who is experiencing disabling apathy as 'being 'dispirited', or not having the 'heart' to deal with life's difficulties - terms that Bergson might well have suggested connote a weakening of the *élan vital*: a lack of psychical energy affecting not only a person's ability to existentially function, but also eroding their very *will* to live. And an Australian aborigine would probably say that the spirit has gone 'walkabout'.... leaving the soul unavailable.

When one reads Mozart's letter to a patron describing how he composes music (Chapter 2), it is obvious that his instantaneous inspiration went beyond the bounds of normal consciousness - becoming what I called an act of 'spontaneous mental combustion'. You may also recall that I described the occasion when I was making drawings of Sir John Barbirolli - the conductor of the famed Hallé Orchestra in Britain - when the Leader of the orchestra suggested that if I stood next to his chair during the afternoon rehearsal of Mahler's First Symphony, I would be in the best position to observe the transformation that overtook Sir John whenever they performed Mahler (his favorite composer) - and so witness 'a musical soul that had just ignited'. Such moments of 'musical vision' present us with a most unusual phenomenon... that of the soul and spirit working in unison to take over consciousness. In Mozart's case bringing complete symphonies to mind as he walks around Vienna - and transporting Barbirolli to exalted heights as the orchestra sounds the first chords of Mahler.

When such transcendent moments have the true visionaries amongst us in thrall, I wonder if the distinctions we have been making throughout these chapters between soul and spirit count for very much? Possibly, the spontaneous arousal of such a combination of creative insight and the accompanying intensity of feeling, may result when the message becomes the messenger - soul and spirit now a unitary transcendent force, invading consciousness in life.... as indeed may be the case in death. *'No man was ever great,'* wrote Cicero in *De Natura Deorum, 'without some portion of divine inspiration.'*

Henry Miller, in his book *The Colossus of Marousi* - a gripping account of his travels among the islands of the ancient Greek world - describes his encounter with a fellow voyager, Katsimbalis, on an overnight Mediterranean ferry. Katsimbalis is a huge man, a veritable colossus who spilled out his thoughts and feelings in a continual stream of words - a flood of opinions voiced in a volume of sound that matched the impressiveness of his bulk. Miller was awed by the performance which went on nonstop for several hours, and only ended when the giant decided it was time to sleep - on the deck, that is - as neither of them had purchased a cabin berth. Within a minute Katsimbalis had gone; the sudden silence startled his listener's ears as the flow of words abruptly stopped. Miller stared wonderingly at the great hulk sprawled unmoving on the boards - now an inanimate, soundless object from which life had suddenly fled. The Katsimbalis of but a few seconds before was no more - no more outpouring of thoughts; no more emotion dancing in the eyes; no more the expressive and extravagant gestures of the orator. All vanquished by sleep in just a few seconds.

Miller walked round the black-coated thing, marveling at the suddenness of transition. And knowing that in the ancient Latin world the word *spiritus* represented the breath of life - together with the resulting vigor of the soul - he pondered the abrupt departure of Katsimbalis' spirit. The barely discernible shallow intake of breath - now totally devoid of words and serving merely to keep the huge bulk in a state of suspended animation, could scarcely qualify as the spirit-force of a *human* being. Where was Katsimbalis - the real one-and-only Katsimbalis who breathed the language of fire, brimstone and honey... while roving over the vast territory of human concerns? His soul had lost its spokesman. The spirit was abroad somewhere? Or was it still communicating in dreams to the sleeping giant at his feet - still carrying the soul's messages to ensure continuity and progression along the Katsimbalis Way when the next day dawns?

> *A slumber did my spirit seal;*
> *I had no human fears;*
> *She seemed a thing that could not feel*
> *The touch of earthly years.*
> William Wordsworth: A Slumber Did My Spirit Seal.

21

CONSCIOUSNESS WITHOUT A BRAIN?
If *not* the Brain... then what?

All matter, living and nonliving, is ultimately an electromagnetic phenomenon. The material world, at least as far as physics has penetrated, is an atomic structure held together by electromagnetic forces. If some people can <u>detect</u> fields from other organisms, why shouldn't some people be able to <u>affect</u> other beings by means of their linked fields? Since the cellular functions of our bodies are controlled by our own DC fields, there's reason to believe that gifted healers generate supportive electromagnetic effects, which they convey to their patients or manipulate to change the sufferer's internal currents <u>directly</u>...*

Dr. Robert O. Becker spent almost thirty years in working on the content of his book *The Body Electric* - a monumental research-study on the role played in life by the biofields of electromagnetic activity that vitalize every aspect of our cellular, physiological structure. In the passage I quote above he discusses how one's vitally active electro-channels may also be capable of affecting the *'internal currents'* of others, as well as being affected in turn by *their* electromagnetic potency. One implication of this theory is that the brain itself may not be the *sole* source for everything of which we become aware; that it may, on occasion, serve as but a conduit to bring forms of cognizance - not available through the operation of the senses, and presented by independent electromagnetic stimuli - to consciousness. For in the final paragraph of the same chapter Becker goes on to suggest that we can experience highly subjective mental images rising directly from the perineurial sheath tissue that encloses bundles of nerve fibers.

> *Over and over again biology has found that the whole is more than the sum of its parts. We should expect that the same is true of bio-electromagnetic fields. All life on earth can be considered a unit, a glaze of sentience spread thinly over the crust. <u>In toto</u>, its field would be a hollow, invisible sphere inscribed with a tracery of all the thoughts*

---
* Robert O. Becker M.D., and Gary Seldon, *The Body Electric* (New York: William Morrow, 1985), p. 269

> *and emotions of all creatures. The Jesuit priest and paleontologist-philosopher Pierre Teilhard de Chardin postulated the same thing, a noosphere, or ocean of mind, arising from the biosphere like a spume. Given a biological communications channel that can circle the whole earth in an instant, possibly based on life's very mode of origin, it would be a wonder if each creature had not retained a link with some such aggregate mind. If so, the perineurial DC system could lead us to the great reservoir of image and dream variously called the collective unconscious, intuition, the pool of archetypes, higher intelligence deific or satanic, the Muse herself.*

One can have little doubt after reading Dr. Becker's book that each one of us is an electromagnetic phenomenon, and the passage I have quoted provides an apt and felicitous introduction to this chapter. For here I wish to discuss two cases of out-of-body experiences that call into question whether or not the brain (as we neurologically know it today), is the instigator of *every* aspect of human knowledge and awareness.

The first of these events presents us with an extraordinary phenomenon: an occasion when the senses completely abdicated their existential, information-providing functions to permit their 'owner' Dr. Victor Frankl - a political prisoner suffering brutally hard physical labor in the German concentration camp at Auschwitz - to 'slip away' and return home to talk with his wife in Vienna... sitting in his old apartment with everything just as tangible and 'real' as it was at the time of his departure three years earlier. The second case would seem to be even more incredible and inexplicable, and involves a woman who survived radical brain surgery and had an amazing story to tell afterwards.

There are certainly many millions of human beings on this earth who exist in terribly harsh environmental circumstances - victims of natural disasters, repressive political regimes, tribal or sectarian violence, extreme poverty and near starvation - who must find it difficult, if not impossible, to believe that life can offer either hope or meaning to comfort the spirit. And yet it is under such appalling conditions of extreme suffering and privation that deep and subliminal psychical resources seem able to take over, providing the kind of mental and bodily 'escape' experienced by Victor Frankl. The shamans and wise men of more ancient societies than our own, have long held the belief that the ability to move from the present into another dimension of being, resides naturally within each one of us.

And, indeed, a few survivors of concentration camps during the Second World War who were subjected to periods of 'solitary' - which took the form of being incarcerated in deep holes covered with planks which were removed only to allow food to be lowered - have described leaving their temporary tombs to wander around outside for a while, aware of a slender silver-like cord connecting them to the body in the hole, and 'knowing' that if it were to break they would indeed die.

Dr. Victor E. Frankl, a highly educated European, recounts an even more perplexing example of such a phenomenon - telling of the 'wondrous strange' (as Shakespeare would have put it) ability to detach himself from the horrors of the German concentration camp at Auschwitz and return home to his wife.... as I have described. He survived the notorious Nazi prison and died in Vienna in 1997, having become an eminent authority on human behavior and the psychology of brain and mind. He was freed from Auschwitz by the Russian army in their advance on Germany from the East. After the war he developed the Third Viennese School of Psychiatry (founded on his therapy known as *Logotherapy*) - his predecessors being Freud and Adler. Shortly after his release Frankl wrote a chilling account of the inhuman cruelties visited upon himself and fellow inmates entitled *Man's Search for Meaning* - (first published in Austria in 1946 and by Pocket Books of Boston in the United States in 1985.) Here he describes the levels through which consciousness moves as suffering is intensified and prolonged. The first reaction is that of disbelief... of the fact that such things can be happening to oneself. This is generally followed by despair and self-pity - the worst kind of apathy which greatly diminishes the chances of survival, and from which one must struggle to rise by accepting the situation... Frankl achieved such transcendence by exorcising all sense of time, place, and happenings from consciousness, and moving into a visionary state of mind in the manner of the Greek stoics. In retrospect he believed that the practice of such stoicism was the only way to become a survivor. It is only when Frankl could attain such 'altered states' of consciousness - when well into three and a half years of incarceration - that he was able to 'leave' Auschwitz.

But such transitory escape was dependent on the body's ability to 'hang on' - cope with terrible malnutrition and constant physical abuse.

Victor Frankl was one of the few who survived for so long in Auschwitz. The only one we know for whom time slipped; the camp was gone. Whether he was out in the bitter cold, breaking frozen ground, dressed only in the rags supplied to prisoners that passed for clothes, or in the starkly barren, overcrowded hut at night... he would move - (it would seem to be quite involuntary) - to go 'walkabout' as the Australian aborigines call it and live a separate existence. In his book he emphasizes the seemingly automatic nature of this phenomenon, leaving the reader in no doubt that such transportations through time and space were perceptually intense and absolutely 'real'. And this despite the fact that his body was going through the motions demanded of it back at Auschwitz. Yet here 'he' was, talking with his wife back home in Vienna - visitations when every sensory aspect was fully realized. His wife was manifestly present; they talked normally to each other; they were sitting in their old well-used chairs in the apartment that been their home in pre-war Vienna. These were not situations that could be described as dreamlike or hallucinatory.

Yet when he was finally liberated he found that his wife was not alive: had been dead from the very beginning of his internment - gassed at Dachau shortly after he was taken to Auschwitz. He talks at length of the great love they had for each other - a love which would seem to be of such transcendent power that it enabled them to be together in spirit during the most unendurable months of his captivity. In re-reading *Man's Search for Meaning* I was reminded, and deeply moved, by Frankl's conviction that the essence of humanness is the capacity to love on the plane of spirit. His experiences, and the words he uses to describe them, support the views held by some philosophers throughout the centuries, that to love deeply is to attain a *duality* of being - a state in which a spiritual essence presents itself - on equal terms with the body - to reveal the immaterial and spiritual constituents of our being. I wonder if Victor Frankl had ever come across a famous poem written by the English poet A.E. Houseman, entitled *A Shropshire Lad* - particularly the verse:

> *If truth in hearts that perish*
> *Could move the powers on high,*
> *I think the love I bear you*
> *Should make you not to die.*

In the summer of 1991, pioneering and risky brain surgery was performed on a woman patient at the Barrow Neurological Institute in

Phoenix, Arizona. The operation was to treat a threatening bulge in a brain artery, and I came across an account - written by Anita Bartholomew - of the extraordinary experiences undergone by the patient during the surgical process in an old issue of the Reader's Digest for August, 2003.

The revelatory nature of the report lent credence to some of the neurological issues I was discussing in the earlier chapters of this book. Consequently, a few days later, I was able to speak to Dr. Robert Spetzler, the neurosurgeon and Director of the Barrow Institute who performed the operation to obtain his view concerning the medical and philosophical implications aroused by his patient's extraordinary account of the surgery. He very graciously talked to me on the telephone for some time, explaining the highly sophisticated medical procedure and the high level of risk to which it subjected the patient, a 35 year-old mother of three children. For it entailed 'cooling' her brain to the point she would have no brain function, draining most of her blood, and putting her on mechanical life-support systems. Dr. Spetzler estimated that this would give the surgeons a 'window' of about 45 minutes to excise the aneurysm without fear of massive hemorrhaging. During this time the patient would be clinically brain-dead. After the surgery her normal brain temperature would be gradually restored, and her blood reintroduced to her body.

However, as doctor and surgeon, he could offer no scientific explanation to account for the extrasensory kind of awareness displayed by the patient when she talked to him in the recovery room many hours later - both concerning the surgical process itself and other out-of-body experiences. And one should bear in mind her condition at the start of the operation. Leads from a machine gave out a clicking sound when they were inserted into her ears to assess brainstem function; other instruments kept track of vital signs; legs and arms were constrained; eyes lubricated before being taped shut. The operation was a success, despite the odds stacked against it. But the surgeon did not expect to be questioned about some of the details. For example, she inquired about the 'toothbrush-like thing' he was holding over her head. Spetzler told her it was the electric saw with which he opened her skull. And she said she got rather worried when they were 'fiddling around' in her groin when the problem was in her head. To which the surgeon responded by explaining that the life-support systems were inserted via the groin. She went on to tell him that as he triggered the electrical saw in order to open her skull, she felt herself fly up from her body to a position just above the surgeon's shoulders from where she looked down on the operation in progress. At the same time she heard a woman's voice complaining that the patient's blood vessels were too small.

Spetzler was rightly concerned, if not somewhat disconcerted, by this inquisition; wondered if she had been talking to someone prior to his visit.... asking questions about how the operation was performed. Apparently not. And she was only quite recently out of the anaesthetic. When I asked him if the event had caused him to wonder about a purely materialistic view of life... he very politely refused to comment, saying that this was psychological or philosophical territory - not that of medical science.

However, he confirmed the accuracy of the following Reader's Digest report:

> But even though her eyes and ears were effectively sealed shut, what she perceived was actually happening. The surgical saw did resemble an electric toothbrush. Surgeons were, indeed, working on her groin: catheters had to be threaded up to her heart to connect to a heart-lung machine
>
> Spetzler gave the order to bring the patient to a 'stand still' - draining the blood from her body. By every reading of every instrument, life left her body. And she found herself travelling down a tunnel towards a light. At its end, she saw her long-dead grandmother, relatives and friends. Time seemed to stop. Then an uncle led her back to her body and instructed her to return. It felt like plunging into a pool of ice water. After she came to, she told Spetzler all that she'd seen and experienced.
> "You are way out of my area of expertise," Spetzler said. And twelve years later, he still doesn't know what to make of it.

The conventional medical explanation for such 'out-of-body' experiences has been to attribute them to hallucinations brought about by drastic changes in brain activity - as when the potent analgesic properties of endorphins 'flood' the brain at times of bodily crisis: seen by some psychologists as nature's way of 'easing' the individual through moments of high stress, and particularly when undergoing the frightening process of dying. Yet this theory obviously cannot hold up in the case we are discussing. For Dr. Spetzler's patient had no brain activity at all - no blood, no vital signs

- yet she observed the medical team working on her. She was in a state of brain death - defined as such when the brainstem, which controls automatic body functions, stops working, and a person can be kept 'alive' only with the help of life-support machines. (It should be noted that in similar circumstances, if a patient was about to become an organ donor, then he or she could be considered legally dead.) When I talked to Dr. Spetzler he emphasized that he would have been unable to repair the artery had there been any blood supply to the brain. The brain had to be out of commission. His 'client' had to suffer brain death.

The author of the Reader's Digest report also introduces the opinion of Dr. Michael Sabom, cardiologist and NDE (Near Death Experience) researcher, who compared what Spetzler's patient said she saw and heard with Spetzler's surgical transcript. Dr. Sabom found that during the period she experienced the tunnel, she had no brain activity at all. Like an unplugged computer, her brain, to all intents and purposes, was dead. And a dead brain can't misfire. Neither can it hallucinate or react to anesthesia or other drugs. "She met all clinical criteria for death," according to Sabom. "She had no blood in her body. She had no vital signs at all. So was this death? And, if it was death, what was this experience that she had while in this state?"

Well, whatever it was... it certainly causes one to ponder whether or not the brain is the exclusive source of awareness.

In talking at length about *Mind and Brain* in Chapter 17, I referred to questions raised by the long-held concept of 'mind over matter' - the psychosomatic factor, as contemporary medicine would call it - and I took the words of the eminent Canadian neurosurgeon Wilder Penfield as my starting point. You may remember that when he was asked how to conceive of 'mind' as opposed to 'brain' (matter), he responded by saying: *'Mind can only be thought of as a non-temporal, non-spatial entity'*. A statement implying - as I wrote then - that Mind is an energy-force not explainable by the laws of physics as we know them: a *'...source of intelligence... distinct from the biological, neurological activity of the brain itself - an as yet unidentifiable energy-force, meta-biological in its constitution, that induces the brain and consciousness to travel in realms of thought and feeling that lie beyond the sensory confines of the world as we know them.'* And I went on to mention that in many cultures throughout history, the imaginative and insightful wanderings of

Mind have been seen as the conduit to consciousness through which the metaphysical forces of Soul and Spirit make themselves known. Aristotle declared that *'Reason is a light...kindled in the Soul'*. Cicero wrote of finding *'... a sort of truth for the soul in cultivating the mind'*. And Joseph Joubert, the 19th century French philosopher stated that *'The mind is the atmosphere of the soul.'* And here I would add that Albert Einstein - while not using the same terminology - would seem to be thinking along somewhat similar lines to Penfield when he wrote:

> *I have no doubt that our thinking goes on for the most part without the use of signs (words), and, furthermore, largely unconsciously. For how, otherwise, should it happen that sometimes we "wonder "quite spontaneously about some experience? This "wondering' appears to occur when an experience comes into conflict with a world of concepts that is already sufficiently fixed within us... The development of the world in thinking is in effect a continual flight from wonder.*
> From *'Autobiographical Notes'* in Schilpp,
> *Albert Einstein: Philosopher-Scientist.*

So what *was* happening to Dr. Spetzler's patient? It is difficult to see how her powers of Mind - however metaphysically inclined - could have much, if anything, to do with it. Even if one theorized that a sudden surge of spiritual insight mentally took over - which would really represent the ultimate case of 'mind over matter' - such an inspired breakthrough would be to little effect... *if there was no brain at the receiving end*. Consequently, it could be argued that some autonomous and absolute metaphysical potency such as that associated with the Soul - having, unlike Mind, no symbiotic relationship with the brain - arbitrarily broke through the temporal and biological nature of her existence when it was hanging by a thread.... to bring about the kind of metamorphosis she experienced.

One can but wonder... if the 'near death' of the patient resulted in the release of a vital quanta of radiant energy that enabled the spiritual nucleus of her being - the Soul - to bring about such a supranatural transformation; thus permitting a disembodied Self (although not fully released as it would be with actual death) to be aware of what was going on in the operating room. However, whatever theory may be advanced to account for the patient's experience, we should certainly reconsider the whole question of just how 'dead' is someone when declared to be 'brain dead', even if only up to one hour - particularly with regard to the moral and legal issues raised in taking organs for transplant from those considered to be brain-deceased

The 'altered' state of consciousness experienced by Dr. Victor Frankl, and the 'out-of-body' condition undergone by Dr. Spetzler's patient, are dissimilar in several ways: the most obvious being that he was 'brain-alive' and she was 'brain-dead'. Yet having said this, there are other significant variances. For while she was hovering invisibly around, she remained within the context of what was taking place... in 'real' time. Dr. Frankl's experiences were not 'out of body' in this sense. Also, *his* brain was functioning - a factor that would allow the influence of Mind to take over completely - become wonderfully authoritative and visionary, as it can in truly desperate situations involving sensory deprivation and acute psychical distress. In which case it might be said that his meetings with his wife - apparently overcoming the physical laws of time and motion - represented the definitive achievement of 'Mind over Matter'. For while his physical body was in one place he was seemingly able to be convincingly present somewhere else.

However, let him speak for himself... as he writes in *Man's Search For Meaning* of one particularly body-breaking night march beyond the camp gates:

> And as we stumbled on for miles, slipping on icy spots, supporting each other time and again, dragging one another up and onward, nothing was said..... Occasionally I looked at the sky where the stars were fading and the pink light of the morning was beginning to spread behind a dark bank of clouds. But my mind clung to my wife's image... with an uncanny acuteness. I heard her answering me, saw her smile, her frank and encouraging look. Real or not, her look was then more luminous than the sun which was beginning to rise.....
>
> Then I grasped the meaning of the greatest secret that human poetry and human thought and belief have to impart: The salvation of man is through love and in love.....
>
> In front of me a man stumbled and those following him fell on top of him. The guard rushed over and used his whip on them all. Thus my thoughts were interrupted for a few minutes. But soon my soul found its way back from the prisoner's existence to another world, and I resumed talk with my loved one. I asked her questions, and she answered; she questioned me in return, and I answered.

In this passage the author uses the words 'mind' and 'soul' as practically synonymous terms - as indeed did Joubert in declaring that *'The mind is the atmosphere of the soul'.* Yet throughout these pages I have suggested that they are different, if related, forces - that Soul is the absolute spiritual power, while Mind plays a hybrid role... in that it conveys the spiritual dictates of Soul to register in the neurophysiological pathways of a material brain. The theory that Mind has a 'foot in both worlds' - so to speak - would account for the difference between the out-of-body experience of Dr. Spetzler's patient, and the 'body-migrations' in time and space recounted by Victor Frankl. For it suggests that she, being brain 'dead', was subjected to an uprush of sheer Soul.... Mind being out of the picture. While he, being brain 'alive', was affected by an exalted state of Mind... inspired, nevertheless, by a certain *quickening* of Soul.

'Omnia vinci Amor' (Love conquers all) wrote Voltaire - a declaration that can be seen as supporting Victor Frankl's conviction that altruistic love stands as the most potent force in the Soul's spiritual arsenal - that it can, as is said, 'move mountains'. And so the man from Auschwitz - whose love for his wife was so profound - overcame the here-and-now obstacles of time and place and was released to travel, in some timeless state of being, to join her. Reading Frankl's accounts of these trysts with a wife he did not know was dead, leave one convinced that he was not delusionary, but living a 'reality' quite inexplicable by the physics of time and motion, or the biomechanical principles governing life. And until science can demonstrate otherwise I would cite Aristotle's dictum that certain phenomena lie beyond the reach of science, beyond physics, and can therefore be termed *meta*physical.

> *Love is the emblem of eternity: it confounds*
> *all notions of time: effaces all memory of a beginning,*
> *all fear of an end.*
> Madame De Stael, *Corinne*. Bk. viii, ch. 2.

The 2007 Nobel Prize in Physics was awarded to Peter Gruenberg from Germany and Albert Fert from France for their 1987 discovery of a particular electromagnetic phenomenon - the application of which has brought about tremendous advances in computer function. The science is complex and my understanding of their work as a layman may be pretty fundamental... but it would seem that they fabricated new material by layering very thin 'slices' of metals comprising magnetic and nonmagnetic atoms, and in so doing created original physical substances having extremely advantageous *electromagnetic* properties in that when outside a magnetic field they would not conduct an electrical current... yet when exposed to such a field became effective conductors.

I mention this latest development in physics because it represents yet another example of the seemingly cardinal importance of electromagnetic energy in the overall scheme of things: one more discovery upholding Dr. Becker's statement in *The Body Electric* that...*'All matter, living and nonliving, is ultimately an electromagnetic phenomenon'* - and in describing the phenomenon he goes on to say:

> *'Electromagnetism can be discussed in two ways - in terms of fields and in terms of radiation, A field is "something" that exists in space around an object that produces it. We know there's a field around a permanent magnet because it can make an iron particle jump through space to the magnet. Obviously there's an invisible entity that exerts a force on the iron, but as to just what it consists of - don't ask! No one knows. A different but analogous something - an electric field - extends outward from electrically charged objects.*

As you can see, there's a certain mystery surrounding both the dynamic nature and the modus operandi of electromagnetism. Just to read in the New Journal of Physics that non-organic galactic dust '... *when held in the form of plasma in zero gravity conditions formed the helical structures found in DNA... the particles being held together by electromagnetic forces...* ' brings one to wonder at the nature and origin of forces that can determine the particular kind of physical characteristics that something will come to possess. Inevitably, one begins to wonder if the matter-altering powers of electromagnetism could play any part in the metamorphosis experienced by Dr. Spetzler's patient.... or in the kind of 'body-time-space-travel' undergone by Dr. Victor Frankl. In his case - as he was in possession of an electronically working brain - I find myself contemplating whether the metaphysical and spiritual energy-force of the love he had for his wife, could in any way transform the physical field of electromagnetic radiation surrounding him - that is, cause his body's particle-structure to become fractured.... allowing him to materialize in a place other than that he occupied in 'real time'... seemingly be in two places at once. But Spetzler's patient, incapable of experiencing any brain excitation at all, could not be 'somewhere in loving' in the Frankl sense. No such quasi-scientific explanation will help here to shed light on her out-of-body wandering. For she was scientifically dead, and so beyond the purview of science to offer any solution exhibiting the methods or principles of science... as Dr. Spetzler rightly pointed out. Therefore in suggesting earlier that when his patient describes herself as 'flying up from her body...' she was experiencing the temporary release of her Soul, I was turning for answers to the side of life's coin not scientifically explainable. To those mystical or spiritual encounters that have invaded consciousness and been described by men and women throughout our history - feelings and thoughts that have given rise to both belief and faith in a personal and incorporeal Self, central to their Being and of spiritual quality. Namely, the Soul.

In times less secular and rational than our own, the 'Frankl experience' would not appear all that fanciful and old-fashioned.

John Donne, a contemporary of Shakespeare, and in my view the greatest of the so-called Metaphysical Poets of the time, was preoccupied with the themes of changeability and death: with the paradoxical nature of the union of spirit and matter that was generally seen to be the human condition. He wrote a wonderful love poem - charged with ardor, romance, and spiritual overtones - entitled *The Extasie*. Here, in his plea to his lady that they should make love, he assures her that their desire is indeed that born of soul mates - however, as their souls occupy their bodies, they have no other way to realize their spiritual affinity than to take to their bodies in the loving. An engaging stratagem in the art of seduction!

> *So must pure lovers soules descend*
> *T'affections, and to faculties,*
> *Which sense may reach and apprehend,*
> *Else a great Prince in prison lies.*
> *To 'our bodies turne we then...*

22

HUMANITY AT LARGE: DECLINE and FALL?

> Though undoubtedly man's genetic nature changed a great deal during the long proto-human stage, there is no evidence that it has in any important way improved since the time of the Aurignacian cave man... Indeed, during this period it is probable that man's nature has degenerated and is still doing so.
> Julian Huxley: *Essays of a Humanist*, (1964)

The French Aurignacian period to which the distinguished English scientist Julian Huxley (1887-1975) refers, attained its highest cultural achievements in Western pre-history during the Upper Paleolithic period - the name given to the last major glacial phase in the Northern Hemisphere. It is the prehistoric period generally regarded by anthropologists as that which saw the emergence of *homo sapiens* - the *fully* modern man more accurately referred to as *homo sapiens, sapiens*. The epoch covers the years from about 40,000 B.C. until the ice began to release its hold some time around 15,000 B.C.

Some neuroscientists say that the human brain was cortically complete - as operationally complex as it is today - between four million and one million years ago, and that during this period consciousness became capable of developing language and art as symbolic forms of self-expression. Others believe that this essentially human faculty was not fully operative until around 200,000 B.C. In either event, the evidence suggesting the existence of one such abstract sensibility as early as 62,000 B.C. was discovered during the excavation of the Neanderthal burial site at Shanidar in Northern Iraq... which we discussed in Chapter 9. Now we have Julian Huxley's somewhat surprising comments concerning Aurignacian man: thoughts which might have been inspired by the archaeological explorations in the caves around Aurignac in Southern France that revealed the presence of artists possessed of great symbolic powers; together with evidence that some form of communal music making went on. And this as early as 32,000 B.C. For here were found small stone sculptures and magnificent wall paintings…. and, surprisingly, a petrified flute-like musical instrument of the same age. Some anthropologists see the Aurignacian culture as extending to the end of the epoch - around 15,000B.C. - when the wonderfully expressive paintings (such as that of the deer reproduced in Chapter 10) that comprise the great 'gallery' of animal art in the Lascaux caves of the Dordogne were created: caves not all that far way from Aurignac.

However, the discovery in 1994 of the Chauvet Cave in the Ardeche Valley, some 100 miles south of Lascaux, revealed what are considered to be the oldest known paintings in the world. Radiocarbon tests found them to be over 30,000 years old, just about *twice* as old as the animal paintings at Lascaux, and more or less contemporaneous with the artifacts and flute found at Aurignac. The Chauvet animal paintings - even though some 15,000 years earlier than those at Lascaux - are just as convincingly observed, as passionally expressive, and as skillfully rendered as the famous works in the Dordogne. They also manage to convey a strength of empathic feeling on the artist's part no less evident than that which impelled the hand of the Lascaux artist: indicative of a similar belief in sympathetic magic and ritual. Sir Julian Huxley died in 1975 and so could not know of the Chauvet Cave findings. Had he been alive they would surely have confirmed his feelings about man living in this region of Europe so many thousands of years ago.

Music, art, and a spirit-sensitivity to animals and life in general are three ways in which I would define *humanness*. And here they are, manifesting in the same general locality; testifying to sensibilities extant more than 30,000 years ago. In his book, *Dark Caves, Bright Visions,* Randall White describes the ancient Aurignacian flute-like instrument as having six holes, two on top and four on the underside: an arrangement allowing complex patterns of sound to be produced, indicating that it was not intended to be used as simply a sophisticated animal-call device. A modern reconstruction of the 'flute' when played, revealed that although the tones that would have reverberated around the Abri Blanchard rock-shelter where the instrument was found, may not be as musically elegant as those produced by a modern flute, they nevertheless possessed an undoubted emotional and ritualistic potency. A second such instrument, unfortunately broken, was later discovered at the Gravettian site in southern France. One remembers Freud's oft-quoted statement that *'music is the royal road to the soul'…*

and wonders if the sounds of the Aurignacian flute affected these ice-age peoples to the depths of their being. (In the Natural History Museum in New York can be seen a similar kind of bone or ivory flute that is a mere 15,000 years old!)

Two of the world's most distinguished anthropologists, the Abbé Henri Breuil and Siegfried Giedion, were impressed not only by the acuity of perception, drawing skill and expressive power displayed by the cave artists of Lascaux and Altamira (in northern Spain) who were working in the later years of the Upper Paleolithic era (20,000 to 15,000 B.C.).... but also by an intangible quality their work conveyed. Both scholars experienced an odd awareness, an impression, that the paintings were infused with a high degree of empathy for the animals depicted: that they were regarded with respect as living creatures sharing the same world, on whom the cave dwellers depended for their own survival. And that one function of the paintings was to indicate to the particular animal 'spirit' that its physical manifestation was not hunted without mercy.... without recognition of the 'sacrifice' the animal was making for the survival of the hunter. If such was the case, it makes the 'modern' hunter - out for 'sport' - an altogether less humane individual. However, something of this very early empathic link with animals has continued to manifest itself over thousands of years. Totemism - the belief in kinship with, or a mystical relationship between humans and animals - may have prehistoric origins, but the ancient civilizations of Egypt, Babylonia, Persia.... produced artists and artisans who displayed extraordinary skill in expressing such feelings in the grace and beauty of the animal forms they created. The Gods in the classical Greek pantheon for example had, in many cases, their symbolic animal counterparts. Nowadays there are societies worldwide dedicated to the protection of animals; yet, many species are also being pushed to extinction by our ruthless exploitation of their environment. In the issue of the London Sunday Times for 26 October, 2008, the headline for an article commenting on a report by The World Wide Fund for Nature, reads: *Humans 'drive biggest mass extinction since dinosaurs...'* Yet it should not be forgotten that the American Indian would drive herds of buffalo over cliffs to their deaths; while even today, in Alaska, the hunting of wolves from low flying aircraft is considered by some to be great sport. And it is evident - given the number of requests for financial help one receives from animal protection societies - that there are many among us who feel no affinity with other forms of life; have no compassion for the creatures whom they subject to the most terrible acts of violence. Such people sadistically enjoy their viciousness: maim and kill not for reasons of self defense, or in order to survive by hunting for food. Commercial fur interests drive the annual seal-pup clubbings in Labrador. While honing shooting skills is given as the reason for bringing thousands of birds tumbling out of the sky in 'shoots' that are also regarded as social occasions.

Even so, in many Western countries it is still said that a dog is 'man's best friend'; in some Veterans Hospitals in the United States, dogs selected for their healing powers - delivered by the 'laying on of paws' - are seen to noticeably raise the spirits of patients through some kind of sympathetic transfer of feeling. No such role for those greyhounds in Britain who - when still comparatively young - are killed when they are no longer capable of performing well on the racetrack. When travelling in parts of Africa and the Far East, it is painful to witness the minimal subsistence so many dogs have to endure. And try not to think about the food industry's slaughterhouses. Or of the ordeals endured in the vivisection laboratories.

Only some 3,500 black rhinos are left in the world, 3,000 of which are to be found in South Africa. In the year 2008 a hundred of them were killed by poachers - just to obtain their single horn which is believed to contain aphrodisiac properties and for which a lucrative market exists in East Asian countries. In the London Sunday Times for February 22nd, 2009, is a description of how the poachers steal the valuable horn by hacking away at the rhino's skull with pangas (machetes) causing terrible injuries. *'It's a terrible thing to come across the poachers' handiwork,'* said Frank Reardon, a wildlife enthusiast. *'To see one lying dead with the carrion feeding off it is an awful sight.'* George Hughes, a former head of the KwaZulu Natal Parks Board, described the white rhino as *' a gentle and friendly animal. They are vegetarians, not predators, and only man preys on them'.*

One has to ask, 'What manner of man?'

All of which brings one to wonder just how many members of contemporary society, worldwide, would respond credulously and sympathetically to the supposition that prehistoric hunters believed a link existed between themselves and *'the animal spirit?'* Or how many would smile cynically and superiorly to themselves at the very idea?

Perhaps Julian Huxley - when he wrote about man's nature having *'degenerated'* since Aurignacian days - had in mind the difference between the obvious expression of the cave artist's feelings for his animal subjects, and the often sadistic, commercial and uncaring attitudes to so many animal species that have accompanied the onward march of *homo sapiens*. Two thousand years ago the Romans, for example, were a sadistic lot: they imported thousands of wild animals annually for blood-letting 'entertainments' of various kinds in the Colosseum. And one should also remember that before the advent of the machine, the more aggressive the commercial interests of a community, the more harshly have animals been exploited as 'beasts of burden'. The following words were written almost twenty years ago, yet I wonder how Huxley would have responded had he been alive to read them.

> *Yet our adult society views animals…… as commodities or resources, as things, objects, and tools. We use them. We eat them. We experiment on them. They no longer enchant and delight us.**

He may have seen them as lending support to the view that the empathic nature of the relationship between man and animal - that mystical sensibility pervading much of Upper Paleolithic cave art - has been steadily eroded throughout the thirty or so following millennia. Seen in this light, Aurignacian man could be seen as naturally having the moral edge on the increasingly more sophisticated generations that succeeded him.

I have no way of knowing if Sir Julian was familiar with John Beston's moving and inspiring words about the animal kingdom in his memorable book, *The Outermost House* - yet, if so, I think he would have smiled approvingly to himself:

> *We need another and a wiser and perhaps a more mystical concept of animals. Remote from universal nature, and living by complicated artifice, man in civilization surveys the creature through the glass of his knowledge and sees thereby a feather magnified and the whole image in distortion. We patronize them for their incompleteness, for their tragic fate of having taken form so far below ourselves. and therein we err, and greatly err. For the animal shall not be measured by man. In a world*

---

* Lines written by H.N. Robbins, author of the Pulitzer-Prize nominated book *Diet For A New America,* in the Foreword to *The Souls of Animals,* written by Gary A. Kowalski.

> *older and more complete than ours they move finished and complete, gifted with extensions of the senses we have lost or never attained, living by voices we shall never hear. They are not brethren, they are not underlings; they are other nations, caught with ourselves in the net of life and time, fellow prisoners of the splendour and travail of the earth.*

However, we can go back to an even earlier prehistoric age than that of the Upper Paleolithic, and find evidence suggesting that some societies embraced the idea of 'spirit' operating as a life force. You may remember that in Chapter 9 I referred to Professor Ralph Solecki's 1968 excavations at a Neanderthal burial site in the Shanidar cave region in northern Iraq - where he discovered that flowers from the region had been ritually used in burial ceremonies. This might have been a symbolic act to support a belief in the 'flowering' of spirit with the death of the body: (a practice we still follow - more or less unconsciously - to this day by placing flowers on both coffin and grave.) For it was at Shanidar that pollen grains from flowers indigenous to the region were found throughout the graves, showing a particular concentration in the fossilized chest cavities of the dead. The date is said to have been about 62,000 B.C.

It is not unreasonable to conclude from Solecki's excavation that these Neanderthals of the Lower Paleolithic period practiced a Rite-of-Passage ceremony: a ritual in accordance with a numinous approach to life and death. (I regard the controversy concerning whether or not Neanderthal man belongs to 'our' species as irrelevant here.... in light of the 'human' way he buried his dead.) Huxley does not disclose whether he knew of Solecki's excavations in 1968, otherwise he might have wondered if some degree of Neanderthal mysticism had not in fact influenced the development of Aurignacian cave art some 30,000 years later. After all, Neanderthal groups existed throughout Southern Europe, including settlements in northern Italy that were relatively close to the Dordogne region and the cave areas around Aurignac.

Consequently, we have instances from two periods, some 30,000 years apart, when it can reasonably be said that a mystical or spiritual sensibility affected first Neanderthal life in ancient Mesopotamia, and then that of Upper Paleolithic cultures in southern Europe. And when Huxley writes that after the Upper Paleolithic Glacial Epoch he considers it *'probable that man's nature has degenerated and is still doing so....'* he is suggesting that

Aurignacian man in particular was strongly influenced by an innate conviction that mystery permeated all of nature's manifestations. And that this, together with the practice of rituals involving the arts of carving, painting and music - enabled him to psychically participate in the mystery of Being. And in so doing come to live a life that could be described as 'naturally religious' - a term denoting a sense of wonder, awe, and respect when connected to nature's world. Here is Gary Kowalski writing in the penultimate chapter of The *Souls of Animals*:

> *What will it mean for the human race if children.... come of age in a world bereft of other living creatures? Their growing years will be immeasurably less vivid and vibrant. Their connection with the earth will be severed, and part of their inborn potential for amazement will go uncultivated. It is not just that animals make the world more scenic or picturesque. The lives of animals are woven into our very being - closer than our own breathing - and our souls will suffer when they are gone.*
>
> *As society becomes increasingly urbanized and animals disappear from our daily lives, and as more and more species slip into the long night of extinction, our humanity will inevitably be diminished. We will become increasingly confused about who we are, and distortions of the self (egos that are chronically over-inflated or under-inflated, having no reference point in nature) may become more common. In spite of our material plenty, our inner world will be impoverished.* *

Perhaps Julian the scientist possessed a 'mystical gene' similar to that exercised by his brother Aldous the visionary poet and novelist. For despite his scientific bent, Julian's words concerning Aurignacian man indicate that he understood the human significance of the ancient, and apparently mystical sensibilities that would appear to have been part of Upper Palaeolithic life: qualities that have been substantially eroded over succeeding millennia.... resulting in the general loss of primitive wisdom as civilizations developed across the globe. Yet I would say that the influence of the times in which they lived also contributed to both Julian's and Aldous' awareness of these lost levels of human sensitivity. Two world wars, and two cataclysmic genocidal events occurred in the 20th Century

---

* From the book *Souls of Animals*. Copyright © 1991, 1999 by Gary A. Kowalski. Reprinted with permission of New World Library, Novato, CA. www.newworldlibrary.com .

to mark their lifetime. Stalin's political purges in Russia after World War I, together with the forced relocations in the 1930's of the peasant farmers during which millions died.... exposed a completely conscienceless regime. While Hitler's persecution of the Jews which preceded the start of World War II and reached its tragic finale during the war itself, reinforced the absolute moral bankruptcy into which human beings can fall. To have lived through those years was not likely to inspire optimism concerning the moral progress of humankind. Particularly when, in addition, you see how arrogantly we have come to regard ourselves as the 'entitled' species on this planet; look on 'the world as our oyster...' free to greedily plunder its treasures and exploit its peoples in the interests of national aggrandizement, and in the gaining of both personal and collective wealth. (As I mentioned in Chapter 16, *Mirror, Mirror, on the Wall,* the evolution of the Classical Ego - as a process leading to the development of *gross*-ego, has a lot to answer for here.)

The Huxleys were not alone in taking this position. Other scientists and philosophers have thought in similar vein. I especially remember the books and poems of the distinguished American archaeologist Loren Eiseley. He writes in *The Star Thrower:*

> .... let us remember man, the self-fabricator, who came across an ice age to look into the mirrors and the magic of science. Surely he did not come to see himself or his wild visage only. He came because he is at heart a listener and a searcher for some transcendent realm beyond himself. This he has worshiped by many names, even in the dismal caves of his beginning. Man, the self-fabricator, is so by reason of gifts he had no part in devising - and so he searches as the single living cell in the beginning must have sought the ghostly creature it was to serve.*

As descendents of Eiseley's *self-fabricator* ice-man, time has certainly delivered us to the contemporary universality and authority of science. Yet there are still parts of the world where shamans exercise naturally transcendent healing powers without recourse to '... *the mirrors and the magic of science*'. As such they may well be described as 'primitive': 'Stone Age stuff...' said a friend of mine when I was recounting an account in the London Sunday Times of 22nd February 2009 - one telling of a feat of

---

* Loren Eiseley, *The Star Thrower* (New York: Harcourt Brace & Company, 1978), p. 121

healing performed by a shaman of the Reindeer people of Mongolia. It involved the autistic and incontinent young son of a well-known horse trainer and animal writer. After a ceremony, the boy's incontinence was apparently cured. From a very early age the boy had exhibited a natural affinity with horses; by being among them and after riding them with his father for three years or so, he no longer suffers from the major dysfunctions occasioned by the mental ravages of autism. And I gather that an important part of this 'horse-therapy' came from riding the tough little horses of Mongolia. I wonder what Julian and Aldous Huxley would have thought of these approaches to healing carried out in the high Mongolian mountains?

If you peruse the writings of the Roman senator and orator Marcus Tullius Cicero, who Octavian (later to become the Emperor Augustus) hounded to his death in 43 B.C. .... you will find that Cicero considered the attainment of *sapienta*, wisdom, to be our main objective in life. In *De Officiis*, Bk. ii, he writes: *Wisdom is the knowledge of things human and divine, and of the causes by which those things are controlled*. So there you have as comprehensive and 'wise' a definition of wisdom as I know: a scientific inquiry into the nature of man and his world, helped along from time to time by the philosophical glimmerings of inspired sagacity.... musings, flashes of insight, regarding ways to account for the *force majeure* responsible for the cosmos, and our particular presence in it.

Unfortunately, Cicero is long gone; and neither Huxley or Eiseley are still with us; yet I find myself wondering just how much 'progress' their shades - if recalled to the world - would consider we had made in the pursuit of wisdom.... as we move through the first decade of the 21st century?

The eminent philosopher William Barrett, when teaching at New York University, wrote a famous and highly influential book entitled *Irrational Man*[*]. The title is intended to emphasize the fact that we are not creatures who operate solely through the much vaunted powers of reason and analysis that developed as the brain's *left* hemisphere grew and matured over the many million years of its evolution. For during this time the

---

[*] William Barrett, *Irrational Man*, (New York: Doubleday Anchor Books, Inc. 1958)

brain's *right* hemisphere was also evolving to present an alternate mode of consciousness: the ability to be taken over by feeling-moods that induce a reverie wherein imagination - 'inspired sagacity' - represents a direct form of 'knowing' without reference to sensory involvement and reason. And so - perhaps around 200,000 years ago as some neuroscientists suggest - with both objective (rational) and imaginative (irrational) modes of consciousness in play - *homo sapiens,* arrived on the scene, bringing with him the ability to both reason and to imagine: hence the mental duality that characterizes our species.

When Oliver Wendell Holmes Sr., the American writer and physician, who died in 1894, said that *"A moment's insight is sometimes worth a life's experience',* he was unknowingly taking the same position that led Professor Barrett, in 1960, to title his book *Irrational Man.* For Barrett stresses the vital role played by intuition, imagination, profound feeling…. in determining the range and depth of our sagacity. And he points out that empirical reasoning and logic do not rule the roost so far as consciousness is concerned. Consequently, the word 'irrational' possesses no derogatory connotations in the context of his book. In fact, one could sum up the author's argument by saying that whether one ascribes our mental evolutionary growth to processes of natural selection, or to unknown forces beyond nature's sphere of influence, this dualistic system of consciousness is obviously meant to function as an alliance - a 'coalition government', so to speak - to promote awareness of two kinds of 'reality'… the sensorily factual and the intuitively envisioned.

Yet in broadly surveying the history of western culture over the last five hundred years or so, Barrett sees the natural balance between the utilization of reason and imagination gradually tipping in favour of a dominantly rational consciousness - concentrating on the material and logistical facts of life as simply the best way to prosper in terms of our existential wellbeing. He tells of an increasing scientific curiosity, thriving at the expense of philosophical speculation as to the 'why's' and 'wherefore's' of existence. The drive to gain wisdom (Cicero's *sapienta*) is steadily losing ground; is, in fact, regarded as something of a diversion - an idealistic philosophical exercise secondary to the pursuit of scientific investigation. Professor Barrett goes on to point out how mediaeval philosophy embraced the duality of matter and *spirit:* a balanced view of life that gave way - as the years of the Renaissance advanced - to a more one-sided and secular belief in straightforward humanistic principles born only of natural experience in a material world. The mediaeval world was left even further behind as the scientific approach to life became ever more consolidated during the Ages of Reason and Enlightenment that followed in the 17th and 18th centuries. Then in the 19th century along came the evolutionary theories of Charles Darwin and Alfred Russell Wallace.... followed by the materialistic ethos of the Industrial Revolution and the rapid acceleration of the natural and physical sciences that ensued and continued apace into the 20th century. And all to be capped by the advent of the Technological Revolution in our own time.

Now, in the first decade of the 21st century, there are those neuroscientists who are suggesting that two important consequences may result from the accelerated dominance of science and technology in our own time.

The first is that the *right* brain, being largely the intuitive source of visionary and creative achievement, may atrophy over time if the *left* brain comes to dominate the way we live our daily life - which it would seem to be well on the way to doing. It's objective and analytical skills are engaged nowadays in dealing with most routine logistical tasks, not to mention that it is the principal mental power serving the 'high-tec', computerized world of modern science. In the long run, this pattern of *left* brain authority may result in their being fewer *right* brain creative geniuses around to bring about those breakthroughs of understanding that advance science and illuminate the life of the human spirit - indeed, further the cause of humanness in general.

The second is that an exclusively material and extroverted way of life contributes to the rise of a rational secularism. This is not a development which - as Barrett points out - is likely to further the growth and spread of wisdom as we have defined it here. Or induce and nurture the reflective musings on life's experiences that the poet Wordsworth saw as contributing to the *'wealth'* of one's psychological treasury. In other words, without benefit of such introspection there may be little awareness of the *'... small mystery'* that is oneself.

Barrett points out that in an age dominated by science, industry, and commerce, the mystical rituals practised by our very early forebears, together with the constantly persisting belief over thousands of years in a life-force called 'spirit', are in trouble. They come to be regarded as mere superstitions brought about by fear of the unknown; or simply as manifestations of ignorance. But then he proceeds to show how the ancient human practice of challenging the absoluteness of time, and of believing the actuality of the material world to be no more, or no less, 'real'.... than the domain of spirit persists to this day - despite the advances of science and technology and the increasingly hedonistic lifestyle that pertains over much of the world. For Professor Barrett sees the creative vitality of art, literature, poetry and music... to be an intuitive drive giving form to the persevering life of spirit and its agent the imagination: an ageless consciousness holding its own as demonstrated especially during the Romantic movements of the 19th century and first half of the 20th century when industry and commerce were well established and prospering. And he sets out to show how truly creative men and women seem always impelled to express and symbolize the *'wealth'* of their innermost thoughts and feelings.... however widespread the prevailing materialistic ethos. I believe the earliest record we have of such a need is a single sentence, hieroglyphically inscribed on a clay tablet from Ancient Egypt (circa 3000 B.C.) belonging to the British Museum and translated as: *'I wish to express from my body, feelings, the words for which are not in existence.'*

Professor Barrett writes:

> *Romantic melancholy, as we have seen in the case of Coleridge, is nothing less than man's discovery of his own estrangement from Being; in Baudelaire this... takes on the dimension of revolt. It is not only a social revolt against the materialism of bourgeois society, but a metaphysical revolt against the kind of world created by the positivism and scientism of the present age.*

And then continues a page later:

> *Rimbaud was thus among the first of the creative artists to announce primitivism as one of the goals of his art and his life. From Gauguin to D.H.Lawrence primitivism has been such a varied and rich source in modern art that academicians or rationalists would be ill advised to dismiss it out of hand as a mere symptom of 'decadence'. One might ask, in any case, whether it is not the civilization itself that has become decadent rather than those creative individuals within it who struggle*

to rediscover the wellsprings of human vitality. With Rimbaud primitivism was far from being a sentimental decor for the spirit, an illicit longing after the South Seas and maidens in sarongs; rather it was a passionate and genuine struggle to get back to the primitive - which is to say, primary - sources of Being and vision... .

Romantic melancholy was no mere matter of languor or the vapors; nor was it an outbreak of personal neurosis, impotence, or sickness among a few individuals; rather it was a revelation to modern man of the human condition into which he had fallen, a condition that is nothing less than the estrangement from Being itself. Once having lost contact with the natural world however, man catches a dizzy and intoxicating glimpse of human possibilities, of what man might become... .

Well.... in this first decade of the 21st century we can certainly see what we are becoming, if not what we have become. The application of science and technology is worldwide, tending to reduce the geographical and cultural diversity that distinguishes individuals and nations. The world's financial markets and institutions, together with government agencies, are electronically interconnected; jet travel shrinks journeys over vast distances to the point where they are viewed as routine weekend jaunts. A contemporary state of affairs that change our perception of Time; diminish the sense of our own Individuality; obscuring any awareness of Purpose in one's progression through life.

'Hitch your wagon to a star...' advised Ralph Waldo Emerson the American essayist and philosopher-poet well over a hundred years ago.

Yet western secularism has been ditching such personal 'wagons' for years: they facilitate the transportation of too much personal psychological baggage for this day and age when the haste to get from A to B, and the constant search for the 'new', leaves little time or inclination to harbor thoughts or feelings of the past. As for stars.... we think of them in the factual terminology of physics (self-luminous gaseous celestial bodies), rather than in the symbolic associations of the poet's (*Th' evening star, Loves harbinger*), as wrote John Milton in *Paradise Lost*. Following the nihilistic legacy of two World Wars, fought within the space of fifty years, there seems little time to waste indulging idealistic feelings or contemplative musings when it comes to constantly living in anticipation of the future. Certainly, by the 1970's in most of the Western world - and increasingly so in the East - it was generally held that success in life was measured in financial terms: money providing the opportunity to spend time pursuing the new and the pleasurable, in the belief that this was the best way to provide a sense of purpose and accomplishment. When John Mortimer, the English novelist and playwright was interviewed by Rosemary Hartill for a book called *Writers Revealed**, she reminds him that he had previously spoken of Britain's *'Bizarre religion of greed...'* and went on to suggest that this represented quite a *'strong attack'* on the state of the country. He replied:

> *'I think it's a pretty disgusting situation, the idea that money is the entire be-all and end-all of existence, that everything is judged now by whether or not you make money. People are meant to be educated now simply in order to make money. The idea that you are educated to be a more tolerable person, or to lead a richer life, when you are sitting by yourself, has totally gone...'*

'When sitting by yourself...' Does anyone in the general population - save perhaps the aged and newly bereaved - sit quietly by themselves any more, television quietened? Do schools and universities teach in a way that requires the student to go beyond the facts as learned; develop a *personal* opinion as to their reliability, comparative worth and applicability to the salient points of the issue under discussion? In an impersonal and computerized educational system is the finger ever pointed, and the question ever asked, *'Now Smith, what do you think about that...?'* In Chapter 5 I used Wordsworth's poem *The Daffodils* to illustrate the benefits of solitude and the richness of the internal soliloquies that follow significant events in life. I reintroduce the first and last verses here because they illustrate so well the revelatory nature of Wordsworth's encounter. Here is the last verse describing the reverie that followed his visual lake-side encounter with..... *'A host, of golden daffodils...':*

> *For oft, when on my couch I lie*
> *In vacant or in pensive mood,*
> *They flash upon that inward eye*
> *Which is the bliss of solitude;*
> *And then my heart with pleasure fills,*
> *And dances with the daffodils.*

---

* Rosemary Harthill, Writers Revealed (New York: Peter Bedrick Books, 1989), p. 58.

As the Technological Age advances with the proliferation of ever more sophisticated personal computers and communication equipment - together with the facts, figures and graphic images they provide - it could be that future generations will come to regard this *electronic* information as more 'real' than the actuality of things and events as perceived in the outside world. And I wonder if those of our heirs who may come to read *The Daffodils* would be able to respond to the eloquence of the opening lines.... and suddenly 'come to their senses' - literally, I mean - and be impelled to realize how evocative is the gift of sight when entranced by the space, light, and forms of the natural (as opposed to electronic) world; how the full range of normal consciousness is diminished when 'virtual reality' supersedes the real thing. Here is Wordsworth's first verse

> **I wandered lonely as a cloud**
> *That floats on high o'er vales and hills,*
> *When all at once I saw a crowd,*
> *A host of golden daffodils;*
> *Beside the lake, beneath the trees.*
> *Fluttering and dancing in the breeze.*

I have always regarded the poetry of Wordsworth - as my reader well knows - to be wonderfully balanced in so movingly conveying the two-sided drama we call consciousness: telling so graphically and knowledgably of the world's presence on the one hand, while revealing the run of his imaginative and contemplative vision on the other. And it is interesting to note how Einstein, the most humane of scientists, regarded this partnership when he said: *'Imagination is more important than knowledge. Knowledge is limited. Imagination circles the world.'*

I have mentioned how William Barrett talks about 'primitivism' in his book *Irrational Man* - and this in terms that convey what I believe Huxley had in mind when he expressed the view that man's nature had likely degenerated since the ice-age-days of Aurignacian man. For Barrett sees Wordsworth as a poet expressing a similar kind of natural, elemental sensibility.... as the most 'serene' of the early Romantic poets: a *'rural' man* rather than a *'city man cut off from nature',* and attaining *'metaphysical'* insights through his visual delight in the drama and splendor of the natural world.

Sumerian documentation, incised on clay tablets, has about 2000 pictographic signs and is the earliest form of 'written' communication known in historic times. The Sumerians - an ancient and non-Semitic people occupying Southern Mesopotamia - settled on the site of the city of Babylon around 4000 B.C. and the civilization they created reached its height about 3500 B.C. The wedge-shaped (cuneiform) *writing* they produced during their rise to power is the earliest known in historic times, and their arrival on the scene is taken to mark the end of the Palaeolithic period along the Mediterranean coastline. I mention this to illustrate the relatively brief span of time that has elapsed since the beginning of literacy (at least of which we have any evidence), to the sophisticated linguistic communication systems of our own day. Some 5000 years in all. And yet more than 15,000 years went by between the cave culture of Aurignacian man and that of his Upper Paleolithic successors barely a hundred miles away at Lascaux. We can have little, if any, idea as to what 'progress' was achieved during this long, long, span of time. For as I said earlier (speaking as a onetime practising painter and writer on the visual arts), the Lascaux cave artists display *no* greater powers of observation or of skill in the rendering of their animal paintings, or in the expressive and symbolic potency their images display.... than had already been accomplished by the cave artists at Chauvet fifteen millennia before.

'Progress' is a term possessing differing implications for different individuals. Yet most people will agree that, in general, it signifies some advance is being made towards a higher and better way of life. Then, of course, one runs into the question as to what actually constitutes a 'higher' and 'better' way of life: the acquisition of a little wealth, professional success, guaranteed good health, finding love.... or all of the above? When asked about my own views, I am always at a loss to find a few incisive one-liners that might provide a ready answer. So I usually bite the bullet, and launch into a psychological and philosophical discourse on the insights gained by trying to penetrate the mysteries of a consciousness that can have us living both a physical and metaphysical state of awareness simultaneously. At which point either my listener waits politely for me to elaborate.... or abruptly changes the subject. It is on such occasions when I think how convenient it would be to have a miniature edition of this book in my pocket and, asking their indulgence, take a few appropriate lines from the relevant Chapters - those advancing theories concerning the function of the psychological forces linking Mind, Spirit and Soul to the operation-center of the Brain and Senses. Of course, I couldn't stop there, but would have to add that this was the way we came to experiencing higher levels of discernment regarding Being in general, and our own in particular - consequently achieving a more serene and meaningful life. However, you can imagine how effectively such a monologue would shut down dinner table conversation.

Yet it becomes increasingly clear that as nuclear weapons proliferate, it is vitally important for the survival of our species that moral progress should keep up with advances in science and technology. In fact, if we take Cicero's definition of wisdom to represent the form our progress *should* take, it is clear he considers that we should continuously seek to advance in metaphysical sensibility, while at the same time pursue all branches of scientific knowledge - thus balancing the existential and the transcendental sides of consciousness' coin. The German philosopher Max Scheler (who died in 1928) echoes Cicero when, writing in his book The *Nature of Sympathy,* he says that we must '…. *press forward into the whole of the external world and the soul, to see and communicate those objective realities within it which rule and convention have hitherto concealed.*'

The chances of discovering the world of the soul - the serenity and meaning it induces - do not seem likely when one reads the melodramatic account of human nature penned by the 17th century French mathematician, physicist and philosopher Blaise Pascal, in *Pensees VIII: What a chimera, then, is man! what a novelty! what a monster, what a chaos, what a contradiction, what a prodigy! Judge of all things, feeble worm of the earth, depository of truth, a sink of uncertainty and error, the glory and the shame of the universe.* The question that follows is whether or not we have become a little more psychologically 'put together' since Pascal's day. (Or Shakespeare's day, for that matter. The Bard certainly made full use of all the human traits Pascal describes.) And I would say that we have progressed a great deal since the 17th century - are less in the dark as to how we are mentally driven. Over the last 150 years innovators in the relatively new sciences of psychiatry and neurology have enabled us to understand, and often control, the myriad of complex mental drives that result in the forming of each individual personality and character. In addition, the contemporary practice of publishing medical and scientific 'news' has enabled many of us to know our own particular inner mix of motivations, and personally impose some psychical order on the kind of *'chaos'* that Pascal describes. Some 300 years post-Pascal, the great 20th century English poet, W.H. Auden, is describing the kind of self-enlightenment - the reconciliation of consciousness with the Unconscious - which both Freud and Jung saw as the purpose of life's journey: Jung in particular saw the soul as the spiritual nucleus of a human life.... manifesting through the agency of the Unconscious. Here is the key verse from Auden's poem, *The Labyrinth*:

> *The centre that I cannot find*
> *Is known to my Unconscious Mind;*
> *I have no reason to despair*
> *Because I am already there.*

When one thinks of the squalor and deprivation endured over the centuries by so many of the 'unfortunates' - otherwise known as the 'common man' - then certainly modern cities (not ravaged by conflict of one sort or another) provide reasonable sanitation and individual living space. That surely represents progress. After all, it's barely 150 years ago since thousands of Londoners lived with little privacy in cramped and dank conditions. Infant mortality was high. It may actually have been relatively more comfortable living in mediaeval townships, in terms of 'personal' space that is, before the migration from the countryside to the city began as land closures reduced the agricultural demand for labor, and the factory system of industry took over the towns. And in turning to the advances made in medicine and health care, the developments here immeasurably improved human existence. Just think of what surgery without anaesthetics would have been like; of how the incredible technologies developed for 'looking' into the body and making it possible to replace its 'parts' has diminished suffering and increased longevity. When it comes to venturing into the cosmos at large, science has had cosmonauts walking on the moon. More personally, the rest of us fly quickly and relatively easily around the planet, thanks to the invention of the jet engine: six hours (depending on headwinds) to cross the north Atlantic from east to west by air. Whereas in the mid-1800s it could take two weeks or more to cross by steamship; two months or more (depending on weather) to venture under sail in the late 1700s. Early in the 19th century, James Watt's invention of the steam engine led to the development of the railways, revolutionizing transportation on land which, together with the later arrival of the internal-combustion engine and automobile, resulted in greater mobility for the general population. This new-found freedom to move from place to place - be it over short distances or long - broke down the isolation and insularity of life in severely local districts, and created regional and ultimately national environments to which one 'belonged'. The result was to promote positive feelings of nationhood, a sense of 'common cause' that politicians were quick to exploit. Patriotism became the *cause célèbre*.

It was also during the 19th century that the practical development of electrical theory led to the creation of a vast range of electrical instruments and machines - a flowering of electronic engineering without which life today would come to a grinding halt. Suppose, for example, that if in the 1960s someone had said to you that one day in the near future - were you to find yourself stuck in Beijing airport - you would be able to contact a travel agent in New York using a small pocket phone programmed to operate via a satellite rocketed into orbit around the earth…. you would most likely have regarded your informant as just a fanatical science fiction enthusiast.

However, all this scientific and technological progress has its more ominous side: the invention and proliferation of what are commonly called, 'weapons of mass destruction'. The first such weapon came into being in 1000 A.D. when the Chinese perfected their invention of gunpowder - a mixture of charcoal, sulfur, and potassium nitrate. Some nine hundred and forty five years later Americans dropped atomic bombs on the Japanese cities of Hiroshima and Nagasaki. Scientific progress. Unfortunately, the science of quantum physics which led us to atomic theory has put weapons in the hands of the most dangerous members of mankind - those who are driven by the need for power, political, religious, or both at once. We have seen in previous chapters how the need to dominate and control both people and events is the most injurious and destructive psychological compulsion in human nature. It is followed by two very strong runners-up, avarice and jealousy…. which often lie behind the compulsion to be controlling and all-powerful. As I write I am reminded of the English statesman Francis Bacon's words on this issue, found in his 17th century treatise entitled, *Of Great Place:*

> *It is a strange desire, to seek power, and to lose liberty; or to seek power over others, and to lose power over a man's self.*

Some psychiatrists see such 'delusions of grandeur' and obsessive self-gratification to be manifestations of a grossly inflated ego - *gross* ego as it was described in Chapter 16: the result of the normal balancing and self-preserving directives of *ego* losing out to the neurotic mental dysfunction of egomania. Nevertheless, such dysfunctional personalities, incapable of seeing and working for the general good, are spread throughout our history. The record of the last 5000 years reveals their presence and responsibility for many of the disasters that have crippled our progress as human beings.

The question that inevitably follows is whether such pathological egos were around some 30,000 years ago in the days of cave and ice? Local rivalries and skirmishes apart, were there any Caesars, Attilas, Napoleons, Hitlers.... making their plays for overall power - albeit on a smaller scale - during the thousands of years of Prehistory? There are no ways of knowing. Although anthropologists seem to agree that Neanderthal man was not territorially aggressive and warlike. Yet whatever degree of egoistical and power-driven characteristics were present in Upper Paleolithic societies, Julian Huxley's remarks suggest that these early Aurignacian ancestors of ours possessed the capacity to be moved by levels of moral, aesthetic, and mystical sensibilities which are not generally found in today's world. And yet there are still a few tribes and ethnic groups living peacefully in extremely isolated parts of the world who do not live a 'modern' way of life, and do not appear to display any of the 'genetic' aggressiveness that characterizes human nature in general. They have their traditional rituals that would seem to enable them to live in a mystical and elemental state of grace with the forces they see as responsible for their existence. A little stretch of the imagination.... and one can think of them as representing a persistence of Aurignacian sensibility - displaying a form of wisdom almost entirely lost to most of contemporary society.

Evolutionary biologists have pretty well determined the time frame during which our general physiological progression took place; while the neuroscientists have concentrated on the 'departmental' growth of the brain responsible for our advancing mental powers. But the neurological specialists cannot ascertain when a particular way of thinking, or a particular quality and range of feeling entered consciousness to endow us with what we think of as human qualities: when, and if, for example, an altruistic sensibility held sway in prehistoric societies. Or when consciousness evolved to create the sense of an individual *self*... unleashing the psychic energies we now lay at the door of ego. When, as is sometimes said, *'the days of innocence were lost.'*

The earliest thoughts on life and 'humanness' - of which we have any 'written' record - were presented by Sumerian thinkers a mere 4000 years ago; were expanded by later Mesopotamian cultures, before being built on by the early Greek philosophers who followed. And it is odd that, given the sum total of their reflections and deliberations, one would have the feeling that these ancient philosophies represent 'human' attitudes more applicable to Huxley's ice-people than to modern man.

If this were so, then what price our achievements in education and literacy.... in scientific and medical discovery, globe-trotting *par avion* and *par internet*; or in the relative luxuries afforded by a materialistic lifestyle? Have we not somewhat anaesthetized the vitalizing power of the human spirit through our appetite for facts - the 'how' of things without considering the 'why'; for a totally existential life in which new sensations and constant entertainment are taken to be the be-all and end-all of life?

An article appeared in *The New York Times* for June 30, 2009. entitled *Tibetan Nuns and Monks Turn Their Minds Towards Science*. It describes a program initiated by Emory University to introduce modern science into Tibetan monasteries in India. The monks and nuns find that science strengthens Buddhist philosophy and thought. The Dalai Lama writes about his confidence in *'critical investigation'* and goes on to say that *'if scientific analysis were conclusively to demonstrate certain claims in Buddhism to be false, then we must accept the findings of science and abandon those claims.'* The article quotes Arri Eisen, a biology professor at Emory, as stating that teaching the nuns and monks helped him consider how *'how to nurture positive thinking. Western education doesn't nurture empathy.'* While Bryce Johnson, an environmental engineer who coordinates the Science for Monks program, comments that *'Science may be far advanced in the West, but a moral vacuum exists... There's something lost in the West. The meeting of science and Buddhism is a healthy exchange that is as much for the scientists.'*

POSTSCRIPT

It will be obvious to my reader that any attempt to grasp the full import of Huxley's proposition must necessarily involve a good deal of guesswork and speculation. After all, we only have hieroglyphic and written records to cover the last 5000 years or so, and the archaeological evidence testifying to human life before then is pretty flimsy - especially with regard to the 'Aurignacians' who were living at a time when it is thought that people were only some five feet tall, and had a life expectancy of but twenty-five years. To reflect on their superiority as human beings might well seem to be an exercise in futility. Nevertheless, in so doing.... a central issue regarding what may be considered to constitute both the *ideal* and the *essential* nature of humanness, becomes evident. And that is the absence of violent and aggressive behavior - either in personal relationships, or in the pursuit of territorial and material gain. Archaeological and anthropological evidence suggests that Neanderthal man displayed such pacifist and altruistic qualities, and the Shanidar cave burials we have discussed date back to about 62,000 B.C. - some 30,000 years prior to whatever was the state of human life in the Upper Paleolithic period. Was human nature in the Aurignacian cave culture still relatively 'post- Neanderthal-tranquil' - which, if so, would lend support to Huxley's reflections? And if so, at what time in our evolving history since those early days did the mindless death-dealing brutality that characterizes human life throughout *recorded* history, become universally endemic to our species?

23

ON THE SIDE OF THE ANGELS: LOVE and COMPASSION

*For the poet is a light and winged
and holy thing, and there is no
invention in him until he has
been inspired and is out of his senses...*
Plato: Dialogues, *Ion*

Some two thousand and four hundred years ago Plato, the Greek philosopher who was the pupil and friend of Socrates - and later became Aristotle's mentor - describes 'the poet' in words that match the age-old concept of angels as ministering spirits or divine messengers: immortal and immaterial beings providing the link between the numinous realms of spirit and the time-bound materiality of life on earth.

The Greek word *angelos* (and its Latin derivative *angelus*) signifies *'messenger'*: not just any old messenger..... rather an ethereal herald bringing the edicts or judgements of the Gods: later to become the 'heavenly angel' performing a similar function in Judaic, Christian, and Islamic monotheistic faiths. *Hermes* - a minor God in his own right - *was* the heavenly messenger for the Greeks, to be re-named *Mercurius* by the Romans. When depicted in illustrations, the 'angel-status' of Hermes was symbolized by the wings on his helmet and sandals. Wings, the flight organs of birds, bats and insects have, throughout our history, been taken out of their natural context and used to signify the airborne journeyings of spirit-beings. In classic myth, Pegasus was the winged horse of the Muses who ultimately made a final ascent to heaven to become the constellation of the same name. While in the traditional paintings of Tibetan Buddhism, the flying horse symbolizes the ability of spirit to rise above the affairs of the world - as does the depiction of those advanced mystics who possess the powers of levitation and are shown in full flight, arms outstretched, keeping them airborne like an aircraft's wings. In ancient Egyptian low-relief and free-standing sculptures, the animal subjects become symbols of spirit-entities simply by the addition of wings. The angels of the Christian faith are depicted as possessing spectacular, furled wings which, if extended in flight, would show a wingspan of at least twelve to fourteen feet.

Throughout our history, the idea of a winged ascent above the earth to reach the 'spirit-zone' has been verbally expressed and artistically depicted. And although our very physicality obviously bears witness to the fact that as creatures of flesh and bone we live but a relatively brief life, the concept of 'the angel' as messenger and guide lends support to the belief that a timeless and immaterial realm of spirit prevails somewhere beyond ourselves. And one to which we have brief access on those occasions when powerfully affected by events of great moment - moved, if only briefly, to have no sense of bodily weight (no longer earthbound) and to lose all sense of linear time. An inevitable consequence of such momentary transformations of consciousness, is to cause one to wonder if the onset of such an experience represents a breakthrough to a transcendent level of awareness - one truly surpassing the day-to-day workings of the senses.

The long-standing records of our ability to experience such shifts of time and place, lends credence to the concept of a *'great chain of being'* - a theory holding that all natural life-forms occupy differing positions in the chain *below* the median line separating the world of nature from the realm of spirit. It goes on to postulate that humankind - as the most advanced form of natural life - occupies the *highest* position of creatures living below this line.... immediately below the *lowest* level of the spirit realm in the chain above the line. However, as we are considered to be hybrids with at least 'part of a foot' in the spirit zone, we are enabled from time to time to move up and mentally interact with the lower ranks of extraterrestrial angels and other spirit beings.

Professor Arthur O. Lovejoy delivered the 'William James Lectures' at Harvard University in 1933. He spoke of the historical background and philosophical implications of this very human and mystical way of thinking about 'being' on a universal scale - and some years later wrote his well known book based on these lectures, entitled: *The Great Chain of Being*. In the chapter, *Conflicts in Medieval Thought*, he quotes from the highly regarded late-medieval theological writer Nicolaus Cusanus, as follows:

> *All things, however different, are linked together. There is in the genera of things such a connection between the higher and the lower that they meet in a common point; such an order obtains among species that the highest species of one genus coincides with the lowest of the next higher genus, in order that the universe may be one, perfect, continuous.*

Lovejoy then takes over and writes:

> *The accepted 'philosophical,' as distinct from the dogmatic, argument for the existence of angels rested upon these assumptions of the necessary plenitude and continuity of the chain of beings; there are manifestly possibilities of existence above the grade represented by man, and there would consequently be links wanting in the chain if such beings did not actually exist. The reality of the heavenly hosts could thus be known a priori by the natural reason, even if a supernatural revelation did not assure us of it. This... continued for many centuries to be the chief reason offered in justification of the belief in 'spiritual creatures...'*

In revelatory and apocalyptic writings, angels are grouped in differing orders or circles. The first circle comprises Seraphims, Cherubims, and Thrones. In the second, are Dominions, Virtues, and Powers. While Principalities, Archangels, and Angels (including Michael and Gabriel), occupy the third and highest circle of the hierarchy. Milton provides a list of the 'fallen' angels in *Paradise Lost*, Bk. 1. (The Muslims say that angels were created from pure, bright gems.... whereas man was formed from clay.) There are guardian angels, recording angels, prophesying angels, avenging angels, miracle-working angels, fallen angels bent on destruction... and then there is the angel of death.

Here is a quotation from Robert Burton's *Anatomy of Melancholy*. Burton was a passionate scholar in every branch of learning who lived between 1577 and 1640; was the vicar of St. Thomas, Oxford and the keeper of his college Library at Oxford. Except for a few minor Latin pieces, he left only one major work, his *Anatomy*.... into which he poured a lifetime's hoard of classical and mystical learning:

> *Every man hath a good and a bad angel attending on him in particular, all his life long.*

I wonder how many of my generation were, as children, told to listen to the good angels, and watch out for the tempting machinations of the bad ones.

On several occasions, and particularly in Chapter 12, *Conscience and Morality...* . we have reflected on the nature of the positive (good) and negative (evil) forces influencing our psychological disposition. Such influences would be considered *psychical* in that they could be said to emanate from metaphysical sources - the supposed territory of soul, spirit, or various angel-types. Whereas our day-to-day existential and sensory mental life is obviously more *physical,* directed by the relatively mechanical, biological and cerebral processes that utilize the brain as such. Consequently, some psychiatrists will treat patients psychically by means of long-term sessions of psychoanalysis, in order to try and reach, and uncover, the spiritually-related problems that are negatively affecting their life. While others, believing that faulty brain-mechanics are more likely to be causing a patient's problems, will treat him or her with drugs that might restore normal functioning of the neurons.

Angels notwithstanding.

Three quotations:

Aleksandr Solzhenitsyn, on accepting the Nobel Prize for literature in 1970: *'Love is the only cure for the world.'* (ii) Erich Fromm, in the Foreword to a book, The Art of Loving: *'Love is the only satisfactory answer to the problem of human existence.'* (iii) Sigmund Freud in a letter to Carl Jung: *Psychoanalysis is in essence a cure through love.*

The ability to love unreservedly and selflessly, requires a sensibility that rises above the automatic self-preservational urges of instinct, and transcends the self-servings desires and necessities of ego. It brings levels of altruism and compassion into play that produce a heightened awareness of the true goal of a human life: an experience, I would suggest, that can best be attributed to a breakthrough of the soul into the regular functioning of consciousness. The triumph of spirit over matter. A feeling of release from the stranglehold of time, and a new-found freedom from the mental limitations of being a player on the world existential stage. In *Pages from the Goncourt Journal,* The Russian writer Ivan Turgenev describes this new self as follows: *I believe that love produces a certain flowering of the whole personality which nothing else can achieve.* His more renowned contemporary, Leo Tolstoy, enlarges on this statement in *Life and Essays on Religion*. (In quoting it here I have substituted the word 'physical' wherever Tolstoy uses 'animal'.)

> *All men know in their very earliest years that beside the good of their physical personality there is another, a better, a good in life, which is not only independent of the gratification of the appetites of the physical personality, but on the contrary, the greater the renunciation of the welfare of the physical personality, the greater the good becomes... This feeling... is known to all.*
> *This feeling is love.*

Turgenev, Tolstoy, Solzhenitsyn, Freud, Jung, Fromm, Rollo May.... all talking throughout these pages about the kind of loving that can be said to pertain more to the life of the soul than to that of the passional senses. Their words brings us back to the notion of angels as immaterial higher beings, with the reminder that the prime characteristic of the good angel - always implied if not specifically stated - is the exercising of the most sublime level of the force of love.

Physical love alone cannot reach this level. *Eros,* the Greek god of love, is primarily the patron of physical love - hence the significance of the adjective *erotic* in the English vocabulary. Yet physical love between 'soul-mates' (as two lovers are sometimes described) may bring about a state of transcendence that moves the heart to a point of being beyond time and place. And the Greek word *agape* was used by the early Christians to denote the more selfless love of the spirit as opposed to that of the body - a love born of ethereal truths rather than worldly ones.

The life of Saint Francis of Assisi was driven by an even more mysterious form of loving and compassion - an extraordinary compound of Merlin-like alchemical knowledge allowing him to miraculously control the recondite forces that govern animal behavior, together with a spiritual intensity that set the seal on his empathic power over man and beast. The legendary account of his encounter with the Wolf of Gubbio is a parable: an allegory symbolizing the manifestation of a supramundane sphere of the life-force.... where the *spirit* of love can be said to 'pass all understanding';

449

induce a state of sympathetic 'fellowship' between a human being and another living creature, inasmuch as they temporarily share a *'common wavelength of being...'* The wolf, who lived in the woods around Gubbio, was terrorizing the local citizenry, both when they ventured into the forest, and when it decided to visit the town. Saint Francis, when asked for his help went into the woods with a group of townspeople and, on meeting the wolf, addressed him. Here is the account of the occasion, written by an anonymous, 14th century hand. Rather than paraphrase, or select extracts from the original Latin translation - powerful in its lucid simplicity - I reproduce it here as initially rendered:

THE VERY HOLY MIRACLE PERFORMED BY FRANCIS WHEN HE CONVERTED THE VERY FIERCE WOLF OF GUBBIO

During the time when Francis was staying in the town of Gubbio, there appeared in the surrounding country, a very large, fierce and terrible wolf, who devoured not only animals but also people. Thus the inhabitants lived in fear, especially when the wolf came near the town. Whoever left the town carried arms as though going to battle. But in spite of this, a person who met the wolf on his own was unable to defend himself from it. Matters reached the point where, out of fear of the wolf, no one any longer dared go outside the confines of the town.

For this reason, Francis, taking pity on the people of this town, decided to go out and meet the wolf, despite the vigorous warnings of the town dwellers. Making the sign of the Holy Cross, he passed through the city gate with his companions, placing his entire trust in God. Although the others were hesitant to go further, Francis took the road that led to the wolf's territory. And then, under the eyes of the town dwellers who had come in great numbers to see this miracle, the wolf rushed at Francis with its jaws agape. Approaching the beast, Francis made the sign of the Cross. He called to it and spoke to it as follows: 'Come here, Brother Wolf! In the name of Christ, I command you to do no harm, either to myself or anyone else'

At this moment a wondrous thing occurred, As soon as Francis made the sign of the Cross, the fearsome wolf closed its mouth and stopped running. And at Francis' command, it came and laid himself, as gently as a lamb, at his feet. Then Francis spoke to him as follows: 'Brother Wolf, you have wrought a great deal of harm in this district and you have committed many misdeeds, injuring and killing the creatures of God without His permission; and not only have you had the audacity to injure and kill human beings made in the image of God. For this reason you deserve to be hung like the worst of robbers or the worst of murderers; everyone here is crying out and grumbling against you, and you are the enemy of the whole country.

But as for me, Brother Wolf, I would seal the peace between you and all these people so that you no longer attack them and they forgive you all your past offenses. Neither men nor dogs will chase you anymore.' When these words had been said, the wolf showed by movements of his body - his tail and his ears - and by lowering his head that he accepted what Francis had said and was willing to abide by it. Then Francis said: 'Brother Wolf, since you are willing to make the peace and respect it, I promise to have the inhabitants of the town provide you regularly with all the food you have need of, to the point where you will no longer suffer hunger. Because I know it has been out of hunger that you have done all this evil. But in consideration of my gaining you this favor, I want you, Brother Wolf, to promise me never to harm any human being or any animal. Do you promise?' By lowering his head, the wolf gave a clear sign of his promise. Francis said to it: 'Brother Wolf, I want you to swear an oath on this promise so that I can have confidence in it.' When Francis held out his hand to receive this oath, the wolf raised his paw and placed it in the hand of Saint Francis, in this manner making the sign of good faith of which he was capable.

Francis said to the people: 'Hear me, my brothers! Brother Wolf, who is here before you, has promised me and sworn an oath to make peace with you and never to trouble you in any way whatsoever if you promise to give him what he needs. As for me, I will stand as his bondsman to guarantee that he will faithfully respect this pact of peace.' Then all the people, with one voice, promised always to feed it... After this the wolf lived for two years in Gubbio. It went from door to door and entered the houses familiarly, without ever doing harm to anyone or having any harm done to it. It was courteously fed by the people, and as it walked through the town and entered the houses, never did any dog bark at it.

Finally, after two years, the wolf died of old age, to the deep sorrow of the town dwellers, because in seeing it move about the town as a peaceful creature, they had been better able to remember the virtue and holiness of Francis.

Like all allegorical stories, this account leaves reason considerably strained. Initially, its fairy-tale-like-whimsicality appeals to the child in us. Yet, on reflection, one starts to ponder whether the spiritual authority possessed by a great saint might *not* encompass an occult power capable of inducing trans-species communication. And, in so doing, bring about an empathic link that can unite man and animal in some form of mutual and sympathetic understanding. In Robinson Jeffers' poem, *The House Dog's Grave* - reproduced in its entirety in Chapter 18 - the intensity of the loving bond between the poet and his dog is no mere romantic idyll about 'a man and his dog'. But rather than have himself - as the man addressing the dog - Jeffers intensifies the poignancy by having Haigh, an English bulldog, talk to him from the grave just outside his front door. Here is the last verse: and if Solzhenitsyn had ever read this poem, I can see him nodding his head and saying to himself, *'This is the kind of love I am talking about...'*:

> *You were never masters, but friends. I was your friend.*
> *I loved you well, and was loved. Deep love endures*
> *To the end and far past the end. If this is my end.*
> *I am not lonely. I am not afraid. I am still yours.*

Also, I wonder if Solzhenitsyn was familiar with the following famous lines from Elizabeth Barrett Browning's *Sonnets from the Portuguese:* this particular sonnet is also to be found in Chapter 18:

> *How do I love thee? Let me count the ways.*
> *I love thee to the depths and breadth and height*
> *My soul can reach, when feeling out of sight*
> *For the ends of Being and ideal Grace.*

Such is the transcendent nature of the love one person can have for another.... a remarkable, if not rather wonderful, aspect of humanness that can transform one's life. However, we often find it difficult to believe that animals can respond to humans with anything like the same degree of feeling and devotion. Yet history provides many stories of animals surpassing their wild nature in responding to men and women with whom they have forged powerful emotional bonds - links that persist despite long periods of separation following their original interaction with a person. (In Chapter 18 I have given several examples of dogs who have devoted their lives to their 'masters'.) One version of the oldest account of such extraordinary affinity as told by the Roman chronicler Aulus Gellius, is that recounting the story of Androcles. He was a Roman Legionnaire who, having completed his service, was on his way home from North Africa when he came across a lion totally incapacitated by a large thorn between its pads. Androcles extracted the thorn and both he and the lion went their separate ways. Some years later, when it would seem that the legionary had converted to Christianity, he was sentenced to die in the arena under attack by wild animals. (The Romans annually imported thousands of dangerous animals from Africa to provide 'entertainment' for the local citizenry, as well as to execute 'criminals'.) When the lion rushed at Androcles the crowd were not prepared for what happened. Instead of attacking Androcles, the lion, on recognizing him, displayed obvious signs of pleasure, pawing him gently, and licking his face. Both were released, although what ultimately happened to the lion is not disclosed. (One likes to think of the lion and Androcles finding some remote country retreat to live out their years together!)

There is a tendency in our overly rational and empirically ordered world to regard such legendary tales as but fanciful and mythic folk-yarns. But the story of the two Australians who bought a lion cub from Harrods in London in 1969 has become an internet sensation of late. It does not exactly parallel the Androcles affair, yet bears testimony to its veracity.... insofar as the demonstration of the grown lion's deep feelings for its former human 'owners' is caught on film. This trio - the cub and its two human companions - became famous in Chelsea as they wandered around.... even playing football together in a park. Yet Christian - as the lion cub was named - was, at the age of 18 months, considered too large for public safety and was transported to the Kenyan wildlife sanctuary featured in the film Born Free. It was a year later that John Rendall and Anthony Bourke visited the sanctuary to observe Christian. George Adamson, the founder, told them to stay well away from the lion as he was now a creature of the wild and had his own pride of lionesses. He said the lion would 'tear them to pieces'. However, a video camera film shows Christian rushing towards them, putting his paws on Rendall's shoulders and licking his face in joy. Then both men pet, stroke and even wrestle with the lion, who nuzzles them gently as it did when a cub. It is a scene of poignant intimacy that has brought tears to the eyes of the millions who have seen the video.

Konrad Lorenz, the ethologist who was awarded the Nobel Prize for Medicine in 1973, writes in his book, *King Solomon's Ring*, of the 'love life' of many higher birds and mammals. Jackdaws, geese, ravens, swans, albatrosses, and some eagles.... apparently mate for life. On the death of one of a pair of jackdaws, the survivor appeared to be grief stricken: it became listless, had no appetite, kept his head and eyes lowered. After a lifetime spent studying the behavior of such advanced members of the animal kingdom, Lorenz concluded that bonds of affection are widespread among them. He is not alone in taking this view. Other scientists and rangers who spend their lives observing and caring for animals, also believe that a number of species have feelings very similar to our own: experiencing emotions to which Lovejoy in *The Great Chain of Being* suggests we are able to respond - allowing us to see them as fellow travellers in the great stream of life. Because - as hybrids occupying the dominant level of natural life - we are sensitized to do so by our immediate proximity to the lowest *supra*natural, spiritual region of Being: that of the lower order of angels. It is but one example of how we can be influenced to transcend the mechanical forces of physics and biology, and the psychological drive of ego. If we can become mentally transformed - if only intermittently - by the spirit-power of love, we can know a transcendent *joy...* An exalted state in which the questing intellect and ego-self shut down and complete psychical serenity brings one to be, as is said, 'in seventh heaven'. (*I* have always thought it an odd coincidence that the author of *The Great Chain...* should be one, 'Lovejoy', by name.)

"*Joy is an elation of spirit...*' wrote the Roman statesman and philosopher Seneca in *Epistulae ad Lucilium*. To which one might add - 'and elation of spirit is the gift that comes with love...' While some 1500 years after Seneca, the great French essayist Montaigne wrote in *Essays* Lxxv: *The most evident token and apparent sign of true wisdom is a constant and unrestrained rejoicing.*

I have just come across a few lines by the English novelist and dramatist John Galsworthy that poetically reinforce the accounts I have given here of the love that can exist between man and dog, and vice versa. Galsworthy died in 1933, one year after receiving the Nobel Prize for Literature. He writes:

> *If a man does not soon pass beyond the thought: 'By what shall this dog profit me?' into the large state of simple gladness to be with the dog, he shall never know the very essence of that companionship which depends, not on the points of a dog, but on some strange and subtle mingling of mute spirits. For it is by muteness that a dog becomes for one so utterly beyond value; with him one is at peace, where words play*

no torturing tricks. When he just sits loving and knows that he is being loved, those are the moments that I think are precious to a dog - when, with his adoring soul coming through his eyes, he feels that you are really thinking of him.

When Aleksandr Solzhenitsyn remarked that 'Love is the only cure for the world', he was surely not suggesting that this remedy for our ills requires that the vast majority of the world's population become saints. For most of the Saints who appear both in the written record and the oral tradition over the last 4000 years or so, have been inspired by some form of direct revelation of spiritual truths: divine insights which they would not renounce even in the face of death. They lived in a state of grace - seemingly disinterested in material needs and worldly wellbeing - vitalized in spirit by an all-embracing form of love, which brought them to live in a state of holiness. But there are also many unsung, caring people among us: men and women who are simply innately good, yet who are not 'holy' in the saintly god-loving way I have just described. Rather, they are driven by moral and compassionate sentiments (which may be regarded as God-centered attributes), and are moved to come to the rescue when faced by the plight of those less fortunate than themselves - needy neighbors and friends, the poor, the sick, the despairing, the dying, the newly bereaved.... Such caring and benevolent men and women are to be found everywhere, often 'putting their lives on the line' to serve the greater good: working either independently or for charitable organizations in regions devastated by war, famine, plague, or other natural disasters. And it is this sense of responsibility for the welfare of others, together with the compulsion to act and relieve suffering, that I believe Solzhenitsyn had in mind when he referred to the reconciling and universal power of 'love'.

Nowadays we live in a world described as a 'global village' where contemporary technology can bring every significant happening, however distant, to the television screen as part of the evening news. One great advantage of such almost 'instant communication' is that millions of us are brought to face the misfortunes and sufferings of others: an exposure without which many of our contemporaries would not become awakened to feel for the plight of fellow human beings; brought to recall the old saying, there *but for the grace of God, go I...*

However, even allowing for all the many ways in which we have materially and scientifically progressed over the last half-century.... is it possible to suggest that we are moving any closer to creating a world society in which differing ethnic, geographical, economic, or religious influences.... are being mediated in order to serve the cause of human brotherhood? This is a cause echoing throughout western history, proclaimed in the writings of Homer, Shakespeare, Emerson, Solzhenitsyn.... And in the words of Richard Aldington, the English poet, novelist and scholar who died in 1962: *A little common sense, goodwill, and a tiny dose of unselfishness, could make this goodly earth an earthly paradise.* (Quoted from the *Colonel's Daughter.*) Yet with 6 billion plus individual human beings on the planet (the number rising steadily), the chances of realizing Aldington's criteria for bringing men and women throughout the world to live their lives respecting and caring for each other, would seem to be - statistically at least - impossible. Throughout the known warlike history of our species the prospects for such a 'coming together' have never looked good; particularly over the last hundred years or so when two world wars, accompanied and followed by the most inhuman politically motivated ethnic atrocities, have resulted in the deaths of millions. It has been left to the spiritual outposts of Christian and Buddhist monastic Orders; the dedicated organisations such as the Quakers, Save the Children, Medecins Sans Frontiéres and Care...; and individuals such as Albert Schweitzer and Mother Teresa.... to keep alive the human virtues of charity and compassion.

Yet we have to remember that in talking about a total world population of 6 billion, we are referring to 6 billion *individuals* - men and women each having a distinctive character and outlook: the result of national and ethnic differences, but also the consequence of each of us being driven by the makeup of our own genes - all factors competing with strength of spirit and the emergence of character in 6 billion plus cases. This is not, however, to say that ultimately some strong, mutually supportive and caring bonds between more and more peoples, driven by an expanding spirit of goodwill, might not further develop to complete the saga of our evolution. In which event we could see the abolition of major international wars in general, and wars of religion in particular; the disappearance of power politics, out-of-date nationalism, international terrorism.... and, at the more personal level, a diminishing of the destructive emotions of jealousy and hate.

However, I see only three ways in which the universal power of love could become the most influential moral and spiritual authority at work in the world. First, by divine intervention: the kind of cosmic phenomenon suggested by Loren Eiseley which propelled us, in the main, to overcome the basic impulsions of instinct, becoming reflective and caring, even at the expense of ego and survival - a development he put down to '... *something happening in the brain, some blinding irradiating thing.* (The complete statement can be found in Chapter 19.) Second, by the *compelling* example that would be set by a few unequivocally moral nations, and the wise counsel offered by their statesmen who would untie Gordian knots in the disputes between warring political factions - aided and abetted by a majority of their citizens who acquiesce in the cause of world peace. And third, by the continuing slow creep of the 'natural' evolutionary process that has been shaping the *humane* part of our being over the last 500 million years.

I feel sure Solzhenitsyn would agree that for the world to become 'healed' through the power of love, it is not necessary to attain a hundred-percent conversion of the world's population at large. Once a working *majority* of benignly disposed humans on the planet can be attained, the 'cure' Solzhenitsyn had in mind could begin to take effect.

There are some neuroscientists and philosophers who wonder if events are not moving us - very, very, gradually - towards achieving such a goal. They note that as early as the 5th century B.C. in the East, Buddhism taught that the reduction, if not the elimination, of the desires of a Self tied solely to a material existence - seeking only aggrandizement and the gratification of worldly desires - would hasten the onset of a peaceful and benevolent world. After the Second World War there was some hope that such a change in the moral climate might be indicated by the creation of world organizations seeking to improve the quality of life where it is most needed. There was the establishment of the United Nations, the World Health Organization, and the World Monetary fund. In addition, Human Rights and the plight of refugees have also become a concern addressed by some countries. And whenever natural disasters strike, both foreign nations and non-governmental philanthropic groups - despite being thousands of miles from the scene - are quick to assist in providing shelter, food, and medical help. (The remarkably humane and caring French medical organization, *Medicine Without Borders,* is made up of doctors and nurses who volunteer their services wherever sickness and disease strikes - often at great risk to themselves.)

It would be difficult to find a writer more 'qualified' by experience - and more morally inspired - than Aleksandr Solzhenitsyn when it comes to extolling the surpassing influence of love and compassion in our lives. The Nobel Prize for Literature awarded in 1970 cited the 'ethical force' his works imparted to contemporary Soviet literature, thus acknowledging the moral strength and conviction that inspired his writing. In February 1945 he had been arrested for referring to Stalin as 'the boss', using the criminal's slang word *pakhan* to denote 'head man'. After eight years in prison camps he was exiled from European Russia and not allowed to return until 1957. In the early 1960s - during the more liberal political atmosphere that prevailed - he was able to publish several major novels, all reflecting the trials and injustices he and many of his contemporaries faced with great courage and strength of spirit. Even in those days much of his work only appeared in the West and charges that he was slandering the Soviet State were being muttered around by the body politic. In 1969, his open letter to the Fourth Congress of Russian Writers denouncing censorship resulted in his expulsion from the Writers' Union. In 1973 the KGB confiscated a manuscript of *The Gulag Archipelago,* and when Solzhenitsyn approved its publication in the West he was arrested. In February 1974 he managed to hastily depart for West Germany, moving on in 1976 to the United States where he lived and continued to write in Vermont.

Few men of his time survived the unspeakable horrors perpetrated by Stalin and his henchmen, and have then gone on to expose them. *The Gulag Archipelago* which unflinchingly describes all the terrible details and suffering of Gulag life, can surely leave no reader unmoved by the lack of any concern or compassion displayed by the faceless communist jailors. Literature, as Solzhenitsyn explained in his Nobel lecture, exists to uncover lies and serve the moral nerve of society in the face of the callous and sadistic terror unleashed by 'bureaucrats and dictators'. So he speaks from both the heart and bitter experience when he declares that only goodwill and love can serve as the saving grace for humanity in terms of what the future might hold.

Looking back over the years, the most generous, self-effacing and naturally *'good'* man I can bring to mind is the Czech soldier and greatest long-distance runner of his time, Emil Zatopek. (I introduce him here for if we ever were to have a majority of 'Zatopek-minded' world citizens, then Solzhenitsyn's hopes for a world 'cure' wise words would be realized.) I remember Zatopek from the late 1950s and early 60s, and hope that recounting a few telling incidents from his life.... will serve to reveal his extraordinary guilelessness and charitableness - both constituents of the wise and caring human being Solzhenitsyn had in mind. I have been greatly helped in this by a wonderfully informative new book entitled *Born To Run*, in which a well-known trainer of professional runners - known as Coach Vigil - exclaimed, *'Such a sense of joy!'* after watching the famed and fast long-distance Tarahumara runners from Mexico's Copper Canyon. It was their smiles apparently that astounded Vigil, *'as if...'* writes the book's author, Christopher McDougall, *'running to the death made them feel more alive'*. The sheer joy of the Tarahumanas - a true celebration of running as the means of evoking a seemingly spiritual transformation, led Vigil to believe that his belief in *'toughness'*, both of physique and character, as the principal factors necessary for great running was less important than the psychical powers of *'Compassion. Kindness. Love'*.

Enter Zatopek.

McDougall writes: *'Emil Zatopek loved running so much that even when he was still a grunt in army boot camp, he used to grab a flash-light and go off on twenty-mile runs through the woods at night... In his combat boots. In winter. After a full day of infantry drills.'* In his first marathon of 26.2 miles, instead of training by running long distances at moderately-testing speeds, Zapotek prepared by doing hundred-yard sprints. Quoting from *Born To Run*:

> *'I already know how to go slow,' he reasoned. I thought the point was to go fast.' His atrocious, death-spasming style was punch-line heaven for track scribes ('The most frightful horror spectacle since Frankenstein'...) 'He runs as if his next step would be his last... He looks like a man wrestling with an octopus on a conveyor belt.'*

He was a runner cast in the Tarahumas mould - running long distances for the joy it gave him. Love of life, love of humanity, can be expressed in many different ways. For Emil Zatopek, running seemed to be the surpassing Way to confirm an elemental and spiritual *joie de vivre*.

Christopher McDougall continues the story - and in words too illuminating to paraphrase - describing the Czech as ....

> '... a bald, self-coached thirty-year old apartment-dweller from a decrepit Eastern European backwater when he arrived for the 1952 Olympics in Helsinki. Since the Czech team was so thin, Zatopek had his choice of distance events, so he chose them all. He lined up for the 5,000 meters and won with a new Olympic record. He then lined up for the 10,000 meters, and won his second gold with another new record. He'd never run a marathon before, but what the hell; with two gold medals already around his neck, he had nothing to lose, so why not finish the job and give it a bash?
>
> Zatopek's inexperience quickly became obvious. It was a hot day, so England's Jim Peters, then the world record holder, decided to use the heat to make Zatopek suffer. By the ten-mile mark, Peters was already ten minutes under his own world-record pace and pulling away from the field. Zatopek wasn't sure if anyone could really sustain such a blistering pace. 'Excuse me,' he said, pulling alongside Peters. 'This is my first marathon. Are we going too fast?'
>
> 'No,' Peters replied. 'Too slow.' If Zatopek was dumb enough to ask, he was dumb enough to deserve any answer he got.
>
> Zatopek was surprised. 'You say too slow,' he asked again. 'Are you sure the pace is too slow?'
>
> 'Yes,' Peters said. Then he got a surprise of his own.
>
> 'Okay. Thanks.' Zatopek took Peters at his word, and took off.
>
> When he burst out of the tunnel and ran into the stadium, he was met with a roar; not only from the fans, but from athletes of every nation who thronged the track to cheer him in. Zatopek snapped the tape with his third Olympic record, but when his teammates charged over to congratulate him, they were too late: the Jamaican sprinters had already hoisted him on their shoulders and were parading him around the infield.'

The author of *Born to Run* goes on to characterize Zatopek as *'one of the most beloved athletes in the world...'* and recounts incidents that tell of the Czech runner's ability to put his feeling for others above his own self-satisfaction at constantly winning. McDougall writes that....*'During a manic stretch in the late '40s, Zatopek raced nearly every other week for three years and never lost, going 69-0'*. He goes on to recount that on one occasion when Zatopek became friendly with an Australian athlete who was running in the 5,000 meters with the hope of breaking the Australian record, Zatopek told him to quit the race for which he was entered, and join him in the 10,000 meter contest... and to keep up with him for the first 5,000 meters. In this way he brought his friend at high speed through the shorter distance to win the Australian record, before speeding away to win the 10,000 meter record for himself. This was not a man given to feelings of superiority, despite his great achievements. No temperamental displays of ego are to be found in the life of this great Czech athlete. On yet another occasion, the sheer *goodness* of his charitable nature was revealed during the Mexico City Games in 1968. A famed - yet it would seem fated - Australian runner, Ron Clarke (who never won a major 'world-class' event) was afflicted by altitude sickness when preparing to run the 10,000 meter finals. It was probably his last chance to fulfill his ambition. He broke his journey home to visit Zatopek in Prague and, as McDougall tells it: *'Towards the end of the visit, Clarke glimpsed Zatopek sneaking something into his suitcase... Zatopek sent him off with a strong embrace. 'Because you deserved it,' he said.*

Only later did Clarke find Zatopek's 1952 Olympic Gold Medal for the 10,000 meters in his luggage.

When the Soviet army invaded Prague in 1968 to crush the democratic-leaning government, they offered Zatopek the post of ambassador of sports if he would join them: otherwise he could be sent to a uranium mine to spend the rest of his life cleaning out toilets. Zatopek made his choice. *'And just like that...'* writes Christopher McDougall, *'one of the most beloved athletes in the world disappeared.*[*]

Why fate should treat a human being of such moral integrity, driven by love and compassion, in such a way.... is one of the great teleological mysteries that has defied philosophical explanation throughout our history. It is the problem that occupied the best minds of Classical Greece, was taken up by Eastern religions and by Christianity. Without any really convincing explanation. All we can do is ponder the transcendent example set by the lives of the Zatopeks of this world. And, in so doing, come to believe - with Solzhenitsyn - that they reveal how humankind can live joyously, at peace with themselves and the rest of the world.

Such has been the purpose of this book: to discuss how the complex human psyche is able, from time to time, to bring soul and its dynamic

---

[*] Christopher McDougall, *Born To Run* (New York: Alfred A. Knopf, 2009) p. 96.

of spirit to deny - as André Malraux put it, the 'nothingness' of our time-determined existential selves.

While writing these last few words it occurred to me that I had no need of all these pages of exposition. Two of the many quotations I have used throughout would have sufficed:

> *There are more things in heaven and earth, Horatio,*
> *Than are dreamt of in your philosophy.*
> Shakespeare: *Hamlet* I, v.

> *Any man's death diminishes me, because I am involved*
> *in Mankinde. And therefore never send to know for whom*
> *the bell tolls; It tolls for thee.*
> John Donne: *Devotions upon Emergent Occasions XVII.*

24

LIFE WITHOUT MYSTERY

The greatest mystery is not that we have been flung at random between the profusion of matter and the stars, but that within this prison we can draw from within ourselves images powerful enough to deny our own nothingness.
André Malraux: from The Walnut Trees of Altenburg

And so, in the final chapter, we conclude with a quotation taken from this book's beginning pages where Malraux's paean to mystery - to the undisclosed secrets of the cosmos and the psychical enigma of ourselves - was first introduced.

The very fact that we can conceive of things - be they cosmic or human – as *'mysterious'*, reveals the limitations of consciousness: our inability - even in this scientifically advanced aged - to explain just 'how' and 'why' a macrocosmic universe, and the microcosm of a single human being, should come to exist. It is a limitation ensuring that we ultimately developed the mental drive of *curiosity* - the urge to *know* that drives our questing and creative life. Science, philosophy, religion, and all the arts - not to mention our introspective search for 'Self' - owe their existence to the sense of mystery surrounding much of natural phenomena.... including ourselves. And with the pursuit of these mysteries comes a sense of wonder: an intensity of feeling, and range of thought leading to our creativity - the expansion of consciousness to which Malraux refers.

The following verse from William Wordsworth's *Tintern Abbey* is as powerful an evocation of the profound depths of feeling and the transcendence of thought in the face of mystery... that I know:

> *And I have felt*
> *A presence that disturbs me with the joy*
> *Of elevated thoughts; a sense sublime*
> *Of something far more deeply interfused,*
> *Whose dwelling is the light of setting suns,*
> *And the round ocean and the living air,*
> *And the blue sky, and in the mind of man;*

> *A motion and a spirit, that impels*
> *All thinking things, all objects of all thought,*
> *And rolls through all things.*

We are not all as finely tuned as Wordsworth in living the kind of 'double life' that takes us from sensory perceptions of the world to an inner reverie serving the more metaphysical aspects of our being. (In Chapter 22 I tell of an unknown Egyptian scribe writing on a clay tablet some 3000 years ago, *I wish to express from my body, feelings, the words for which are not in existence.* There he was, struggling with the paradox of knowing himself to be physically in the world, yet psychically not entirely of it.) This is the enigma which Alfred North Whitehead, English philosopher and mathematician, was addressing in his *Dialogues* when he wrote, *There are no whole truths, only half truths.* A contemporary neuroscientist might elaborate on that statement by saying that the right-hand half of the brain delivers the more visionary, idealistic and subjective 'truths' resulting *from* an act of perception - whereas the left-hand side presents the material factual 'truths' as presented *by* the act of perception itself. Wordsworth would likely respond by saying that it is then up to us, individually, to reconcile the two - see them as parts of one *whole* truth.

In the opening pages I quoted the following passage from Albert Camus' *Myth of Sisyphus*. Again, his words speak eloquently to the 'halves' of consciousness. They also indicate that while his perceptions of the world were rendered more meaningful by the objective, explicatory role of science.... the surges of feeling and contemplative musings aroused by nature's forms and moods brought him, like Wordsworth, to experience internal states best described as feelings of spiritual equanimity:

> *And here are trees and I know their gnarled surface, water and I feel its taste. These scents of grass and stars at night, certain evenings when the heart relaxes -how shall I negate this world whose power and strength I feel... The soft lines of these hills and the hand of evening on this troubled heart teach me... that if through science I can seize phenomena and enumerate them, I cannot, for all that, apprehend the world.*

To return to the healing achievements of Carl Gustav Jung for a moment, it was his belief in the need to bring patients to realize that as human beings they live a double life - physically in the world and psychologically in terms of the inner drama of which both Wordsworth and Camus speak. *'A motion and a spirit...'* writes Wordsworth - words which Jung would regard as the poet's way of referring to the active life of soul and spirit. The whole thrust of Jung's analytical therapy was to enable patients to build a bridge over the divide between existence in the world and living in spirit; come to recognize the 'reality' of both - make of two 'half truths' a 'whole' truth... and so become a whole and individuated Self.

And so we have to live a riddle - an uncertainty wondering where the ultimate reality' of one's existence lies.... as consciousness automatically takes one in and out of the time-bound material world. Yet.... can you imagine living a life knowing only a sensory involvement with outside things and events; having no reflective inner life such as that on *'...certain evenings when the heart relaxes...'*: moments of solitude when those contemplative felt-thought's whisper as to *'the meaning of it all...'*

It is extremely difficult when being pulled in two such opposite directions to experience any convincing and persisting insight as to the purpose and meaning of human existence in general, and one's own in particular. Jung's psychoanalytical methods proved their worth for those most alienated from their inner selves. Yet we all, to greater or lesser degree, suffer from this psychical uncertainty - the kind of dilemma-cum-mystery that the Welsh poet Dylan Thomas had in mind when he coined the phrase, *The Wire Dangled Human Race*.

But once again, here is my 'old friend' William Wordsworth to put the enigma in its poetic place:

> *Those obstinate questionings*
> *Of sense, and outward things,*
> *Fallings from us, vanishings;*
> *Blank misgivings of a Creature*
> *Moving about in worlds not realized.*
> *Intimations of Immortality*, 1. p, 145.

And just the other day I came across the following lines penned by the German philosopher Nietzsche:

> *Everywhere the wasteland grows; woe*
> *To him whose wasteland is within.*

Gertrude Stein, the American writer who for many years lived in Paris presiding over a notable literary and artistic circle - and whose advice was solicited by Hemingway, Fitzgerald and other authors - was well known for her aphoristic, witty, and philosophical declarations. On one occasion she posed the query, *'What would the question be if we knew the answer?'* It is an astute and provocative 'one-liner', implying that both the need and capacity to question are 'built-in' to the human condition; and that even if we *did* ultimately have all the answers available at the time we would - because we are human - still have to find some question to ask if we were not to become zombies. Stein is suggesting that one should try to imagine our life if the physicists had discovered all there was to know about the origin of time, matter and energy; if philosophers, shamans and priest-mystics had managed to demonstrate that the soul was indeed a spiritual entity, and our link to a metaphysical dimension of existence. And if neuroscientists finally came to understand exactly how the trillions of neural links in the brain work.... enabling the psychologists to move in and manipulate every motivation and behavioral drive in the human psyche.... *'Aye, and what then...?'* any philosopher-poet would have to ask.

Were such prospects ever realized, and there were no questions left - no ambiguities, paradoxes, enigmas waiting to be resolved - one might well wonder what would happen to the right-brain neurons responsible for the *imaginative* and *creative* workings of the mind-brain link. Would they atrophy to a degree that would leave consciousness to be basically a left-brain affair, simply registering information supplied by the senses? If so, it is doubtful we could then say with Shakespeare, *'We are such stuff/ As dreams are made on...'* Such a questionless consciousness would result in the loss if that inner life which largely defines our individuality. 'Slim pickings' at this juncture for 'thinkers' of whatever stripe. Or perhaps, without the need to ask 'how' or 'why', we would enjoy a state of bliss... a form of nirvana?

Gertrude Stein's question acts as a reminder that it is given to us - more than to any other creatures on the planet - to be driven by curiosity: about the mysteries presented by the material world as one phenomenon, and those that shroud our psychologic selves as another. In Chapter 22 I quoted a statement made by the philosopher Max Scheler (who Heidegger described as 'the greatest philosopher from modern Germany), which I repeat here to sum up this final chapter. For Scheler believed that it is our purpose in life to '....*press forward into the whole of the external world and the soul, to see and communicate those objective realities within them which rule and convention have hitherto concealed.*'

We are 'built' to question. Such is the nature of the vital force driving our life. When, and if, all is ever known, one mystery.... one question will remain: what kind of journey will it be through the Valley of Death?

BIBLIOGRAPHY

Aurelius, Marcus (new Introduction Gregory Hays, 2002) *Meditations*, New York: Random House.

Barrett, William (1962) *Irrational Man*, New York: Doubleday Anchor Books.

Becker, Robert O & Gary Selden (1985) *The Body Electric*, New York: William Morrow.

Bettelheim, Bruno (1983) *Freud And Man's Soul*, New York: Alfred A.Knopf.

Bewley, Maurice (1966) *The Selected Poetry Of John Donne*, New York: The New American Library.

Bronowski, J. (1978) *Magic, Science, and Civilization*, New York: Columbia University Press.

Calaprice, Alice (2005) *The New Quotable Einstein*, Princeton: Princeton University Press

Camus, Albert (1955) *The Myth of Sisyphus*, New York: Alfred A. Knopf, Inc.

Chauvet, Jean Marie (1996) *Dawn Of Art, The Chauvet Cave*, London: Thames and Hudson Ltd.

Collier, Graham (1972) *Art and the Creative Consciousness*, Englewood Cliffs: Prentice-Hall Inc.

Collier, Graham (1999) *Antarctic Odyssey*, London: Robinson Publishing: New York: Carrol & Graf.

Collins, Francis, S. (2006) *The Language Of God*, New York: Free Press.

Delacroix, Eugéne - written between 1822 and 1863 - *The Journal*: (1961) translated by Walter Pach, NewYork: Grove Press Inc. (First Evergreen Edition).

Dole, George F. & Robert Kirven (1997) *A Scientist Explores Spirit*, New York: Swedenborg Foundation.

Dubos, René (1968) *So Human an Animal*, New York: Charles Scribner's Sons.

Ehrlich, Gretel (1994) *A match to the heart*, New York: Penguin Books USA Inc.

Eiseley, Loren (1960) *The Firmament Of Time*, New York: Atheneum.

Eiseley, Loren (1973) *The Man Who Saw Through Time*, New York: Charles Scribner's Sons.

Eiseley, Loren (1978) *The Star Thrower* (Introduction by W.H.Auden), New York: Harcourt Brace & Co.
Evans, Richard L. (1966) *Dialogue with Eric Fromm,* New York: Harper & Row.
Everitt, Anthony (2001) *Cicero,* New York: Random House.
Fordham, Frieda (1963) *An Introduction To Jung's Psychology,* Baltimore: Penguin Books.
Frankl, Viktor (1959) *Man's Search For Meaning,* New York: Pocket Books, Simon & Schuster Inc.
Fromm, Eric (1956) *The Art of Loving,* New York: Bantam Books, Harper & Row Inc.
Fromm, Eric (1959) *Psychoanalysis & Religion,* New Haven, Yale University Press.
Furlong, Monica (1980) *Merton: A Biography,* New York: Harper & Row Inc.
Giedion S. (1962) *The Eternal Present: The Beginning of Art,* Princeton: Princeton University Press.
Goethe (1961 Trans,) *Faust,* New York: Doubleday & Co, Anchor Books Edition 1963.
Goldbrunner, Josef (1964) *Individuation: A study of the Depth Psychology of Carl Gustav Jung,* Indiana: University of Notre Dame Press.
Goodall, Jane & Phillip Berman (1999) *Reason For Hope,* New York: Warner Books.
Highet, Gilbert (1954) *Man's Unconquerable Mind,* New York: Columbia University Press.
Isaacson, Rupert (2009) *The Horse Boy,* New York: Hachette / Little Brown.
Jeffers, Robinson (1965) *Selected Poems,* New York: Vintage Books, Random House.
Jens, Inge (1987) *At the heart of the White Rose: Letters and diaries of Hans and Sophie Scholl,* New York: Harper & Row.
Jung C.G. (1961) *Psychological Reflections,* New York: Bollingen Library, Harper & Row.
Jung C.G. (1961) *Memories, Dreams, Reflections,* New York: Pantheon Books, Random House.
Jung C.G. (1969) *Four Archetypes,* Princeton: Princeton University Press.
Kaku, Michio (1997) *Visions,* New York: Anchor Books, Random House.
Kowalski, Gary (1991) *The Souls of Animals,* Walpole N.H., Stillpoint Publishing.

Lovejoy, Arthur O. (1942) *The Great Chain Of Being,* Cambridge: Harvard University Press.

McDougall, Christopher (2010) *Born to Run,* New York: Alfred A. Knopf.

Maritain, Jacques (1962) *A Preface to Metaphysics,* New York: The New American Library.

May, Rollo (1967) *Psychology and the Human Dilemma,* New York: W.W. Norton & Company.

Moussaieff, Masson & Susan McCarthy (1995) *When Elephants Weep,* New York: Delacorte Press.

Newberg, Andrew & Eugene D' Aquili (2001) *Why God Won't Go Away,* New York: Ballantine Books.

Ornstein, Robert, & Richard Thompson, *The Amazing Brain,* Boston: Houghton Mifflin Company.

Read, Herbert (1960) *The Form Of Things Unknown,* New York: Horizon Press.

Scholl, Inge (1970) *The White Rose,* Hanover N.H.: The University Press of New England.

Tomkins, Peter & Christopher Bird (1973) *The Secret Life of Plants,* New York: Harper & Row.

Verschuur, Gerrit (1993) *Hidden Attraction: The History and Mystery of Magnetism,* New York, O.U.P.

ABOUT THE AUTHOR

GRAHAM COLLIER served with Bomber Command, R.A.F., during World War II. He was Professor of the Philosophy of the Arts at the University of Georgia and is now Professor Emeritus there; he is also an Associate Fellow of Davenport College, Yale University.

Collier's previous books include *Form, Space and Vision* (in print through four editions from 1963 to 1995), *Art and the Creative Consciousness,* and *War Night Berlin* (1993), described by the *London Weekend Telegraph* as "a rare and rewarding book indeed . . . ".

His most recent book, *Antarctic Odyssey*—an account of several voyages to circumnavigate Antarctica—received the following comment from *Publishers Weekly:* "Collier's crystalline account of his several recent trips to the bottom of the world . . . is a wondrous, serendipitous adventure . . . an eloquently expressed romantic view of the continent and the human encounter with it."

INDEX

A

Adams, Dr. John, xi
Adamson, George, 453
Ad animan (Hadrian), 330
Ad Atticum (Cicero), 241
Adler, Alfred, 26
Ad Lucilium (Seneca), 221
The Advancement of Learning (F. Bacon), 279–280
Aeschylus, ix, 300, 396–397
Affection, 42
Age of Chivalry, 375
Age of Enlightenment, 147–148, 168, 369
Age of Faith, 160–162, 168
Age of Reason, 53–54, 330–331
Age of Science, 381
Aion (Jung), 305
Aircraft, 113–126
Air Ministry, 8
Aldington, Richard, *Colonel's Daughter*, 456
Aldred, Cyrilin, *The Egyptians*, 133
Aldrich, Thomas Bailey, *Identity*, 408
Alexander the Great, 274, 285
Alice's Adventures in Wonderland (Carroll), xxvi, 95–98, 102, 285
Allah, 174–175
The Amazing Brain (Ornstein and Thompson), ix, xxviii, 22, 108–109, 152
Amiel, Henry Frederic, *Journal*, 265
Anatomy of Melancholy (Robert Burton), 447–448
Andrewes, Lancelot, 265

Androcles, 452–453
Angels, 445–462
Anima Poetae (Coleridge), 374
Aniseed balls, 341–344
Antarctic Odyssey (Collier), xxix, xxxiii
Apartheid, xxxi
The Apprenticeship of a Mathematician (Weil), 35
Aquinas, Thomas, 380
Archeology, 134
Architecture
    Gothic, 293–295
    Greek, 275–276
    monastic, 294–295
    Roman, 277
Areopagitica (Milton), 387
Aristotle, ix, 10–12, 24, 106, 127, 130, 149, 174–175, 187, 207, 274, 297, 329, 386, 418, 420
    Nicomachean Ethics, 330
Armistice Day (1948), 372
Art, 27
Art and the Creative Consciousness (Collier), 26
Arthur, King, 158, 285
The Art of Loving (Fromm), 399, 448
Astronomy, 400
As You Like It (Shakespeare), 73
Atkinson, Johnny, 247
At the Heart of the White Rose, 38–39
Auden, W. H., xx–xxi
    The Labyrinth, 138, 440
Auguries of Inocence (Blake), 17
Augustine

The City of God, 103–105
The Confessions of St. Augustine, 103–105
Aulus Gellius, 452–453
Auschwitz, xxv, 412, 413
Autobiographical Notes (Einstein), 418

B

Bach, Johann Sebastian, xxx, 406
Bacon, Roger, *Opus Majus*, 329
Bacon, Sir Francis, 218, 265, 279
    The Advancement of Learning, 279–280
    Of Atheism, 9
    Of Great Place, 442
    New Atlantis, 232
Baraka, 384–387
Barbirolli, Sir John, xxxi, 309–310, 324, 409
Barker, Jack, 101
Barrett, William, 434–435, 437
    Irrational Man, 171–172, 431–432
Barrow, John, *Theories of Everything*, 5
Barrow Neurological Institute, 414–415
Barth, Karl, 33
Baudelaire, Charles, 304
B.B.C., xxxi
Beauchamp, Squadron Leader, 197
Beaumont, Squadron Leader, 355
The Beautiful Years (Thompson), xxi
Becker, Dr. Robert O., 411
    The Body Electric, x, 110–111, 334–335, 411–422, 421
Beethoven, Ludwig van, 61–62, 325, 332
Behavior, 42
Benedictine Order, 162–166, 368

Bergson, Henri, 314, 407
Bernstein, Leonard, 309
Bettelheim, Bruno, *Freud and Mans' Soul*, 333–334
Beyond Good and Evil (Nietzsche), 180–181
Bible, 147, 230–231
Bierce, Ambrose, *The Devil's Dictionary*, 5
Bishop Blougram's Apology (R. Browning), 284
Blackmore, Richard, xxviii
Blake, William, 159, 314–315
    Auguries of Innocence, 17
    The Marriage of Heaven and Hell, 313–314
Blessedness, 378
The Body Electric (Becker and Seldon), x, 110–111, 334–335, 411–422
Bohrs, Niels, 9n
Bold, Samuel, 322
Booth, David, 101
Born to Run (McDougall), x, 459
Bourdillon, Francis William, 347
    Light, 144
Boyle, Robert, 18
Bradley, General Omar, 372
Brain, xiv, xxvii–xxx, 147–167, 433. *See also* Consciousness; Neurology
    cerebellum, 195
    cerebrum, 150–159
    consciousness without a, 411–422
    cortex, 150–159
    description of, 22–23
    drawings of, 154
    function, 19–41
    limbic system, 195

mapping, 140–141
memory, 150, 216
mind and, xix–xxi, 316–325
physiological complexity, 193
physiological differences, 21–22
weight of, 191
Brave New World (A. Huxley), 221–224, 228, 232
Breuil, Abbé Henri, 425
Brill, A. A., *Psychoanalysis: Its Theories and Practical Application*, 308–309
Broakes-Carter, Richard, xi
Bronowski, Jacob, 23–24, 26–27
Brooke, Rupert, 209
Broom, Dr. Robert, 184–185
Browne, Sir Thomas, 71, 73, 203
  Religio Medici, 201–202
Browning, Elizabeth Barrett,
  Sonnets from the Portuguese, 43, 347–348, 452
Browning, Robert
  Bishop Blougram's Apology, 284
  Paracelsus (Part i), 264–265
Bucky, Peter, *The Private Albert Einstein*, 400–401
Buddha, 260
Buddhism, 457
Bulletin Academie Medical, Vol. XCI, 191
Burgess, Anthony, 306
Burke, Edmund, *On a Regicide Peace*, 223–224
Burns, Robert, 258
  A Man's a Man for A' That, 257–258
Burnt Norton (T. Eliot), 107
Burton, Robert, *Anatomy of Melancholy*, 447–448
Burton, Sir Richard, 285
  The Kasidah, 203–204
Butler, Samuel, *Erewhon*, 232
Byron, Lord George Gordon
  Child Harolde, 349
  Don Juan, 264

C

Cabanis, George, 98–99
Caesar, *Gallic Wars*, 178
Caird, James, 67–68
Calaprice, Alice, *The New Quotable Einstein*, x
Campion, Thomas, 209–210
Camus, Albert, *The Myth of Sisyphus*, xxxiii–xxxiv, 464
Care, 456
Carlyle, Thomas, 70
  Sir Walter Scott, 288
Carroll, Lewis (Charles Lutwidge Dodgson), *Alice's Adventures in Wonderland*, xxvi, 95–98, 102, 285
Carruth, W. H., *Each in His Own Tongue*, 207–208
Carse, Duncan, 68
Cathedral of Beauvais, xxxii
Cathedral of Bourges, 164–165
Catholicism, 318
Cerebellum, 195
Cerebrum, 150–159
  drawing of, 154
Cézanne, Paul, 288–289
Character, 180–220
  ego and, 299–313
Charcot, Jean-Baptiste, xxxiii, 86
Charlemagne, King, 160, 166
Chatwin, Bruce, *Notes*, 408
Cheshire cat, 98
Chesterton, G. K., *Tremendous Trifles*, 220
Child Harolde (Byron), 349

Chinn, Austin, xi
Chou Li, 171
Christ, 106, 260, 274, 398–399
Christianity, 120, 138–139, 392–393
Churchill, Sir Winston, 116–117, 241
Church of England, 7
Cicero (Marcus Tullius), ix, 161, 192, 224, 274, 339–340, 370
   Ad Atticum, 241
   De Natura Dorum, 409
   De Officiis, Book ii, 370, 431
   Philippicae, 200
   Tusculanarum Disputationum, 396
The City of God (Augustine), 103–105
Clarke, Ron, 461
Claudel, Paul, 38
Cloetta, Yvonne, 99
Club Reps, 219
Cognition, 42
Coleridge, Samuel Taylor, 81, 82, 144
   Anima Poetae, 374
Collected Works (Jung), 69
Collier, Graham
   Antarctic Odyssey, xxix, xxxiii
   Art and the Creative Consciousness, 26
Collins, Francis, 394, 395
Colonel's Daughter (Aldington), 456
Colosseum "Games," 274
The Colossus of Marousi (Miller), 409–410
Compassion, 445–462
Conation, 42
Concept of wholeness, 170
Concerning the Spiritual in Art (Kandinsky), 393
Condition of Man (Mumford), 292
The Condition of Man (Mumford), 71
The Confessions of St. Augustine (Augustine), 103–105

Confucius, 260, 263–264, 297, 303
Conrad, Joseph
   Heart of Darkness, 196–197, 199, 253–254
   Under Western Eyes, 199–200
Conscience, 180–220, 210
Consciousness, xiii, 1–6, 49–70, 326–374. *See also* Neurology; Psychiatry
   choices and, xxvii
   diagram of, 50
   ego and, 288–289
   levels of, xiv
   neurology and, 20–41
   without a brain, 411–422
Constantine, 160, 276–277
Conway, Moncure D., *Dogma and Science*, 232–233
Corinne (De Stael), 420
Cortex, 150–159
   drawing of, 154
Crimean War, 379
Cromwell (Hugo), 141–142
Crowther-Smith, Ruth, xi
cummings, e. e., 343
   Times One, 228
The Curate Thinks (St. John Lucas), 359
Curiosity, 97–98
Cusanus, Nicolaus, 446–447
Cymberline (Shakespeare), 230–231

D

Dachau, xxv, 395
The Daffodils (Wordsworth), 81–82, 238, 436–437
Dalai Lama, 158, 443–444
Daniel Deronda (G. Eliot), 296–297
D'Aquili, Eugene, 139–140

Dark Caves, Bright Visions (White), 424–425
Darwin, Charles, 207, 219–220, 433
   The Descent of Man, 205
   On the Origin of Species by Means of Natural Selection, xvii, 10–11
Davenport, Guy, 194
De Animi Tranquillitate (Seneca), 214
Death, 1, 63–70, 183, 467. *See also* Out-of-body experience
Death in the Afternoon (Hemingway), 205–206
De Cervantes, Miguel, *Don Quixote*, 375–377
De Chardin, Pierre Teilhard, 412
De cometis (Kepler), 400
Deer of Lascaux, 159, 160
Delacroix, Eugène, *The Journal*, 203
De la Mare, Walter, *Go Far; Come Near*, xvii, xxiii
De l'Amour (Stendahl), 346
Democritus of Abdera, ix, 276, 319
De Natura Deorum (Cicero), 409
De Officiis (Cicero), 202–203, 370, 431
De Rachewiltz, Boris, 193–194
De Rerum Natura (Lucretius), 233
De Saint-Exupéry, Antoine, 347
   Flight to Arras, 231
   Night Flight, 231
   Wind, Sand and Stars, 232
Désastres de la Guerre (Goya), 389
Descartes, René, 317–318, 369
   Discourses on Method, 49
   Traite des Passions de l'ame (Treatise on the Ardor of the Soul), 49
The Descent of Man (Darwin), 205
De Stael, Madame, *Corinne*, 420
Destiny, 113–126

The Devil's Dictionary (Bierce), 5
Devotions upon Emergent Occasions XVII (Donne), 462
De Waal, Frans, 206
Dialogues (Whitehead), 464
Dickinson, Emily, 81, 341, 350
   Poems, 28
Dickinson, Lowes, *The Greek View of Life*, 278–279
Diogenes, 274
Discourses on Livy (Machiavelli), 210–211
Discourses on Method (Descartes), 49
Disraeli, Benjamin, xvii, xx, 190, 380–381
Distinguished Service Order, xxx
Docker, Dudley, 68
Dodgson, Charles Lutwidge. *See* Carroll, Lewis
Dogma and Science (Conway), 232–233
Don Juan (Byron), 264
Donne, John, 188, 369
   Devotions upon Emergent Occasions XVII, 462
   The Extasie, 354, 422
Don Quixote (Cervantes), 375–377
"Double life," xiii, xxi–xxiv
Dowding, Lord, *Many Mansions*, 100
Draper, John William, *History of the Intellectual Development of Europe*, 102
Dreams, 43–48, 87–101
Dualistic approach to psychology, 25
Dubos, René, 190
Du Fay, Carles François, 128
Durant, Will, *The Story of Philosophy*, 200–201
Dure, Professor Leon, 10, 12–13

E

Each in His Own Tongue (Carruth), 207–208
Easter Island, 42–48, 99
Edison, Thomas Alva, 328
Education, 237
Ego, 282–315
    character and self, 299–313
    consciousness and, 288–289
    identity and, 293–299
    supra-ego, 307–309
Egocentrism, xxvii
The Egyptians (Aldred), 133
Ehrlich, Gretel, 30
    A Match to the Heart, ix, 19
Einstein, Albert, xv, 98, 106, 158, 394, 395, 400, 437
    Autobiographical Notes, 418
    Ideas and Opinions, 329
    On Science, 62, 329
    The World as I See It, 212, 291
Eiseley, Loren, xix, xx–xxi, 394, 457
    The Firmament of Time, 177–178, 312–313
    The Man Who Saw Through Time, 279–280
    Science and the Sense of the Holy, 387–389
    The Star Thrower, ix–x, xxi, 430–431
    The Unexpected Universe, 301–302
Eisen, Arri, 443–444
Élan Vital, 407–410
The Electrical Life of the Cell, 285
Electricity, 6–9
Electromagnetic energy, x, 6–9, 127, 411–422
Electronic Age, 218–219, 234–240
Electrons, 109–110

Eliot, George, *Daniel Deronda*, 296–297
Eliot, T. S.
    Burnt Norton, 107
    The Waste Land, 84
Ellis, Flight Lieutenant, 246
Emerson, Ralph Waldo, 435
Emotions, 212–220, 235
Endurance (ship), 55, 63–70, 73, 83, 209, 319
Endurance (Lansing), 62–63
Energy, 174. *See also* Yin/Yang
English Book of Common Prayer, 257–258
English Proverbs (Ray), 53
Epicurus, 274
    Fragments, Physics 58, 139
The Epistles (Horace), 278
Epistulae ad Lucillium (Seneca), 40, 258
Epistulae Morales (Seneca), 303
Erewhon (Butler), 232
An Essay on Man (Pope), 168
Essays II (Montaigne), 282
Essays III (Montaigne), 87
Essays LXXVI (Montaigne), 454
Essays of a Humanist (J. Huxley), 423
The Eternal Present (Giedion), 358–359
Ethos, 297
Euclid, 274, 276
Euripides, ix, 274, 300, 396–397
    Medea, 200
Everson, William, 60
Evil, 241–272, 286
    opposing forces of good and, 304
Evolutionary theory, xvii–xxxiv, 148, 175
The Excursion (Wordsworth), 39
Expostulation and Reply (Wordsworth), 298–299
The Extasie (Donne), 354, 422

F

A Fable for Critics (Lowell), 324
Face to Face, 145
Fact, versus imagination, xiv
Fatal Curiosity (Lillo), 56
Faust (Von Goethe), 14
Felltham, Owen, *Of Dreams*, 101
Ferrier, Kathleen, xxxi
Fert, Albert, 420–421
The Firmament of time (Eiseley), 177–178, 312–313
The First Flower People (Solecki), 133–134
First International Congress of Psychoanalysis (1908), 25
Fitts, Dudly, 352
Flight to Arras (de Saint-Exupéry), 231
Flying Bishop, 57, 113–126, 149–150, 182–183, 209–210, 268, 368
Force majeure, xviii, 407–410
Fordyce, Rev. J., 11
The Forms of Things Unknown (Read), 54–55, 59–60
Four Quartets (Eliot), 107
Fragments, Physics 58 (Epicurus), 139
France, Anatole, 191
France, Peter, *Hermits*, 321
Frankl, Dr. Viktor E., 412–414
  Man's Search for Meaning, xxv, 413, 419–420
Franklin, Benjamin, 98–99, 128, 224
Frans Liszt-The Final Years (Walker), 359
Freeman, John, 145
Freud, Sigmund, xxxi, 25, 51, 54, 70, 89, 332, 397, 448
  Group Psychology and the Analysis of the Ego, 306
  The Interpretation of Dreams, 334
  Lectures in Psychoanalysis, 302
Freud and Mans Soul (Bettelheim), 333–334
Fromm, Eric, 204, 210, 304, 312, 332, 350
  The Art of Loving, 399, 448
  Saturday Review, 180

G

Galileo, 106, 400
Gallic Wars (Caesar), 178
Galsworthy, John, 454–455
Gang of Four, 341, 343
Garden of Eden, 168–169, 232
Gargantua and Pantagruel (Rabelais), 232
Gauguin, Paul, *Whence Come We? What Are We? Wither Do We Go?*, xiii–xiv, xxii, xxiii
The Gay Science (Nietzsche), 366
Gaythorpe, Elizabeth, *Somewhere In Loving*, x–xi
Genesis, 147
Genetics, 381–384
Ghandi, 158
Giedion, Professor Siegfried, 425
  The Eternal Present, 358–359
God, 113–126, 174–175
  concept of, 130
  human beings and, 119–120
  reliance on, 136–138
Goddard, Victor, 100
Goethe, 392
  Spruche in Prosa, 275
Go Far; Come Near (De la Mare), xvii, xxiii
Goncourt, Edmond, 142
Goncourt, Jules, 142
Goncourt Journal, 142

Goodall, Jane, 124–125, 166–167, 183, 184
Good and evil, 304
Gothic architecture, 293–295
Goya, Francisco, *Désastres de la Guerre*, 389
Graves (death), 134
Graves, Robert, 300, 384–385
Gravitational force, 129
The Great Chains of Being (Lovejoy), 446–447, 454
Great Depression, 286
The Great Round, 93, 168–179
Great Spirit, 174–175
Greece, 273–281
    architecture, 275–276
    classical, 276, 286–287
    Olympic Games, 274, 277
    philosophy, 22–23, 24, 26, 84, 131, 278, 283
The Greek Anthology, 352
The Greek View of Life (L. Dickinson), 278–279
The Greek Way (Hamilton), 341
The Green Child (Read), xxx
Greene, Flight Lieutenant, 245
Greene, Flying Officer, 355
Greene, Graham, 209
    A World of My Own, 99
Group Psychology and the Analysis of the Ego (Freud), 306
Gruenberg, Peter, 420–421
Gryphon, 98
Guderian, General Hans, 263
Guillon, L., 191
Guinevere (Tennyson), 229
The Gulag Archipelago (Solzhenitsyn), 458
Guterman, Sergeant Richard, 241–249, 244, 268, 305–306

H

Hadrian, *Ad animan*, 330
Hallé Concerts Society, xxxi
Hallé Orchestra, xxi, 324, 409
Hamilton, Edith, *The Greek Way*, 341
Hamlet (Shakespeare), 42, 174, 215, 238, 462
Handel, George Frideric, xxx
Haralson, Carol, xi
Harmonice mundi (Kepler), 400
Harris, "Bomber," 241
Hartill, Rosemary, *Writers Revealed*, 436
Hartnagel, Fritz, 201
Hayden, Joseph, xx
Heart of Darkness (Conrad), 196–197, 199, 253–254
Heisenberg, Werner, 187, 367
Hemingway, Ernest, *Death in the Afternoon*, 205–206
Henry VIII, 238
Henry VIII (Shakespeare), 212
Herbert, George, 79
Hermes, 445
Hermits (P. France), 321
The Heroic Encounter (Norman), 113
Heydrich, Reinhard, 253
Hilberg, John, xi
Hilton, James, *Lost Horizon*, 232
Himmler, Heinrich, 190
History of the Intellectual Development of Europe (Draper), 102
Hitler, Adolf, xv, 190, 201, 222, 224, 306–307
Holiness, 387–391
Holmes, Oliver Wendell, Sr., 432
Holocaust, 336–337
Holy Roman Emperor, 160, 166
Homer, 274
    Odyssey, 144, 264
Homo neanderthalensis, 179

Homo sapiens, xxviii, 179, 186–187, 423
    development, xiv
Hopi Indians, 170
Horace, 274
    The Epistles, 278
The Horse Boy (Isaacson), x
The House Dog's Grave (Jeffers), 352–353, 452
Housman, A. E., *A Shropshire Lad*, 142–143
Huddleston, Father Trevor C.R., 198, 208–209, 268
    Naught for Your Comfort, xxxi, 209
Hughes, George, 426
Hughes, Squadron Leader, 248–253
Hugo, Victor, *Cromwell*, 141–142
Human beings, 190–212, 423–444
    God and, 119–120
"Humanness," xv
Human order, 185–190, 195
Human rights, 457–458
Human spirit, xviii
Hunt, Flying Officer, 245
Huxley, Aldous, xvi, xvii
    Brave New World, 221–224, 228, 232
    Proper Studies, 264
    Themes and Variations, 367–368
    Time Must Have a Stop, 105
    Views of Holland, 222
Huxley, Sir Julian, xvi, 442
    Essays of a Humanist, 423
    The Outermost House, 427–428
Huxley, T. H., 156–157
Hyperspace (Kaku), 5

I

Idealism, 193–194
The Idea of the Holy (Otto), 345–346
Ideas and Opinions (Einstein), 329
Identity (Aldrich), 408
Identity, ego and, 293–299
Imagination, xviii, 24
    versus fact, xiv
Imitations of Horace (Pope), 299–300
Information technology, 216–217
In Memoriam (Tennyson), 184
Innocence, 241–272
Inspiration, 4
Instinct, 199
The Interpretation of Dreams (Freud), 334
Intimations of Immortality (Wordsworth), 326, 465
Intimations of Mortality (Wordsworth), 16
Intuition, xviii, 52, 54
Inventions, 216–217
Ion (Plato), 30–31, 445
Irrational Man (Barrett), 171–172, 431–432
Isaacson, Rupert, *The Horse Boy*, x

J

Jack the Ripper, xv, 379–380, 384
Jacobi, Jolande, x, 397–398
James, William, *The Principles of Psychology*, 198–199
James Caird, 84
Jeans, Sir James Hopwood, 8–9, 10, 127, 356, 382–383
Jeffers, Robinson, *The House Dog's Grave*, 352–353, 452
Joan of Arc, 158
Jock, Auld, 356–357
John Bull's Other Island (Shaw), 264
Johnson, Bryce, 443–444
Johnson, Samuel, *The Rambler*, 97–98

Jones, David, 13
Jones, "Taff," 8
Joubert, Joseph, 418
    Pensées, 324
Journal (Amiel), 265
The Journal (Delacroix), 203
Journal (Thoreau), 401
Jung, Carl Gustav, 25, 33, 51, 54, 70, 89, 106, 145, 158, 283, 321, 332–333, 335–341, 372, 373, 397, 448, 464–465
    Aion, 305
    biography of, 335
    Collected Works, 69
    Man and his Symbols, 146
    Memories, Dreams, Reflections, 254–256, 292
    Modern Man in Search of a Soul, 132, 338
    Psychological Reflections, x, 310, 350, 373, 397–398
    Psychology of the Unconscious, 287
Juvenal, *Satires*, 225

K

Kaku, Michio, *Hyperspace*, 5
Kandinsky, Wassily, *Concerning the Spiritual in Art*, 393
The Kasidah (Richard Burton), 203–204
Kazantzakis, Nikos, 132–133
Keats, John, 73
    Ode on a Grecian Urn, 406
Kendrick, Keith, 358
Kepler, Johannes, 406
    De cometis, 400
    Harmonice mundi, 400
Kerkule, F. A., 98
Key, Thomas Hewitt, *Punch*, xx, 316
KGB, 458

King Henry VI (Shakespeare), 257
King Solomon's Ring (Lorenz), 453–454
Knight, India, 219
Kowalski, Gary, *The Souls of Animals*, 429
Kramer, Josef, 390–391

L

The Labyrinth (Auden), 138, 440
Lacey, Robert, *The Year 1000*, 292–293
Language, 350–351. *See also* Linguistics
Lansing, Edward, *Endurance*, 62–63
La Rochefoucauld, François de, *Maximes Posthumes*, 204
Lawrence, D. H., 27, 153, 347
Leake, Jonathan, 357–358
Lectures in Psychoanalysis (Freud), 302
Leibniz, Gottfried, 362
Lenin, Vladimir, 190, 222
Leonardo de Vinci, 158, 330, 369
Lerner, Rita G., 315
Leucippus, 274, 276, 329
Levin, Bernard, 311
Levine, Amala, xi
Levine, Eric, xi
Lewis, Damien, 262
Life and Essays on Religion (Tolstoy), 449
The Life of a Cell, 6–9, 361
Light (Bourdillon), 144
Lillo, George, *Fatal Curiosity*, 56
Limbic system, 195
The Limitations of a Scientific Philosophy (Read), 54–55
Lincoln, Abraham, 243–244
Linguistics, 299–301, 319–320. *See also* Language

Liszt, Anna, 359
Liszt, Frans, 359
Little Man and Little Soul (Moore), 375, 377
The Living Bread (Merton), 370
Locke, John, 322
Logos, 297
Logotherapy, 413
London Sunday Times, 219, 292–293, 357–358, 394, 425, 426
Longden, Jack, 342
Longfellow, Henry Wadsworth, *The Spanish Student*, 407
Lorenz, Konrad, *King Solomon's Ring*, 453–454
Los Angeles Times, 140
Lost Horizon (Hilton), 232
The Lousiad (Wolcot), 208
Love, xxvii–xxviii, 141–146, 445–462
Lovejoy, Professor Arthur O., *The Great Chains of Being*, 446–447, 454
Lowell, J. R., *A Fable for Critics*, 324
Lucretius, *De Rerum Natura*, 233
Lyne, G. M., 178, 286–287

M

Ma'at, 133
Macaulay, David, *The Amazing Brain*, ix
Machiavelli, Niccolo, *Discourses on Livy*, 210–211
The Magic Mountain (Mann), 346
Mahler, Gustav, xxxi
Mahler's First Symphony, 409
Malraux, André, 462
  The Walnut Trees of Altenburg, xxxiv, 463
Man and his Symbols (Jung), 146

Mandala, Nelson, 158
Mann, Thomas, 144
  The Magic Mountain, 346
A Man's a Man for A' That (Burns), 257–258
Man's Search for Meaning (Frankl), xxv, 413, 419–420
The Man Who Saw Through Time (Eiseley), 279–280
Many Mansions (Dowding), 100
Mao Tse-tung, 190
Mapping, brain, 140–141
Marcus Antoninus Aurelius, ix, 73, 74, 157, 158, 161, 194, 274, 285
  Meditations, 74, 192, 260, 261, 263, 329–330
Marcus Tullius Cicero, *De Officiis*, 202–203
Marlowe, Christopher, 188, 204
Marquis de Sade, 210
The Marriage of Heaven and Hell (Blake), 313–314
Mars Global Surveyor, 405–406
Mars Pathfinder, 405–406
Marvell, Andrew, 188
Marx, Groucho, 127
A Match to the Heart (Ehrlich), ix, 19
Materialism, 15
Maximes Posthumes (La Rochefoucauld), 204
Maxims of Ptahotep, 194–195, 263
May, Rollo, 332
McClellan, Michael, xi
McDougall, Christopher, 459–461
  Born to Run, x
Measure for Measure (Shakespeare), 197–198, 303
Measurement of Adult Intelligence (Wechsler), 191
Medea (Euripides), 200

485

Médecins Sans Frontières, 456
Medicine Without Borders, 457–458
Meditations (Marcus Aurelius), ix, 74, 192, 260, 261, 263, 329–330
Memories, Dreams, Reflections (Jung), 254–256, 292
Memory, 150, 216
Menos, xxi, 23, 316
    description of, 24
Mens, 316
Mental processes, 42
Merchant Marines, 72–74
Meredith, George, *The Ordeal of Richard Feverell*, 142
Merton, Thomas, 321
    biography of, 322
    The Living Bread, 370
    Vow of Conversation, 321–322
"Metaphysical anxiety," xiv
Metaphysical poets, 188–189
Metaphysics, xxix, 149
Michelangelo, 35, 106, 362–363, 369, 401–402
    Pietà, 364
Military Cross, xxx
Mill, John Stuart, *Principles of Political Economy*, 74–75
Miller, Henry, *The Colossus of Marousi*, 409–410
Miller, Stanley L., 11
Milton, John
    Areopagitica, 387
    Paradise Lost, 402–403, 435, 447
Mind
    brain and, xix–xxi, 316–325
    definition, xxi
    over matter, 319
Mock Turtle, 98
Modern Man in Search of a Soul (Jung), 132, 338

A Modern Utopia (Wells), 232
Mohammed, 260
Montaigne, 295, 298, 303
    Essays II, 282
    Essays III, 87
    Essays LXXV, 454
Monte Cassino, 162–166
Moore, Thomas, 400–401
    Little Man and Little Soul, 375, 377
Morals, 29–30, 180–220, 210, 221–240
    principles of, ix
    war and, 242–243
More, Sir Thomas, 239
    Utopia, 232
Morse Code, 331–332
Mortality, 1–2, 257–258
Moses, 224, 260
Mother Teresa, xv, 158, 456
Mozart, Wolfgang Amadeus, xi, xv, xxv–xxvi, xxx–xxxi, 26, 28–29, 31–34, 35, 40, 43, 57, 61, 76, 106, 295, 400, 409
Much Ado About Nothing (Shakespeare), 28–29
Mumford, Lewis, *The Condition of Man*, 71, 292
Murdoch, Iris, 306
Music, xxx–xxxiv, 61, 309–310, 332, 375–407
    musical therapy, 35–36
    of the spheres, 400–406
Music of the Spheres, 402
My Life with the Microbes (Waksman), 129
Mysterium conunctionum, 372
The Myth of Sisyphus (Camus), xxxiii–xxxiv, 464

N

Nagele, Rose, 39
Napoleonic Wars, 269–270
Native American Indians, 338, 425–426
Native Son (Wright), 289
Natural disasters, 119, 226, 308
Natural order, 182–185, 189
Nature, versus nurture, 265–267
The Nature of Sympathy (Scheler), 439
Naught for Your Comfort (Huddleston), xxxi, 209
Neanderthal man, 428–429
Neurology, xix–xxi, 19–41.
  See also Brain; Consciousness; Psychiatry
  consciousness and, 20–41, 27–28
Neurons, 42–48
  diagram, 21
Neuroscience, 24, 59
New Atlantis (F. Bacon), 232
Newberg, Professor Andrew, 139–140, 141, 259–260
The New Quotable Einstein (Calaprice), x
The New Yorker, 58
The New York Times, 33, 443–444
Nicol, Christine, 357–358
Nicomachean Ethics (Aristotle), 330
Nietzsche, Friedrich, 465
  Beyond Good and Evil, 180–181
  The Gay Science, 366
Night Flight (de Saint-Exupéry), 231
Nightingale, Florence, xv, 158, 379–380
911, 297–298, 308
Nobel Prize in Literature
  1921, 191
  1925, 264
  1927, 314
  1933, 454
  1970, 399, 448, 458
Nobel Prize in Medicine, 1973, 453
Nobel Prize in Physics, 2007, 420–421
Norman, Dorothy, *The Heroic Encounter*, 113
Notes (Chatwin), 408
Nurture, versus nature, 265–267

O

Ode on a Grecian Urn (Keats), 406
Odyssey (Homer), 144, 264
Oedipus complex, 333
Of Atheism (F. Bacon), 9
Of Dreams (Felltham), 101
Of Great Place (F. Bacon), 442
Old Stone Age, 169
Ollivier, Blandine, 359
Ollivier, Emile, 359
Olympic Games, 274, 277, 461
On a Regicide Peace (Burke), 223–224
On Hearing a Symphony of Beethoven (St. Vincent Millay), 31
On Science (Einstein), 62, 329
On the Origin of Species by Means of Natural Selection (Darwin), xvii
On Tranquility of Mind (Seneca), 316
Opposing forces, 127–141
Opus Majus (R. Bacon), 329
Oracle of Delphi, 396
The Ordeal of Richard Feverell (Meredith), 142
Orego, Iris, 373–374
  The Vagabond Path, 373–374
Organ Symphony (Saint-Saens), 31
Ornstein, Robert, *The Amazing Brain*, ix, xxviii, 22, 108–109, 152, 153
Otto, Rudolph, *The Idea of the Holy*, 345–346

The Outermost House (H. Beston), 427–428
Out-of-body experience, 416–417. *See also* Death
Ovid, ix, 30, 105, 161, 192, 274
Owen, Wilfred, 209
Oxford Union, xvii

P

Pages from the Goncourt Journal (Turgenev), 448–449
Paine, Thomas, *Rights of Man*, 258
Palaeolithic art, 152–153, 195
Paracelsus (Part i) (R. Browning), 264–265
Paradise Lost (Milton), 402–403, 435, 447
Parmenides, 24
Pascal, Blaise, 168, 331, 369
  Pensées, xvii, 240, 439–440
Pathos, 297
Patience, 215–216
Penfield, Wilder, 316–317, 381–382, 417–418
Pensées (Joubert), 324
Pensées (Pascal), xvii, 240, 439–440
The People of the Forest (Van Lawick), 124–125
Per Ardua ad Astra, 232, 247, 393
Periclean Age of Athens, 157, 158
Pericles, 274, 277
Peter the First, xxxii–xxxiii
Phaedrus (Plato), xxix, 211–212
Philippicae (Cicero), 200
Philosophy
  Eastern, 23
  Egyptian, 133, 408
  Greek, 22–23, 24, 26, 54, 131, 278, 283
  Roman, 161–162

Pietà (Michelangelo), 364
Planck, Max, 104, 394, 395
Planets, 400–406
Plato, ix, 11, 69, 106, 171, 274, 291, 332, 407–408
  Ion, 30–31, 445
  Phaedrus, xxix, 211–212
  The Republic, 232, 401
  Symposium, 348–349
Playwrights, 131
Pliny the Elder, ix
Poe, Edgar Allen, *To Helen*, 273
Poems (E. Dickinson), 28
Poetry, 73, 81–82, 84, 341, 370–371, 436–437, 440, 447–448, 452, 463–464
  metaphysical, 188–189
A Poet's Epitaph (Wordsworth), xxxiii
Pol Pot, xv, 190
Pontius Pilate, 265
Pope, Alexander, 369
  An Essay on Man, 168
  Imitations of Horace, 299–300
Pope John Paul the Second, 373
Pope Pius XI, 238
Possessions, 15
Poultney, Dr. Joan, xi
Prayer, 141–146
Preface to the Lyrical Ballads (Wordsworth), 235
Priestly, Sir Raymond, 69, 72
Principles of Political Economy (Mill), 74–75
The Principles of Psychology (James), 198–199
The Private Albert Einstein (Bucky), 400–401
Probability theory, 175
Prometheus Unbound (Shelley), 36
Promiscuity blues (Knight), 219

Proper Studies (A. Huxley), 264
Psyche, 22–23
Psychiatry, xix–xxi
Psychoanalysis: Its Theories and Practical Application (Brill), 308–309
Psychological Reflections (Jung), x, 310, 350, 373, 397–398
Psychology of the Unconscious (Jung), 287
Ptatahotep, 200
Punch (Key), xx, 316
Pyramidal structures, 391–400
Pyramid of souls, 382
Pythagoras, 274, 276, 400, 406

Q

QED, 7–9
Quakers, 456
Quantum physics, 60, 104, 188, 314–315, 367
Queen of Hearts, 98
Quintus Horatius Flaccus, *Satires*, 211

R

Rabelais, François, *Gargantua and Pantagruel*, 232
Radio Times, xxxi
The Radio Times, 309–310
The Rainbow (Wordsworth), 272
Raleigh, Walter, 188
The Rambler (Johnson), 97–98
Rapahango, Benito, 46–48
Rause, Vince, 140, 141
Ray, John, *English Proverbs*, 53
Read, Sir Herbert, 14, 198, 204, 208–210, 268
  The Forms of Things Unknown, 54–55, 59–60
  The Green Child, xxx
  The Limitations of a Scientific Philosophy, 54–55
Reason, 136
Religio Medici (Browne), 201–202
Religion, 187–188, 241–272
Renk, Professor Elizabeth, xi
The Republic (Plato), 232, 401
Rights of Man (Paine), 258
Rilke, Rainer Maria, 38
Rites of passage, 271–272
Robinson, Edwin Arlington, 291
Rome, 273–281
  architecture, 277
  Colosseum "Games," 274
  infrastructure of, 277
  philosophy, 161–162
Rommel, Field Marshal Erwin, 263
Royal Air Force, xxx–xxxi, 8, 72, 100, 127, 232, 241, 356, 393

S

Sabom, Dr. Michael, 417
Saddam Hussein, xv, 198, 223
Saint-Saens, Camille, *Organ Symphony*, 31
Salk, Jonas, 158
The Saracen's Head, 243–244, 247
Satires (Flaccus), 211
Satires (Juvenal), 225
Saturday Review (Fromm), 180
Save the Children, 456
Scheler, Max, 373, 466–467
  The Nature of Sympathy, 439
Scholl, Hans, 29–30, 36–41, 254, 307
Scholl, Sophie, 29–30, 36–41, 200–201, 254, 307

Schweitzer, Albert, 285, 456
Science, 147–149
Science and the Sense of the Holy (Eiseley), 387–389
Scott, Sir Walter, 70
Seacole, Mary, 379
Séances, 369
The Second Coming (Yeats), 279, 371–372
Secularization, 180–181
Selden, Gary, *The Body Electric*, x, 110–111, 334–335
Self
    ego and, 299–313
    scrutiny, 14
Seneca, Lucius Annaeus, 179, 192, 210, 274, 378
    Ad Lucilium, 221
    De Animi Tranquillitate, 214
    Epistulae ad Lucillium, 40, 258
    Epistulae Morales, 303
    On Tranquility of Mind, 316
Senses, xviii, 52, 74–82
Shackleton, Sir Ernest, xxix, xxxiii, 62–69, 63–70, 72–74, 84–85, 284, 319
Shakespeare, William, x, xiii, 120, 188, 331, 369, 374, 380
    Cymberline, 230–231
    Hamlet, 42, 174, 215, 238, 462
    Henry VIII, 212
    King Henry VI, 257
    Measure for Measure, 197–198, 303
    Much Ado About Nothing, 28–29
    Sonnets CVII, 188–189, 383
    The Tempest, 1–3, 226
    Troilus and Cressida, 296
    As You Like It, 73
Shamans, 43–48, 412
Shanidar cave burials, 135

Shaw, George Bernard, *John Bull's Other Island*, 264
Shelley, Percy Bysshe, xxi
    Prometheus Unbound, 36
A Shropshire Lad (Housman), 142–143
Sir Walter Scott (Carlyle), 288
Sistene Chapel, 362
A Slumber Did My Spirit Seal (Wordsworth), 410
Smith, Logan Pearsall, *Trivia*, "The Spider," 325
Snake story, 168–179
"Society of thyself," 73
Socrates, ix, 274, 300, 317, 332
Solecki, Professor Ralph, 428
    The First Flower People, 133–134
Solitude, 71–86
    ultimate, 82–86
Solon, 274
Solzhenitsyn, Aleksandr, 399, 448, 455, 457, 458
    The Gulag Archipelago, 458
Somewhere In Loving (Gaythorpe), x–xi
Somewhere in Loving (Tate), 344–349
Sonnets CVII (Shakespeare), 188–189, 383
Sonnets from the Portuguese, 43 (E. Browning), 347–348, 452
Sophocles, 300, 396–397
Soul, xviii, xxix, 23, 326–374, 375–407, 407–410
The Souls of Animals (Kowalski), 429
The Spanish Student (Longfellow), 407
Spetzler, Dr. Robert, 415–416
Spirit, xxix, 4, 130, 407–410
Spiritualism, 369
Spirituality, 241–272
The Spiritual Teachings of Marcus Aurelius, 260–261

Spruche in Prosa (Goethe), 275
St. Benedict, 162–164, 368
St. Bernard, 294
St. Francis of Assisi, 158, 285, 347, 449–450
St. John Lucas, *The Curate Thinks,* 359
St. Mark's Gospel, 158–159, 327
St. Thomas Aquinas, 327
St. Vincent Millay, Edna, *On Hearing a Symphony of Beethoven,* 31
Stalin, Joseph, xv, 190, 222, 310, 430
The Star Thrower (Eiseley), ix–x, xxi, 430–431
Stein, Gertrude, 466–467
Steiner, Professor George, xxvi, 112
Stendahl, Marie-Henri, *De l'Amour,* 346
Stevenson, Robert Louis, 109
The Story of Philosophy (Durant), 200–201
Stratas, Teresa, 58
Stress, 319–320
Sullivan, J.W.N., 61–62
Supra-ego, 307–309. *See also* Ego
Sutherland, Graham, 76–80, 88
Swedenborg, Emanuel, 398
Swift, Graham, *Waterland,* 220
Symbolism, 168–179
Symposium (Plato), 348–349

T

Tate, Elizabeth, *Somewhere in Loving,* 344–349
Technology, 148, 212–220, 233–240, 437
The Tempest (Shakespeare), 1–3, 226
Ten Commandments, 224

Tennyson, Lord Alfred
  Guinevere, 229
  In Memoriam, 184
Terra Nova, 277
Teutonic Knights, 403
Themes and Variations (A. Huxley), 367–368
Theories of Everything (Barrow), 5
Third Millennium, 312–313
Thomas, Dylan, xviii–xix, xxiv–xxviii, 78, 212, 230
Thompson, Francis, *The Beautiful Years,* xxi, xxviii
Thompson, Richard F., 19
  The Amazing Brain, ix, 22, 108–109
Thoreau, Henry David, xxv, 132
  Journal, 401
Thorn Tree series (Sutherland), 78–80
Time, 87–101, 413–414
  linear, 104
  nature of, 103
  passage of, 102–112
  sacred, 103
  secular, 104
  timelessness, 105–106
Time Must Have a Stop (A. Huxley), 105
Times One (cummings), 228
Tintern Abbey (Wordsworth), 79, 463–464
To Helen (Poe), 273
Tolstoy, Leo, *Life and Essays on Religion,* 449
Toynbee, Sir Arnold, 160, 161, 368
Tragedies, xxiv–xxv
The Transparent Self, 397
Trappist Monks, 322
Treatise on the Ardor of the Soul (*Traite des Passions de l'ame;* Descartes), 49

491

Tremendous Trifles (Chesterton), 220
Trivia, "The Spider," 325
Troilus and Cressida (Shakespeare), 296
Truth, 265, 463–467
Turgenev, Ivan, 191
    Pages from the Goncourt Journal, 448–449
Turner, Laurence, 310
Tusculanarum Disputationum (Cicero), 396

U

Unamuno, Miguel, 351
Uncertainty principle, 367
Unconscious, 56–60, 70, 95
    as magma chamber, 60–62
The Unconscious Before Freud (Law), 51
Understanding, 54
Under Western Eyes (Conrad), 199–200
The Unexpected Universe (Eiseley), 301–302
United Nations, 262, 457
Uroburos, 169–170, 172
U.S. National Human Genome Research Institute, 394
Utopia, 221–240
Utopia (More), 232, 239

V

The Vagabond Path (Orego), 373–374
Valentini, Livio, 301
Valéry, Paul, 16, 19, 217, 235–236
Van Gogh, Vincent, 27, 76, 153, 347
Van Lawick, Hugo, *The People of the Forest*, 124–125
Verlaine, Paul, 38
Victoria, Queen of England, xvii, 369

Views of Holland (A. Huxley), 222
Virgil, 319
Voltaire, 158, 304, 305, 369, 420
Von Arnim, Bettina, 332
Von Bellingshausen, Thaddeus, xxxii
Von Goethe, Johann Wolfgang, 50
    Faust, 14
Von Runstedt, Karl Rudolf Gerd, 261
Vow of Conversation (Merton), 321–322

W

Waiting, 214–215
Waksman, Selman A., *My Life with the Microbes*, 129
Walker, Alan, *Frans Liszt-The Final Years*, 359
Wallace, Alfred Russell, 433
The Walnut Trees of Altenburg (Malraux), xxxiv, 463
War, 241–272. *See also* World War I; World War II
    morality and, 242–243
The Waste Land (T. Eliot), 84
Waterland (Swift), 220
Watt, James, 441
Weapons systems, 223
Webster, John, 358
Wechsler, David, *Measurement of Adult Intelligence*, 191
Wechsler-Bellevue Intelligence Scale, 191
Wedding ring, 169
Weil, André, *The Apprenticeship of a Mathematician*, 35
Wells, H. G., *A Modern Utopia*, 232
Whence Come We? What Are We? Wither Do We Go? (Gauguin), xiii–xiv, xxii, xxiii

White, Flight Sergeant, 355
White, Randall, *Dark Caves, Bright Visions*, 424–425
Whitehead, Alfred North, *Dialogues*, 464
White Rabbit, 96, 97
"White Rose," xi, 41
Whitman, Walt, 269, 292
Whyte, Lancelot Law, *The Unconscious Before Freud*, 51
Wild, Commander Frank, xxxiii, 75, 84, 307–308
Willpower, 130
Wills, Stancombe, 68
Wind, Sand and Stars (De Saint-Exupéry), 232
The Winds of Mars, 405–406
Wisdom, 54, 155
Wittgenstein, Ludwig, 332
Wolcot, John, *The Lousiad*, 208
Wordsworth, William, x, 284, 370–371, 433
  The Daffodils, 81–82, 238, 436–437
  The Excursion, 39
  Expostulation and Reply, 298–299
  Intimations of Immortality, 326, 465
  Intimations of Mortality, 16
  A Poet's Epitaph, xxxiii
  Preface to the Lyrical Ballads, 235
  The Rainbow, 272
  A Slumber Did My Spirit Seal, 410
  Tintern Abbey, 79, 463–464
The World as I See It (Einstein), 212, 291
World Health Organization, 457
World Monetary Fund, 457
A World of My Own (Greene), 99
World War I, 209, 222–223, 336. *See also* War
World War II, xxv, 8, 72, 76, 100, 113–126, 222–223, 241, 336, 413, 457. *See also* War
Worsley, Captain Frank, 55, 64–70
Wright, Richard, *Native Son*, 289
Writers Revealed (Hartill), 436
Writers' Union, 458
Writing, 438

X

Xenophon, 285
Xerxes, 277

Y

The Year 1000 (Lacey), 292–293
Yeats, John Butler, 41
  The Second Coming, 279, 371–372
Yeats, W. B., 41, 260, 373
Yin/Yang, 170–171, 173

Z

Zatopek, Emil, 459
Zeus, 274